CAMBRIDGE STUDIES IN ECOLOGY

Physiological ecology of lichens

Also in the series:
Hugh G. Gauch, Jr *Multivariate analysis in community ecology*
Robert Henry Peters *The ecological implications of body size*
C. S. Reynolds *The ecology of freshwater phytoplankton*

Physiological ecology of lichens

KENNETH A. KERSHAW, FRSC

Professor of Biology, McMaster University, Ontario

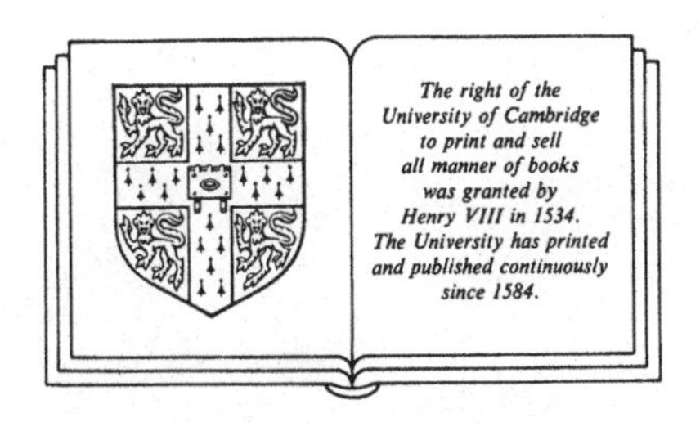

CAMBRIDGE UNIVERSITY PRESS

Cambridge

London New York New Rochelle

Melbourne Sydney

CAMBRIDGE UNIVERSITY PRESS
Cambridge, New York, Melbourne, Madrid, Cape Town, Singapore,
São Paulo, Delhi, Dubai, Tokyo

Cambridge University Press
The Edinburgh Building, Cambridge CB2 8RU, UK

Published in the United States of America by Cambridge University Press, New York

www.cambridge.org
Information on this title: www.cambridge.org/9780521283496

First published 1985
Re-issued in this digitally printed version 2010

A catalogue record for this publication is available from the British Library

Library of Congress Catalogue Card Number: 83-18870

ISBN 978-0-521-23925-7 Hardback
ISBN 978-0-521-28349-6 Paperback

To Peter Greig-Smith and Jack Rutter –
one taught me to recognise a pattern of distribution,
the other the corresponding environmental
and physiological patterns

Contents

Preface

Lichenology has remained a fairly obscure branch of biology until relatively recently when it became evident that lichens were particularly good indicators of atmospheric pollution and could be used as experimental systems for monitoring ambient levels of sulphur dioxide or even particulate emissions from mining and smelting operations. During this same period there was an increasing awareness in plant ecology that the examination of physiological adaptations to specific environmental situations perhaps offered a more constructive research approach than was then apparent from the results obtained from the quantitative or energy-flow studies that had dominated ecology in the 1960s and early 1970s. Lichens particularly, played an important role in this development of our understanding of physiological adaptations in plants.

Lichens as terrestrial plants have a number of attributes which are particularly appealing when examining the physiological ecology of plants. They have no roots-organs, which present some formidable experimental problems in studies of higher plants. Equally they do not flower or have stomata, and as a result, examination of seasonal photosynthetic rates is considerably easier. Lastly, they can be contained in small and relatively simple cuvettes without undue stress to the thallus. Accordingly my initial interest in lichens was in terms of their robust qualities as experimental systems in the field of physiological ecology. Subsequently, however, as lichens are also a dominant component of much of the vegetation of northern Canada, our work on the ecology of low arctic ecosystems in general necessarily involved an understanding of the ecology of the more abundant lichen species.

The publication of *The Lichens* by Ahmadjian and Hale in 1973 provided a rounded summary of our knowledge at that time but also provided considerable stimulation to examine some of the gaps. The intent here is to provide an equivalent critical summary of our current understanding of the physiology of lichens, but only where it relates specifically to their ecology.

Since lichens are often in very close proximity to their substratum I have also dealt extensively with their thermal and ionic environments and particularly with their water relations. The seasonal metabolic responses of a lichen to temperature, moisture and light in terms of nitrogenase activity, net photosynthesis and respiration, form the central section of the book. Many of the examples covered in detail inevitably originate from our own work, particularly that in northern Canada. For this I should apologise, but I hope that it may stimulate a critical examination of some tropical species about which we know little. In the final chapter I have introduced a discussion on ecotypic and population differences and have attempted a provisional summary of the range of strategies which appear to be available to lichens, both in terms of their morphological and their photosynthetic attributes. This section is in part speculative but again it may serve to stimulate further work and discussion in this area.

It is with considerable pleasure I thank my colleagues in the lab for their stimulation, their criticisms and help, throughout much of the work that is included here, but especially during the preparation of the manuscript.

McMaster University, 1983 K.A.K.

1

The lichen environment: temperature

Standard meteorological temperature data are always measured under well-ventilated, diffuse radiation conditions, at approximately 1 m above the ground. Consequently the values obtained have little relevance to the actual temperature experienced by a lichen on the ground, although they may approximate to the thallus temperatures of pendulous arboreal lichens. To appreciate the operating temperature environment, particularly of crustaceous or adpressed soil and rock lichens, it is essential to understand the continuous energy exchange processes and the control of these processes, which take place at all surfaces.

1.1 Surface energy balance

Natural surfaces are never flat and always have some degree of roughness, but for most purposes it is convenient to take a 'principal plane' or 'active surface' to describe, for example, a soil surface with its numerous small irregularities. During the daytime, solar energy arrives at this surface usually at a rate which exceeds the output, resulting in an accumulation of energy. This in turn causes the temperature at the surface to rise. At night the converse is true with the soil surface being the site of radiant emission and hence energy deficit, leading to the development of lowered surface temperatures. This continuous diurnal temperature change at the air/soil interface diminishes with distance away from the interface, resulting in a vertical temperature profile both above and below the soil surface (Fig. 1). The temperature profile below the soil does not concern us here although it has considerable implications for the physiological environment of the roots of higher plants. What are of importance, however, are the values of the surface temperature maxima and minima and the actual temperature gradient above the surface. It is evident from Fig. 1 that the gradient immediately above the surface is extremely steep and it is this environment which is largely utilised by crustaceous as well as many foliose and fruticose lichens.

The steepness of this gradient is due to the presence of a thin layer of air adhering to the surface within which any motion is parallel to the surface

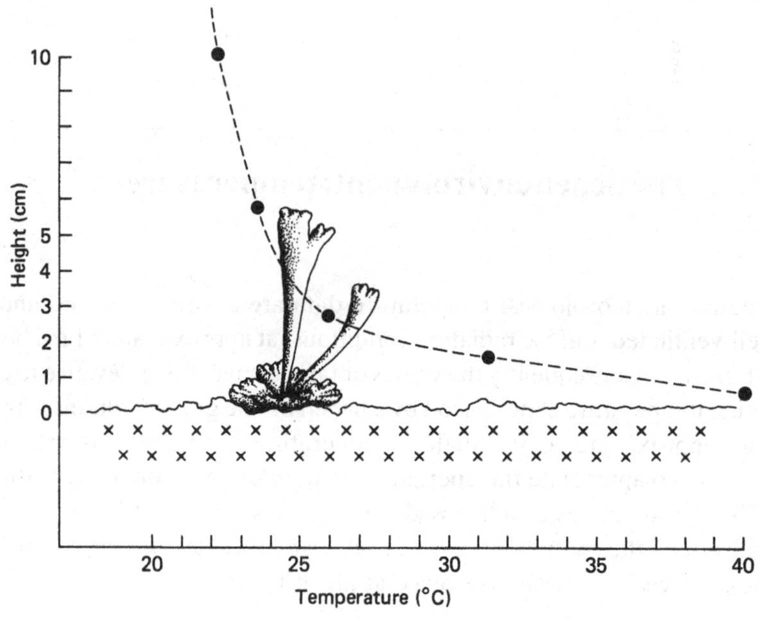

Fig. 1. Diagrammatic representation of the steep temperature gradient adjacent to the soil surface.

(Fig. 2). This layer is called the laminar boundary layer. Its actual thickness is dependent on wind speed and even under still air conditions does not exceed more than a few millimetres. However, because of the laminar flow within the boundary layer the only energy transfer processes across this layer are by radiation and molecular diffusion. As a direct result of the low molecular diffusivity of air, the steep temperature gradient within and adjacent to the boundary layer develops. Above the boundary layer the laminar flow breaks down into a complex of swirling eddies, allowing a much more rapid, turbulent transfer of energy with a correspondingly rapid decrease in the temperature gradient.

The magnitude of the surface temperature values is also dependent partially on the thickness of the laminar boundary layer, and thus under still air conditions surface temperatures will be higher than under windy conditions. However, a number of other factors affect surface temperature to a much greater extent. For example, the level of solar radiation is the primary factor controlling the amount of energy arriving at the surface and indeed will determine the net energy available and thus the final limits of surface temperature.

The net surface energy is usually termed the net all-wave radiation (Q^*), and during the daytime

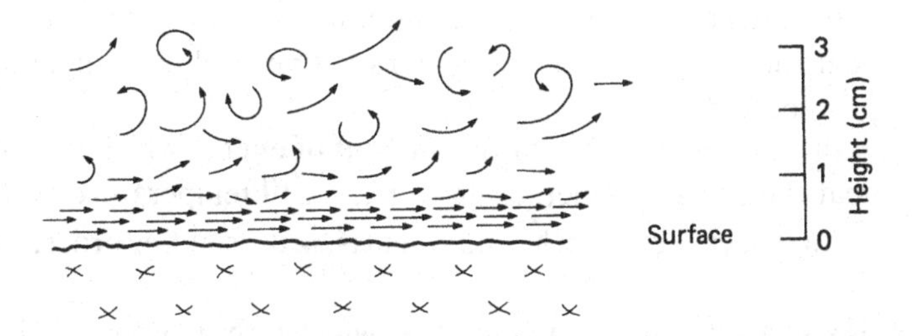

Fig. 2. Diagrammatic representation of the laminar flow in the boundary layer adjacent to the soil surface.

$$Q^* = K\downarrow - K\uparrow + L\downarrow - L\uparrow \quad (1)$$

where $K\downarrow$ and $L\downarrow$ are incoming short-wave and long-wave radiation respectively and $K\uparrow$ and $L\uparrow$ are outgoing short-wave and long-wave radiation. At any given location $K\downarrow$ and $L\downarrow$ are both controlled by latitude, the time of day and angle of slope, with maximal values at solar noon, in mid-summer at the equator. Conversely $K\uparrow$ and $L\uparrow$ are controlled by site-specific factors: $K\uparrow$ by albedo (a) and $L\uparrow$ by the surface temperature (T_0). Albedo is a measure of the amount of the incoming short-wave radiation which is reflected by a surface, expressed as a ratio

$$a = \frac{K\uparrow}{K\downarrow} \quad (2)$$

A surface with high reflectivity has a high albedo approaching a value of 1. Thus, fallen snow has an albedo of *c*. 0.95, in contrast to a wet black soil which has an albedo of *c*. 0.15. A white surface reflects a large proportion of the incoming short-wave radiation, with resultant surface temperatures potentially cooler than those of an equivalent but dark surface, where most of the incoming radiant energy is available at the surface. However, the net radiation Q^* is partitioned into a number of components, only one of which (H) is responsible for heating the surface and developing a vertical temperature profile above the ground. Thus

$$Q^* = LE + H + G \quad (3)$$

where

Q^* = net all-wave radiation or the total amount of energy available at the surface,

LE = the latent heat flux or the amount of energy which is used to evaporate water,

H = the sensible heat flux or the amount of energy which is used to heat the active surface and generate a vertical air temperature profile, and

G = the ground heat flux or the amount of energy which is used to heat the soil and generate the vertical soil temperature profile.

The proportion of energy in each component is, in descending order, $LE > H > G$.

There is also a very small component, P, which is the energy used in the photosynthetic activity of a vegetated surface, but it represents such a small proportion of Q^* that it can be conveniently ignored here. The size of H determines the thermal environment of all corticolous, terricolous and epiphyllous lichens on an hour-to-hour basis at the local scale. The local environment is further modified by the position of the lichen in relation to the active surface. This in turn is seasonally and latitudinally modified by variations in $K\downarrow$ and $L\downarrow$ (Equation 1). On the local scale, however, since Equation 3 has always to balance, any increase in LE will automatically mean an equivalent decrease in H and G. Thus, the size of the sensible heat flux is very dependent on the moisture availability at the active surface. If the soil, for example, is at field capacity, water will be freely evaporated, LE will be large and accordingly H will be proportionally much less than for an equivalent but dry soil surface. For a dry soil, with a limited water supply, LE therefore will be small and H, as a direct result, very large. Thus in a sandy desert the radiation budget (Equation 1) has typically a large initial value for incoming radiation as there is little or no water vapour in the atmosphere and cloud cover is totally absent. Since there is no soil moisture at all, $LE = 0$ and all of the energy must be dissipated as sensible and ground heat fluxes. As a result, surface temperatures in extreme deserts can reach 70°C, with more typical values falling in the range 50-60°C, (Lange 1953, 1965). Lange (1954) similarly reports thallus temperatures as high as 54°C for *Cladonia rangiformis* on sandy heaths in Germany. These values are almost certainly due to the very high sensible heat flux over the dry heathland.

In summary, the lichen ground surface environment is considerably more extreme than meteorological air temperature data might indicate. Not only are temperature maxima much higher but the diurnal range is also considerably greater. Although the overall control of surface temperature is a function of latitude and time of day, local values are profoundly modified by both albedo and soil moisture status. (See also Monteith (1973, 1975); and Geiger (1971) for more detailed treatments.) As a result direct measurement of the thallus operating temperature at suitable time

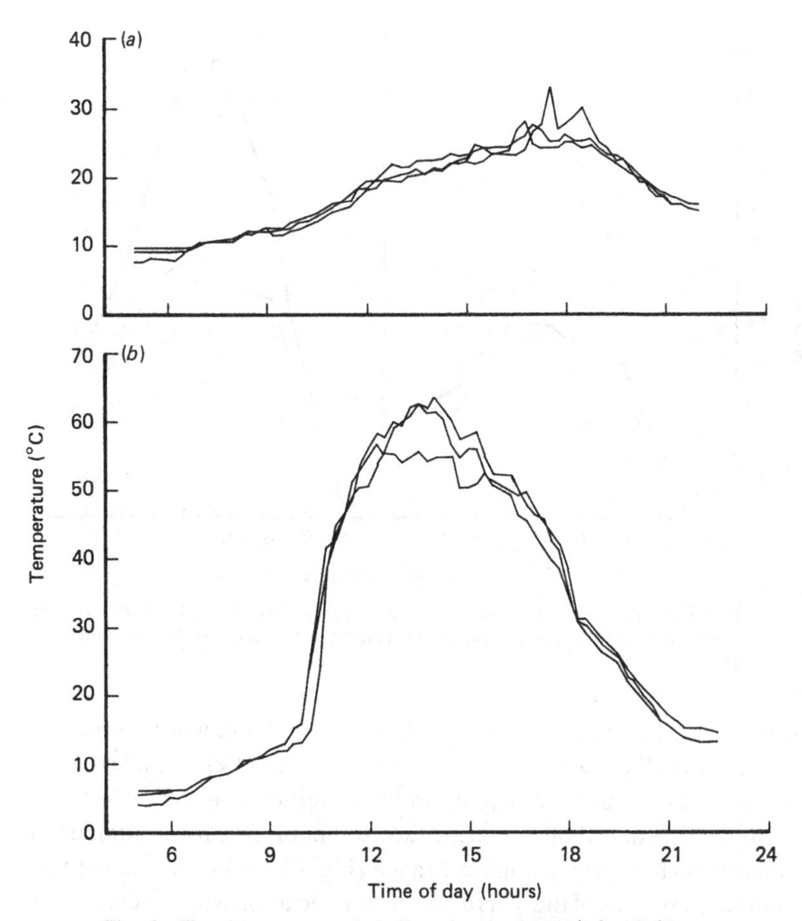

Fig. 3. The time course of thallus temperature (*a*) for *Peltigera praetextata* growing in closed-canopy deciduous woodland, and (*b*) for *P. rufescens* growing on an adjacent exposed roadside. (From MacFarlane and Kershaw 1980*a*.)

periods throughout the year is essential for an understanding of the ecology and the physiological responses of lichens.

1.2 Examples of boundary layer thallus temperatures

MacFarlane and Kershaw (1980*a*) have documented the contrasting thallus temperature environments of two species of *Peltigera* during mid-summer in North Michigan, USA. *P. praetextata* was found in closed-canopy woodland, whilst *P. rufescens* grew 30 m away on an exposed sandy roadside. Under maximal solar radiation conditions, following a rain-free period, thallus temperatures reached 60°C in the open (Fig. 3), a value which is probably typical for boundary layer conditions on windless days.

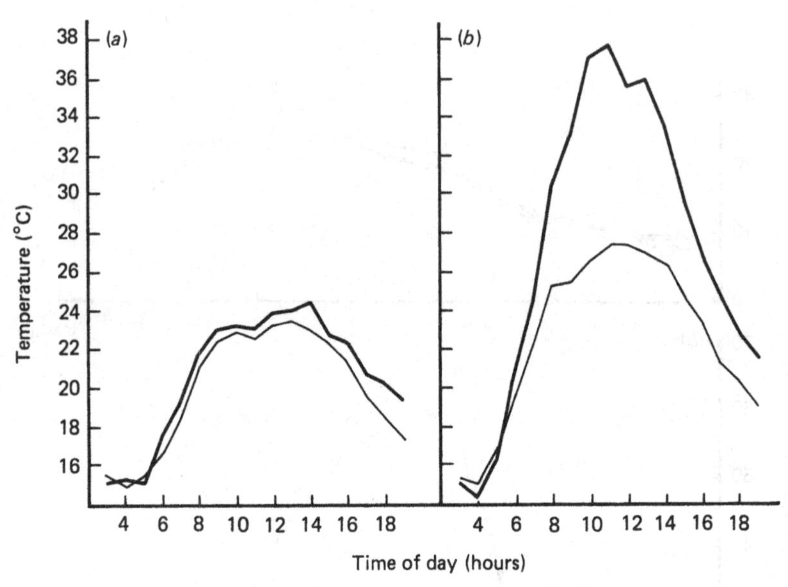

Fig. 4. Thallus (**—**) and air (—) temperatures for (*a*) the shade Verrucarietum cazzae and (*b*) the open Aspicilietum calcareae associations. (From Kershaw, 1983.)

This contrasted markedly with the adjacent woodland where some soil moisture was still evident and thallus temperatures briefly reached 30 °C during periods of sunfleck activity in late afternoon. Roux (1979) also presents contrasting thallus and surface temperatures for a number of crustaceous lichen associations in France (Fig. 4). In *Verrucaria cazzae*, the dominant species of the Verrucarietum association which is characteristically found on shaded rock surfaces, thallus temperature was close to air temperature throughout the day. However, in *Aspicilia calcarea*, the dominant species of the Aspicilietum association which is always found on open and fully exposed rock surfaces, thallus temperatures in June reached 38 °C even at 430 m elevation and when the air temperature was only 27 °C. Coxson and Kershaw (1983*a*) similarly reported that thallus temperatures for *Rhizocarpon superficiale* at 2500 m in Alberta reached 35 °C under full radiation conditions, and with artificial sheltering from the wind, values rapidly climbed to 44 °C (Fig. 5). Persistent windy conditions – which prevent such high thallus temperatures by reducing the thickness of the boundary layer and hence allowing more rapid transfer of energy from the thallus and rock surface to the surroundings – are among the more important environmental parameters governing the ecology of this species.

Correspondingly high temperatures for lichen thalli can be expected for

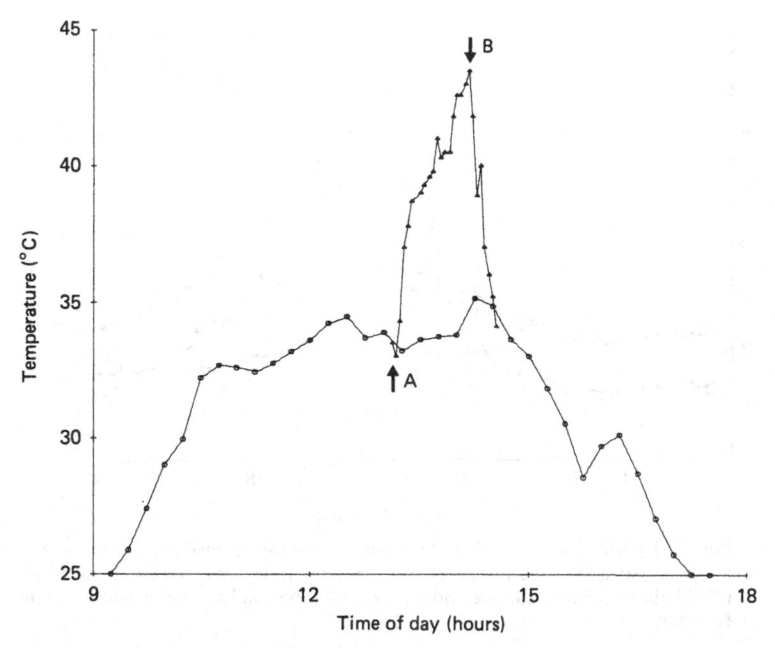

Fig. 5. Normal thallus temperatures in early July for *Rhizocarpon superficiale* under advective cooling (○–○); thallus temperatures screened from the wind for 1 h (A) and after removal of the screen (B). Values are 15 min averages. (From Coxson and Kershaw 1983*a*.)

both arctic and antarctic surfaces: thus, the data of Lewis and Callaghan (1976) for tundra surfaces show that temperatures at the surface exceed 23°C under full radiation conditions with a concurrent maximum air temperature of *c*. 13°C. Similarly Warren Wilson (1957), Bliss (1956, 1962), Mayo, Despain and Van Zinderen Bakker (1973); and Rudolph (1966) have shown that leaf temperatures in the Arctic and Antarctic are closely correlated with radiation levels, aspect and angle of orientation as well as position in relation to the surface boundary layer of the substrate. In general, lichen thallus temperatures, particularly of species growing in the boundary layer, will show similar or even more pronounced correlations with radiation.

For example, *Parmelia disjuncta*, a small adpressed foliose lichen growing on exposed rocks adjacent to Hudson Bay at Churchill, Manitoba, can show considerable elevation of thallus over air temperature (Kershaw and Watson 1983). The diurnal range of thallus temperature on 1 August is given in Fig. 6. The maximum air temperature recorded at 1 m was 9.2°C under good radiation conditions but with a northerly wind. In direct contrast, thallus temperatures reached *c*. 26°C shortly after solar noon

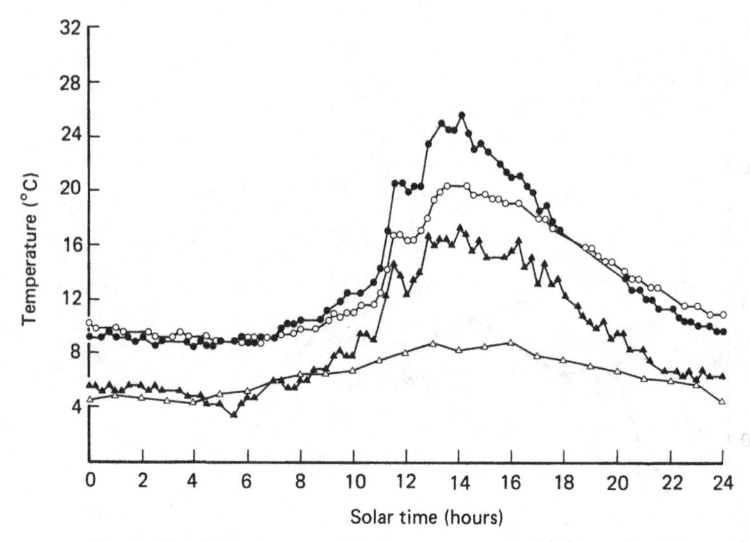

Fig. 6. The diurnal range of thallus temperature for *Parmelia disjuncta* (●–●), temperature at the rock surface (○–○), air temperature at 2 mm (▲–▲) and 1 m (△–△) above the rock surface under good radiation conditions in August. (From Kershaw 1983.)

whilst the temperature at the rock surface was *c.* 20 °C and the air temperature 2 mm above the rock surface was 18 °C. During the night, temperatures at the rock surface were consistently warmer than those of the thallus, which gradually cooled to approximately 4 °C. Under poor radiation conditions, and particularly on northern aspects, the pattern of thallus temperature is quite different. Under full cloud cover on 2 August and with a maximum recorded air temperature of 7.4 °C, thallus temperatures with a south-east exposure were still consistently higher than ambient temperatures, reaching 17.8 °C around 1400 hours (Fig. 7). However, replicate thalli with a northerly exposure were some 5 deg C cooler, reaching a maximum of only 12.8 °C at 1400 hours. The thermal environment of *Parmelia disjuncta* during the summer months is thus characterised by very modest thallus temperatures during periods of rain. Under average summer radiation conditions temperatures rise to 35 °C and probably reach over 45 °C under maximum radiation conditions, especially in the absence of the strong prevailing northerly winds. These data are representative of much of the low-Arctic and emphasise the quite specific thermal operating environments that many lichens have irrespective of the latitude of their geographical distribution. Kappen, Friedmann and Garty (1981) similarly document temperature profiles of rocks with contrasting southerly and northerly aspects in Antarctica. Under high radiation conditions, rock

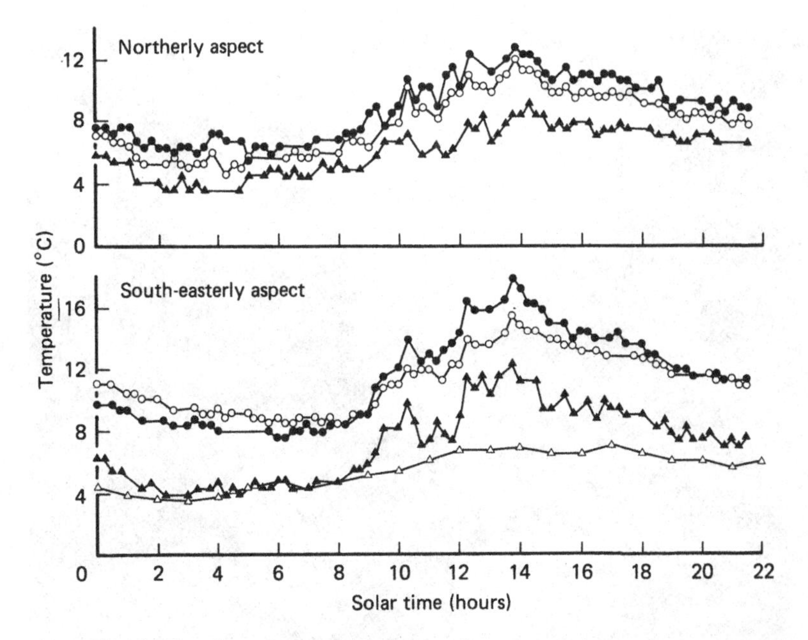

Fig. 7. The diurnal range of thallus, rock surface and air temperatures under poor radiation conditions in August. Symbols as in Fig. 6. (From Kershaw 1983.)

temperatures 15 mm below the surface are 10-15 deg C higher on northerly sunny aspects than the corresponding measurements from rocks with a southerly aspect. It is probable that these higher temperatures in the cryptoendolithic lichen zone are significantly more favourable for metabolism than the temperatures on the rock surface.

1.3 The importance of boundary layer temperatures and energy balances in the recovery succession of lichen woodland following fire

The importance of surface microclimate and its complex interactions with lichen ecology have been examined in detail in relation to the successional sequence following fire in *Stereocaulon* woodland, an extensive vegetation type in the Northwest Territories of Canada (Plate 1; Maikawa and Kershaw 1976). Johnson and Rowe (1975) have examined fire reports prepared by the Northwest Forestry Service for a region of subarctic forest which lies to the east of Great Slave Lake. They show that fire is a frequently occurring but natural event, with almost all of the fires caused by lightning. Four phases of recovery are recognised:
Phase 1, the *Polytrichum* phase with *P. piliferum*, *Biatora granulosa* and

Plate 1. *Stereocaulon* woodland, a characteristic vegetational type of dry eskers, drumlins and ridges in the Northwest Territories, Canada.

Lecidea uliginosa forming the major cover over the recently burnt surfaces.

Phase 2, the *Cladonia* phase developing after 15-20 years with the establishment of a number of *Cladonia* species, in particular *C. gonecha, C. uncialis* and *C. stellaris*.

Phase 3, the *Picea mariana-Stereocaulon paschale* woodland phase, developing after 60-70 years; this is the mature woodland phase and dominates large areas of the region.

Phase 4, apparent after 130-150 years, by which time the canopy closes and the luxurious carpet of *Stereocaulon paschale* is replaced by a carpet of mosses dominated by *Hylocomium splendens, Pleurozium schreberi* and *Dicranum* species. Only rarely does this final closed canopy spruce-moss woodland develop, since the *Stereocaulon* woodland is extremely fire-susceptible and is usually reburnt before the final phase of the succession is initiated.

During the vegetational succession there is a concurrent and equally marked development of surface microclimate. Kershaw, Rouse and Bunting (1975), Kershaw and Rouse (1976), Rouse (1976) and Kershaw (1978) have described the surface microclimate of freshly burnt plots and contrasted this with the microclimate of 23- and 80-year-old recovery phases, representative of the *Cladonia* and *Stereocaulon* phases of the recovery sequence respectively. Considerable differences develop in surface albedo following fire, with values dropping from 20% to 5% on freshly burnt surfaces (Fig. 8*b*). The subsequent recolonisation by *Cladonia* species and *Polytrichum piliferum* produces a marked increase in albedo on the 23-year burn. Within the first 25 years of recovery after fire, two-thirds of the albedo increase has already occurred. The outgoing long-wave fluxes vary largely as a function of surface temperature and during the daytime period the 1 year burn, with the highest surface temperature, shows the strongest outgoing long-wave loss. The summation of these two effects is seen in the surface net radiation of the sites (Fig. 8*c*). The *Cladonia* phase, which has both a high albedo and a marked infrared radiation loss, thus has a lower amount of energy available at the surface than the *Stereocaulon* woodland. The long-wave radiation losses are simply a function of surface temperature of the sites and there are, of course, corresponding soil temperature patterns (Fig. 9). For a typical summer period the soil temperature at the surface of the 1 year burn shows very high diurnal fluctations which are heavily damped in the 80-year-old woodland. The soils in the lichen woodland are colder at all depths, but of more significance to the developing lichen cover, surface maxima and minima calculated from the long-

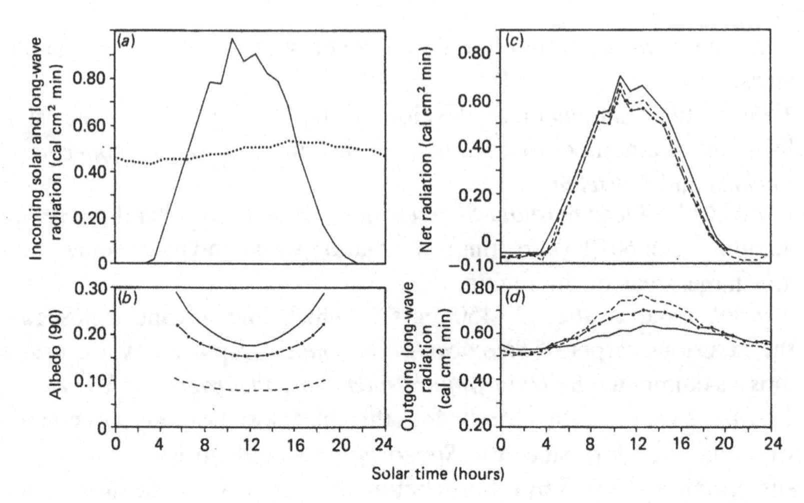

Fig. 8. Albedo and component radiation fluxes over three recovery phases following fire in *Stereocaulon* woodland, Abitau Lake region, Northwest Territories. (*a*) Incoming solar (—) and long-wave (····) radiation; (*b*) albedo; (*c*) net radiation; (*d*) outgoing long-wave radiation. In (*b*), (*c*) and (*d*): ----, 1 year burn; ●–●, 23 year burn; —, 80 year burn. Data cover the period from June to August 1975. (From Kershaw, Rouse and Bunting 1975.)

wave radiation data show extreme values on the 1 year and 23-year-burnt surfaces. Maximum surface temperatures of 65 °C probably could occur during high radiation periods on the 1 year burn, with a diurnal range of up to 46 deg C. On the 23 year burn, surface temperatures up to 45 °C are likely with early morning ground frost even in July, as a result of extreme long-wave radiation cooling.

These high surface temperatures occurring during the early successional stages following fire reflect the very xeric conditions typical of these initial phases. Under these dry conditions of the *Cladonia* recovery phase, the majority of the net radiation is dissipated as a sensible heat flux (see Equation 3). This is visually evident as 'dust devils' which form frequently over the burnt surfaces on hot summer days. These severe conditions may form one of the essential limitations to growth of higher plants and allow the development of the lichen surface without competition from such potentially faster growing species. Surface temperatures are ameliorated in the *Stereocaulon* woodland by the effective mulching properties of the lichen mat (see below) and the resultant maintenance of soil moisture at field capacity. This allows evaporative cooling with a diminished sensible heat flux and results in lower surface temperatures.

These results are confirmed by direct measurement of surface, thallus and air temperatures. Data comparing the 25-year-old *Cladonia* phase

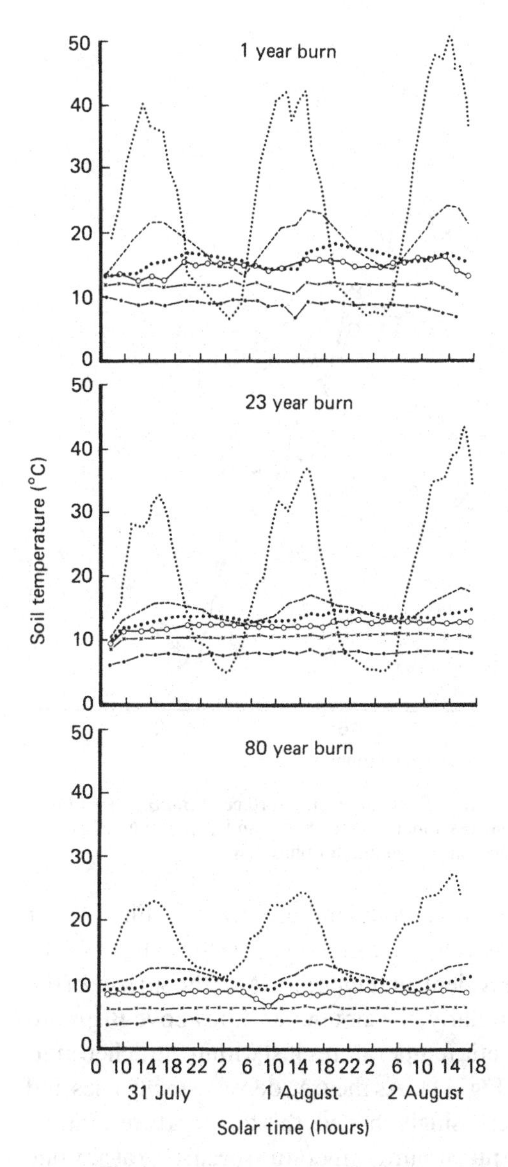

Fig. 9. Mean hourly soil temperatures for 5 (····), 10 (----), 20 (····), 40 (○–○), 80 (×–×) and 160 (•–•) cm depths on 31 July-2 August 1975 for the first three recovery phases following fire in *Stereocaulon* woodland, Abitau Lake region, Northwest Territories. (From Kershaw *et al.* 1975.)

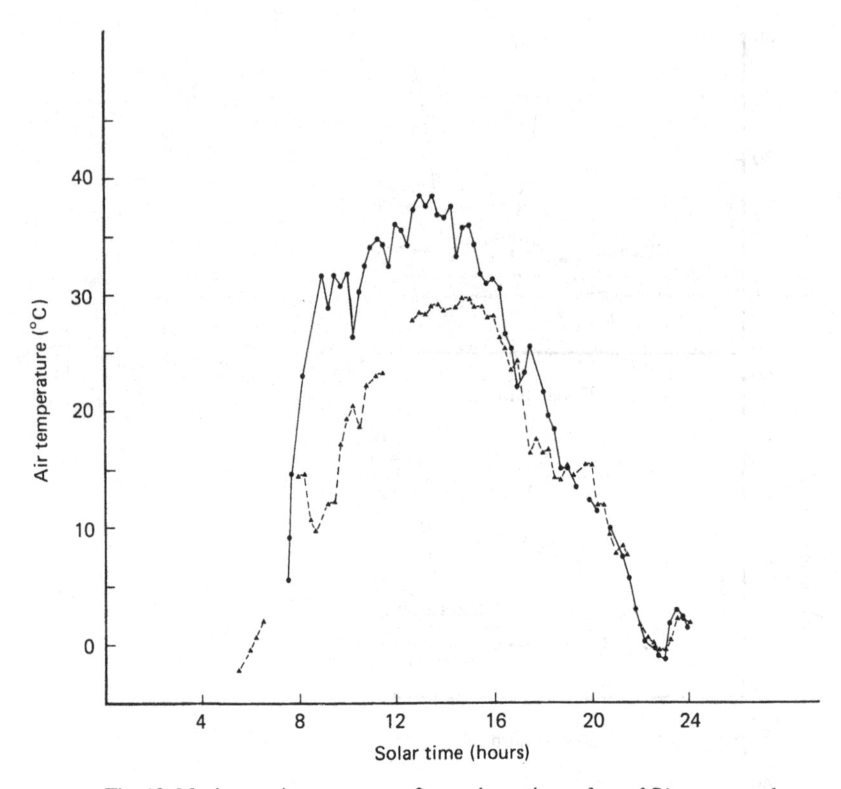

Fig. 10. Maximum air temperature 2 mm above the surface of *Biatora granulosa*, characteristic of the early successional phases (●–●), and at the mat surface of *Stereocaulon paschale* in the mature woodland phase (▲–▲).

with 80-year-old open *Stereocaulon* woodland are given in Fig. 10; air temperature 2 mm above *Biatora granulosa* crusts reaches about 48 °C, contrasting markedly with equivalent air temperatures at the surface of the *Stereocaulon paschale* mat. Similarly, the air temperature above *B. granulosa* exceeds *Stereocaulon* podetia temperatures just within the lichen mat, even in very open woodland (Fig. 11). As the tree density slowly rises and the lichen surface becomes increasingly shaded, this temperature contrast is even more marked, with the maximum temperature occurring at any one site being of very restricted duration (Fig. 11). The interaction of the boundary layer with surface temperature throughout the early successional sequence is very evident. Air temperature 1 mm above *B. granulosa* on 6 June reached 38°C, contrasting with thallus temperatures of 32°C in *S. paschale*, 30 °C at 1 cm on the podetium of *Cladonia gonecha* and only 25 °C at 2 cm on the same *Cladonia* podetium (Fig. 12). This emphasises particularly the potential interaction between propagule establishment and

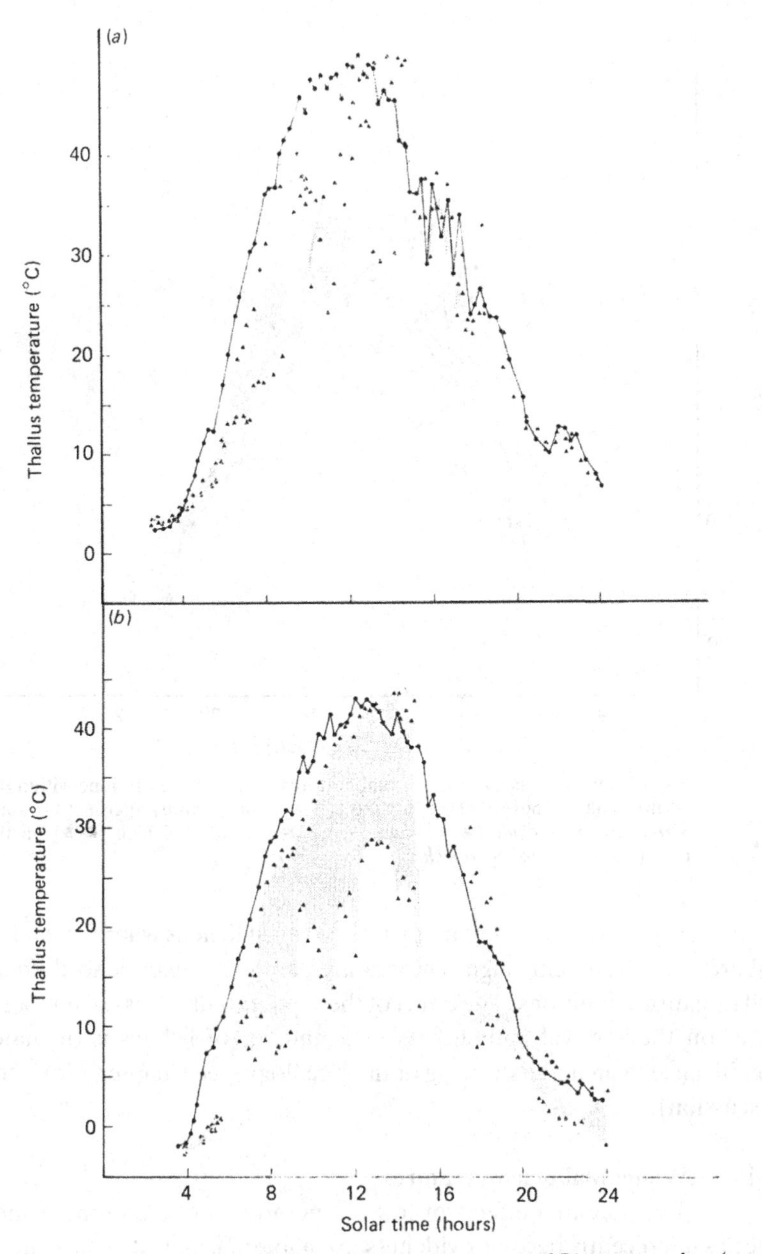

Fig. 11. Comparative maximum thallus temperatures of *Biatora granulosa* (●–●) and *Stereocaulon paschale* (▲–▲) together with the range of values from the experimental woodland plot as tree shadows pass across replicate thermocouples (stippled area). (*a*) 28 June, (*b*) 20 July.

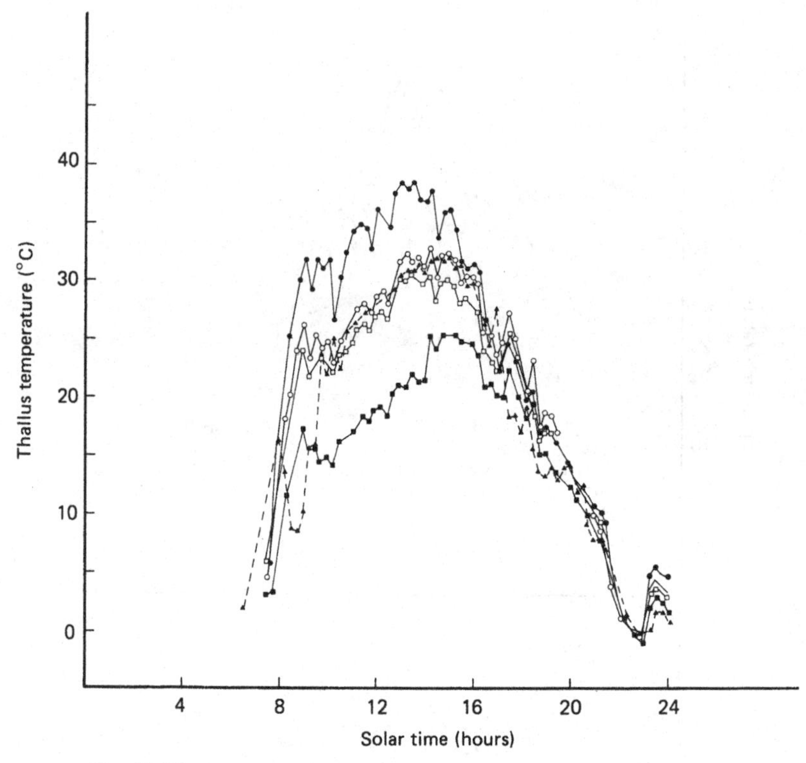

Fig. 12. The sequence of thallus temperatures recorded in early June within the boundary layer above *Biatora granulosa* (●–●), *Polytrichum piliferum* (○–○) and *Stereocaulon paschale* (▲–▲), and 1 cm (□–□) and 3 cm high (■–■) on the podetium of *Cladonia gonecha*.

boundary layer conditions in many fruticose soil lichens where, even in a subarctic environment, high temperatures are often experienced under full radiation conditions. The effect of these potentially stressful temperatures on the survival and success of a number of lichens is of major significance in an understanding of their ecology (see Chapter 7 for a full discussion).

1.4 Winter thallus temperatures

The potential effects of low temperature stress on some lichen species has recently become evident (see Chapter 7), and any treatment of the thermal environment of lichens must include their low-temperature environment during winter. For example, there is an interaction between the thallus thermal environment and the deep snow accumulation in open lichen woodlands. On average 85-100 cm of snow accumulates during the

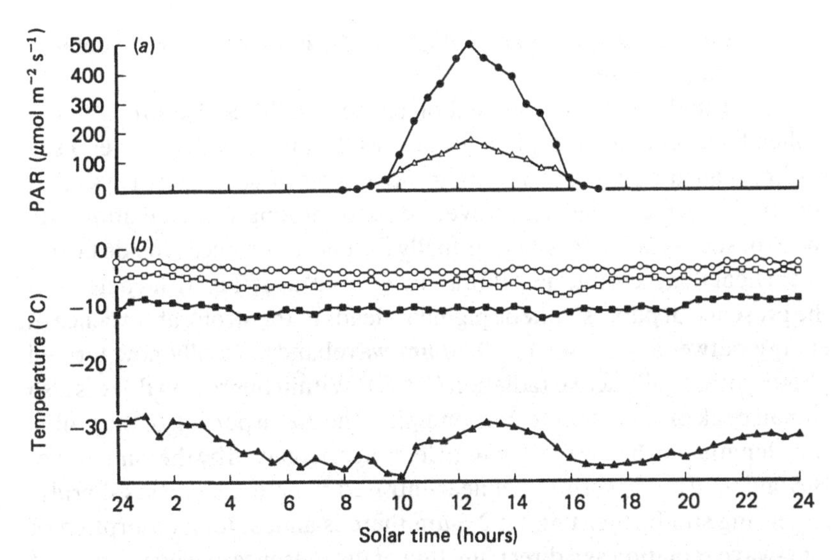

Fig. 13. (*a*) The winter pattern of photosynthetically active radiation (PAR) in the open (●–●) and in shade (△–△). (*b*) Air temperature at 1 m (▲–▲), thallus temperature of *Cladonia stellaris* (○–○), and temperatures within the snow pack (□,■), in low-arctic spruce-lichen woodland in northern Ontario. (From Tegler and Kershaw 1980.)

winter, very effectively insulating the lichen surface from low ambient temperatures. Typical data show that when ambient temperatures above the snow-pack are below −30 °C, lichen temperatures under 85 cm of snow are virtually constant at around −2°C (Fig. 13).

Analogous data for snow-cover over raised beaches in northern Ontario (Larson and Kershaw 1975*a*) show that there is a similar level of protection from low winter temperatures for the *Cladonia stellaris* association (Kershaw and Rouse 1973; Neal and Kershaw 1973). This appears to be one of the ecological parameters that limits the geographical distribution of this particular lichen. It is implicit here that most of the metabolic activity of the lichen takes place during the summer months, but an alternative strategy is seen in, for example, *Bryoria nitidula*, which is a species restricted to exposed ridge tops adjacent to the *C. stellaris* association (Kershaw and Rouse 1973). During mid-winter these ridge tops are virtually blown clear of snow and *B. nitidula* will certainly frequently be exposed to temperatures of −45 °C. However, during snow-melt periods in the spring and autumn, the sheltered environment of snow-melt pockets which develop around the dark thallus will provide a favourable situation for very active photosynthesis (see below) and it is probable that a considerable amount of carbon gain is achieved during these times.

1.5 Lichen colour and morphology, and their interaction with thallus temperature

The thermal environment of a lichen thallus is also strongly controlled by its detailed morphology. Just as the energy balance over a soil surface is an integration of incoming and outgoing fluxes, there is a similar summation of incoming short-wave and outgoing long-wave radiation over the exposed surface of a lichen. Equally the energy balance is made up of a latent heat flux, sensible heat flux and a 'ground' heat flux. In higher plants the presence of photosynthetic pigments leads to the strong absorbence of energy between the 0.40 and 0.70 μm wavebands, usually now termed photosynthetically active radiation (PAR). Within this region there is also a small peak of reflection corresponding to the green portion of the visible wavelengths and hence the colour of leaves as perceived by the human eye. Beyond 0.70 μm absorption of near-infrared radiation increases sharply, increasing still further until at 2.5 μm there is almost total absorption of long-wave radiation as a direct function of the presence of water in the leaf cells. At these wavelengths the leaf behaves almost as a perfect 'black body' with virtually complete absorbence and re-radiation of energy.

A fully hydrated lichen thallus will show a closely similar radiation budget. However, when dehydrated to air-dryness a lichen will be unable to re-radiate its infrared heat load as effectively as a leaf. The energy balance of a leaf includes a level of temperature control by maintenance of the latent heat term through evapotranspiration. The sensible heat term is, as a result, low and the 'ground' heat flux extremely small. These low values are also influenced partly by the internal architecture of the leaf. In lichens the same is true only during periods of hydration. When air-dry, however, the sensible heat term is large; the 'ground' heat flux or storage term equally is maximal in an air-dry lichen (see Fig. 5 for example). As for any surface, radiative transfer processes are dependent on the thickness of the thallus laminar boundary layer which in turn is controlled by wind speed. The effective radiation of energy is completely a *surface* exchange process and accordingly the surface area to weight ratio of a lichen will strongly affect the rate of this process. A fruticose lichen which is very finely branched, for example, will radiate energy much more effectively than a more massive, foliose lichen and their resultant thermal metabolic environments can be very different. Morphological control of the thermal metabolic environment may also interact with the albedo of the thallus.

A number of authors have commented on the possible adaptive significance of thallus colour in lichens. Laudi, Bonatti and Trovotelli (1969) attribute high levels of yellow, red, brown or black pigmentation to an

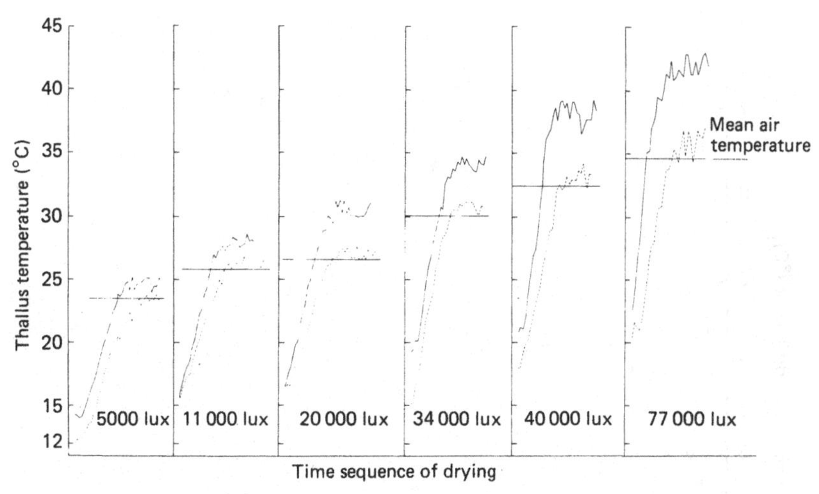

Fig. 14. The extent of evaporative cooling of the thallus under a range of radiation conditions, using natural white *Thamnolia vermicularis* (----) and replicate thalli painted black (—). (From Kershaw 1975*a*.)

adaptive and protective response of the lichen to high levels of irradiation and they suggest, although they do not present any evidence, that there is a general correlation of heavily pigmented lichens with open habitats. Thus Ertl (1951) measured the light absorption of strongly pigmented lichens under bright illumination and showed that almost half the light at the thallus surface was absorbed by the cortex. Similarly Billings and Mooney (1968) comment on pigmentation as an adaptation in alpine lichens to protect them from ultraviolet radiation (see also Kappen 1973, for discussion). These interpretations may be correct for selected species, but certainly many exceptions to the suggested correlation between heavy pigmentation and high illumination occur and it is unlikely that a general overall interpretation of thallus colour as a protective mechanism against radiation stress is correct.

Lange (1954) examined the thallus temperature of *Cladonia rangiformis* and compared the maximum value found of 54.9 °C with the temperature of similar podetia painted black (maximum 58.9 °C) and with the temperature of one painted white (maximum 50.1 °C), at an air temperature of about 27°C. Kershaw (1975*a*) presents data obtained under low wind speed conditions with an ambient air temperature of 18°C and using podetia of *Thamnolia* either in their natural white state or painted black. Under a range of illumination levels, during drying of the thallus, there is initially a very large degree of cooling as a result of free evaporation of water. This may be as great as 20 deg C under high radiation levels in the

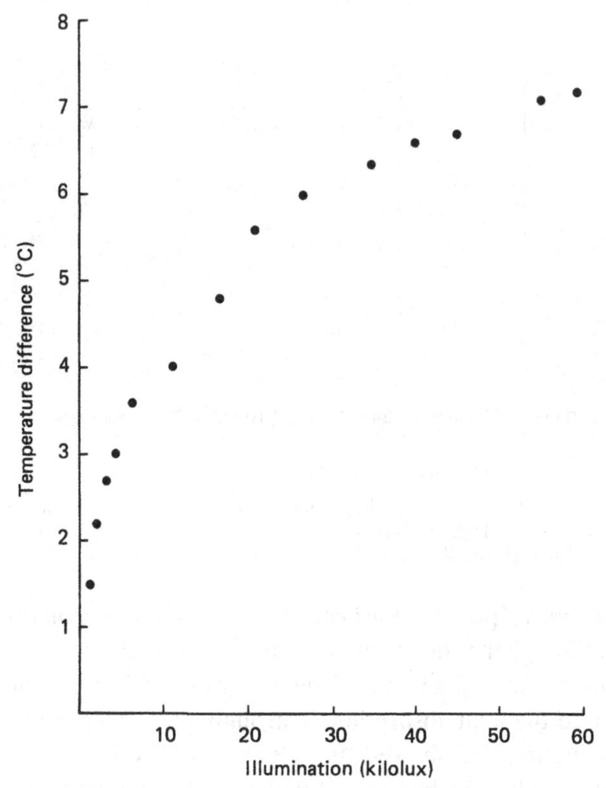

Fig. 15. The mean temperature difference, under a range of radiation conditions, between natural dry podetia of *Thamnolia vermicularis* and those painted black (From Kershaw 1975*a*.)

black podetia (Fig. 14). Air-dry thalli show a consistent variation in thallus temperature between black and white replicates, with a 7 deg C difference apparent under strong levels of illumination (Fig. 15).

Under more natural conditions in the field, the comparisons are not as extreme. Experiments were conducted in late May with about 80 000 lux illumination (corresponding to a range of 1.06 ly min^{-1} to 1.32 ly min^{-1} on the solarimeter throughout the experimental period) and an ambient air temperature of about 16 °C. Wind conditions were fairly variable, ranging from infrequent calm periods of 0.2 m s^{-1} up to strong gusts of 3.6 m s^{-1}, with an average wind speed of 1.0 m s^{-1}. *Thamnolia* as well as *Bryoria nitidula* and *Alectoria ochroleuca* were used for the thallus colour comparison. The results for *Thamnolia* and *Bryoria* are very similar (Fig. 16), showing a 2-3 deg C temperature difference between dark- and light-coloured thalli and an overall temperature drift from saturation to dryness of 7-8 deg C.

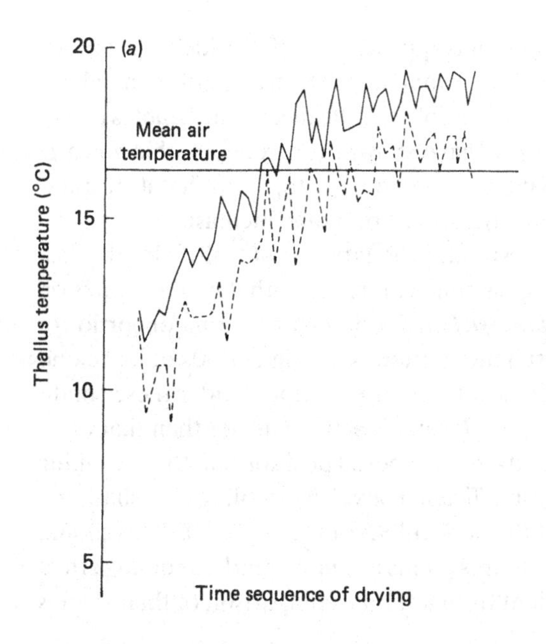

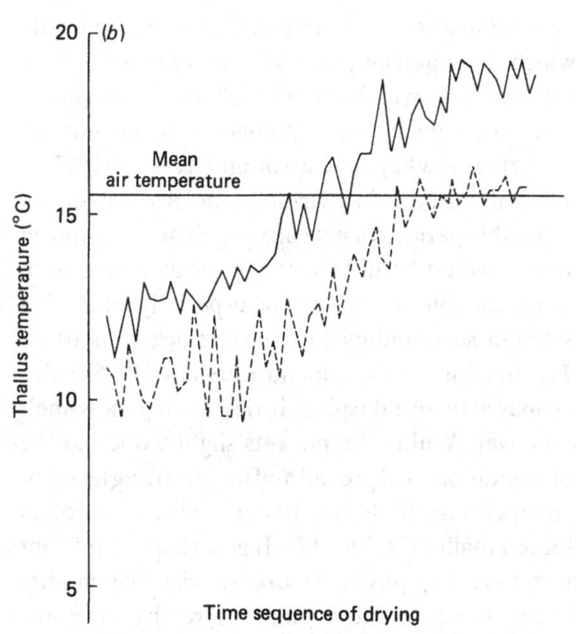

Fig. 16. The contrast between the temperatures of dark and light thalli under natural drying conditions in the field: (*a*) natural (----) and black-painted (—) podetia of *Thamnolia vermicularis*; (*b*) *Bryoria nitidula* (—) and *Alectoria ochroleuca* (----). (From Kershaw 1975*a*.)

The albedo of a lichen thallus thus appears to be of considerable importance in regulating the thermal environment in which it operates metabolically, and accordingly is of considerable significance to the lichen's ecology. Kershaw and Larson (1974) have shown, for example, that *Bryoria nitidula* is especially abundant on exposed ridge tops and that its respiration rate is particularly sensitive to temperature increase. During the summer months excessive respiration is largely controlled by the fast evaporation rates on the ridge summit compared with the slower rates on the lower slope, where *Cladonia stellaris* is dominant. The major portion of carbon assimilation in the summer period is during periods of low cloud after frontal activity and with low levels of radiation. Under these conditions the thallus temperature of *Bryoria* is actually lower than that of *C. stellaris*, simply as a result of its more exposed position and the prevailing higher wind speed resulting in efficient convective cooling (Kershaw and Larson 1974). The fact that the dark colour of the *Bryoria* thallus should induce *higher* thallus temperatures points to a more marked cooling effect due to position than is evident from a simple comparison of their thallus temperature data.

Since the rates of evaporation also vary with position, over the whole summer, *B. nitidula*, which dries quickly, will have a deficit of carbon assimilation relative to the more slowly drying *C. stellaris*. However, in winter the ridge summits have only a very thin cover of snow and are exposed rapidly when the spring thaw begins (Larson and Kershaw 1975*a*). The higher thallus temperatures attained by virtue of the black thallus of *Bryoria* will enable considerable periods of net carbon gain to be achieved when ambient temperatures would be unfavourable for assimilation in those species with a pale thallus colour. This will be especially marked in the snow-melt pockets found surrounding the dense black tufts of *B. nitidula* (cf. Geiger 1971). In addition Larson and Kershaw (1975*a*) also indicate that assimilation may actually take place in developing snow-melt pockets under thin snow cover. Within the pockets slightly open to the surface, saturated still-air conditions will prevail and under strong levels of solar radiation, thallus temperature could rise 10 deg C above that of an equivalent but pale-coloured thallus (cf. Fig. 17). It is perhaps significant that many of the northern boreal epiphytic *Bryoria* species may modify significantly otherwise low ambient winter temperatures by their low thallus albedo. There are several noteworthy examples of lichens which are restricted to a northerly aspect and which are black in colour, and these may similarly utilise snow-melt periods for maximal carbon assimilation.

Coxson and Kershaw (1983*a*) have documented the thallus tempera-

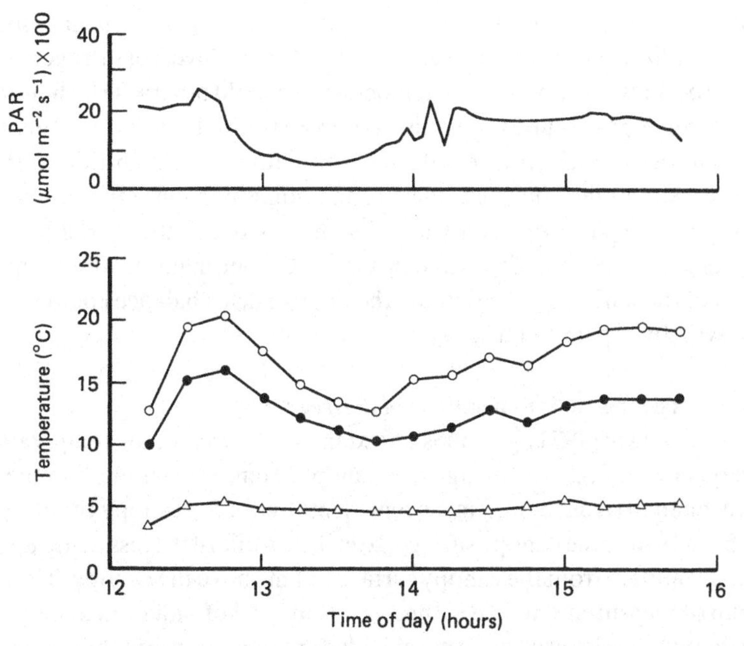

Fig. 17. The pattern of thallus temperature in *Rhizocarpon superficiale* during spring snow-melt conditions under good solar radiation. Time course of photosynthetically active radiation (PAR: —); dry thallus temperature in an open snow-melt pocket (○–○); wet thallus temperature in a partially open snow-melt pocket (●–●); air temperature at 2 m (△–△). (From Coxson and Kershaw 1983*a*.)

tures which actually prevail in *Rhizocarpon superficiale* under active snow-melt conditions at an altitude of 2160 m in Alberta. Under intermittent cloud and full radiation conditions the air temperature at 2 m is approximately 5 °C (Fig. 17), with wet thalli under 20 cm of undisturbed snow being approximately 2-3 deg C cooler. Within the sheltered confines of a snow-melt pocket, dry thallus temperatures rise to 20 °C under full radiation conditions, in contrast to saturated and freely evaporating thalli in a smaller and still-active pocket which are at a temperature of approximately 15 °C. Under these active melt conditions it is obvious that there is considerable potential for carbon gain, and in a similar way *Bryoria nitidula* with its black thallus can be expected to be metabolically active in snow-melt pockets on the ridge tops when ambient temperatures are well below 0 °C.

Similar interpretations may be equally relevant in either alpine or high arctic situations and the dark colours of *Umbilicaria rigida* or *Cetraria sepincola*, for example, may allow quite significant carbon gains during thaw periods. The black cephalodia of *Peltigera aphthosa* and some *Stereocaulon* species may also enable considerable levels of nitrogen fixation to occur when low ambient temperatures would otherwise be limiting. The inverse possibilities may also be ecologically important. Thus the *Lichina* species that form very distinct zones just above high-tide mark in exposed situations, may be achieving an optimum thallus temperature in relation to respiratory activity as a 'balance' between the strong cooling effects of evaporation of thallus moisture on the one hand and the warming effect of the dark thallus colour on the other. Such a balance could not be achieved by a pale thallus.

1.6 The thermal environment of the tree canopy

Geiger (1971) provides an extensive data set of air temperature divergence through a 24 m high tree canopy. Temperature measurements were taken (*a*) from 3 to 19 m, as representative of full canopy conditions; (*b*) from 19 m to the canopy surface, to give a profile of the less dense upper canopy; and (*c*) from the canopy surface to 1 m above the canopy. The last group of measurements are subject to some level of radiation error since the thermocouples used were not shielded and aspirated, but the pattern of events is clear-cut. The maximum temperature divergence between 3 and 19 m is only 3-4 deg C during full radiation conditions, falling to less than 1 deg C during the night and early morning. The upper canopy heats up rapidly in the morning and is then maintained at 2-3 deg C above air temperature until late afternoon. Denmead (1964) and Allen and Lemon (1976) report similar profiles in the canopy of a pine forest and for a 40 m rain forest in Costa Rica, respectively, where in the latter there is a 5 deg C divergence from the forest floor to the lower canopy at 35 m, with a further 5 deg C divergence in the upper canopy. From both case studies it is evident that there are only small daily temperature fluctuations throughout much of a canopy profile, coupled to quite restricted elevations of the canopy surface temperature.

A more detailed examination of the canopy sensible heat depends on the partitioning of individual leaf and branch components, an analysis which has rarely been attempted (Jarvis, James and Landsberg 1976). Rutter (1967) found that needle temperatures in a wet Scots pine canopy were 0.3-1.0 deg C below adjacent air temperatures and under dry conditions were only 0-0.2 deg C above air temperature. These small differences between needle and air temperature result from the small boundary-layer resistance of needles and the consequent rapid exchange of sensible heat which leads to a very close coupling of needle and air temperature. Jarvis *et al.* (1976) have calculated probable leaf-air temperature differences (Table 1) and under normal conditions of canopy ventilation, needle temperatures of more than 2 deg C above air temperatures are most unlikely, even with closed stomata and strong radiation on the needles. Conversely broad-leaved species may exhibit 10-15 deg C temperature elevations under full radiation conditions. Accordingly it seems probable that a pendulous fruticose lichen with its characteristic finely branched morphology and hence a large surface area to weight ratio and a resultant efficient rate of energy transfer, will have a thallus temperature which corresponds fairly closely to air temperature at the equivalent canopy height. The upper canopy, where the majority of arboreal lichens are usually found, is thus characterised by good levels of illumination but moderate thallus temperatures, in marked contrast to the microenvironment of lichens growing on exposed soil or rock surfaces under standard boundary-layer conditions (see Section 1.1).

The thermal environment of bark, however, is quite different. Geiger (1971) presents the temperatures of Sitka spruce bark in Denmark at 1.5 m from the ground and with contrasting aspects; temperature maxima were 38-42 °C on the southerly and south-westerly aspects, contrasting with 26 °C on the northerly aspect when the air temperature was 23 °C. Barkman (1958) presents very similar data. Inclined branches at right angles to the incident radiation achieve even higher temperatures. Presumably, however, with increased ventilation in the upper canopy, bark temperatures are reduced despite increased solar radiation levels.

Table 1. *Calculated leaf-air temperature differences (°C) for Scots pine needles with a range of stomatal and boundary-layer resistances*

	r_{aH} (s m^{-1})[a]			
r_s (s m^{-1})[b]	1	10	50	100
0	−6.83	−6.69	−6.07	−5.30
250	−0.03	−0.27	−0.76	−0.72
500	0.00	0.05	0.36	0.86
1000	0.00	0.23	1.11	2.14
2000	0.03	0.32	1.55	3.01
4000	0.04	0.36	1.80	3.53

Net radiation (R_n) = 500 W m^{-2}; saturation deficit (δe) = 12 mbar.

[a] r_{aH}, boundary-layer resistance to sensible heat transfer.

[b] r_s, stomatal resistance.

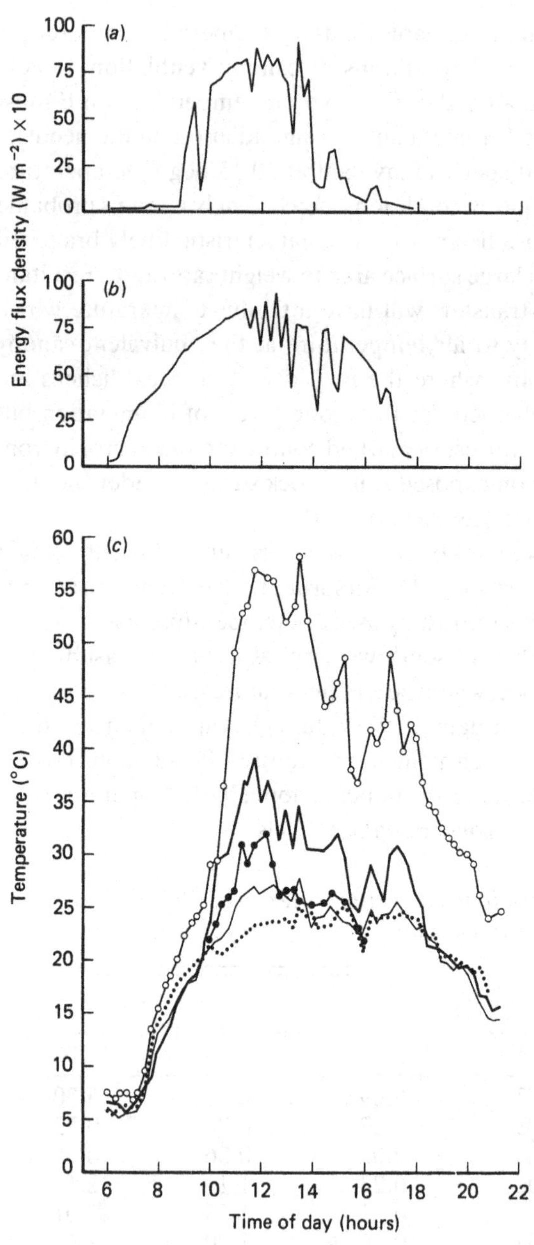

Fig. 18. The thermal environment of corticolous and pendulous tree lichens. (*a*) and (*b*) Solar radiation adjacent to the experimental thalli and above the canopy respectively. (*c*) The range of temperatures for air (····); the thalli of *Usnea fulvoreagens* (—); *Letharia vulpina* (●–●); *Hypogymnia physodes* in the canopy (**—**), and at ground level (○–○). (From Kershaw 1983.)

The thermal regime of corticolous fruticose lichens has been examined by Coxson, Webber and Kershaw (1984) and their findings fully confirm the theoretical expectations for both foliose and fruticose lichens. The temperature data for *Hypogymnia physodes* (at a height of 3 m and on a fallen branch on the ground) and *Usnea fulvoreagens* and *Letharia vulpina* (at 3 m) are given, together with the corresponding radiation data, in Fig. 18. Under good radiation conditions, despite some cumulus activity in the afternoon, air temperatures reached a maximum of 25 °C. *Hypogymnia* on a south facing branch at 3 m reached 38.5°C around noon despite good convective cooling. In marked contrast, replicate thalli of *Hypogymnia* on bark on a fallen branch simultaneously reached 57 °C, typical of boundary-layer conditions in the absence of convection. *Usnea fulvoreagens* at noon was 3 deg C above air temperature although this measurement was for the thicker podetia required for embedding microthermocouples. *Letharia* with its thicker branches and caespitose habit gave an intermediate result, reaching 32°C around noon. The data suggest strongly that the very fine branches of pendulous lichens will remain close to air temperatures during full solar radiation conditions when there is an average convectional level of cooling. Caespitose epiphytes, particularly if the thallus is at all massive, will achieve slightly higher temperatures and only foliose lichens will reach any substantial level of thallus heating (although its amplitude will be considerably reduced by the normal convective cooling which is usually present in a canopy).

1.7 Evaporative cooling in the boundary layer

It is evident from the discussions above that the thallus temperature of an air-dry lichen growing in the boundary layer is considerably higher than the equivalent thallus temperatures experienced by a fruticose species which projects well above this still-air zone. Indeed such differences are clearly documented in Fig. 12 and reflect the impedance of the boundary layer to the transfer of energy to the atmosphere as a sensible heat flux (H). In an analogous way the transfer of latent heat (LE) from a hydrated thallus is impeded by the boundary layer, resulting in diminished levels of evaporative cooling compared with the thallus cooling experienced in the zone above, where turbulent transfer processes operate.

Coxson and Kershaw (1983*a*) present data for the thallus temperature of *Rhizocarpon superficiale*, a crustaceous species of exposed rock surfaces, at 2500 m in the Alberta Rockies. During typical summer conditions dry thallus temperatures reach 35°C under full levels of radiation (Fig. 5). However, in characteristic periods of shower activity with alternating high

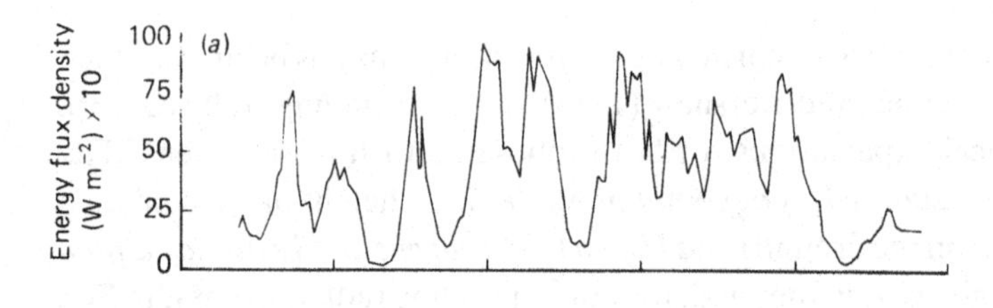

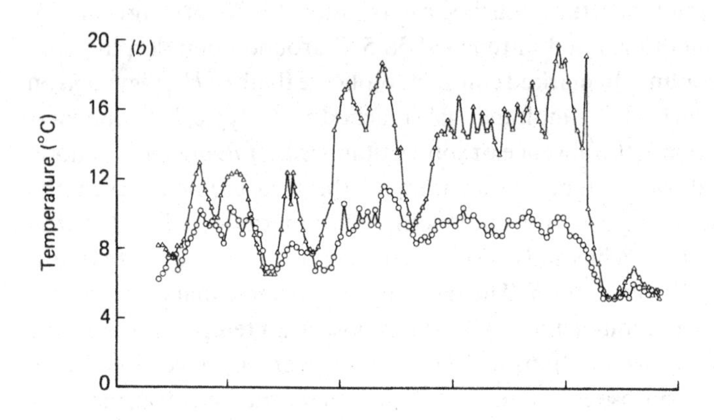

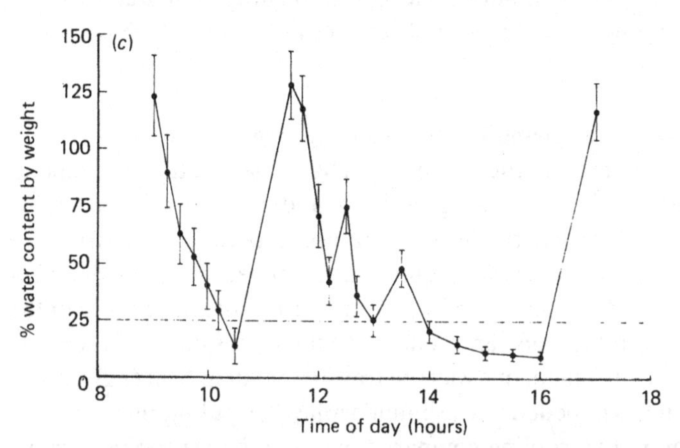

Fig. 19. The interaction of *Rhizocarpon* thallus temperature and hydration in the boundary layer in July. (*a*) Solar radiation at 2100 m over the experimental site; (*b*) the time course of changes in thallus temperature during shower activity throughout the day; (△) and air temperature at 2 m (○–○); (*c*) thallus hydration levels during alternating periods of rain and sunshine.

radiation conditions, temperatures of fully hydrated thalli frequently fall in the 20-25 °C range (Fig. 19) despite the continuously windy conditions. Under conditions of free evaporation outside the boundary layer, thallus cooling would be expected to be around 20 °C (cf. Fig. 15). As a direct result the net photosynthetic response matrix of this species shows a very broad response range to temperature. High rates of photosynthetic activity are generated at low temperatures during periods of snow-melt (cf. Fig. 17) but also at 25 °C in the summer during the alternating showers and sunny periods typical of this alpine environment.

1.8 The concept of the thermal operating environment

It thus becomes very evident that the thermal operating environment of a lichen is extremely specific and can moderate to a considerable extent the overall temperature pattern induced by latitude or elevation. It is not valid to make the simple assumption that it is cold in the Arctic, Antarctic or at altitude. The specific niche utilised by one species may indeed be cold but other spatially adjacent species may utilise a boundary-layer niche which is quite 'temperate' in character throughout the photosynthetically active year. As a result the thermal operating environment of each lichen has to be examined in detail during actual maximum thermal loads experienced under full solar radiation conditions, and during metabolically active periods whilst the thallus is fully or partially hydrated. The thermal operating environment of the dry thallus will define the thermal stress limits which the species must survive, whilst the thermal range during hydration will potentially define the required optimum for net photosynthesis. In the latter case high, medium or low temperature optima will thus correlate fully with the thallus operating environment and not necessarily with the overall meteorological daily average. In some instances, however, the latter average may happen to be closely correlated with the actual thallus environment. Alternative strategies are also available to the lichen, developed in response to the constraints of the physical environment. They may be expressed as, for example, colour or morphological adaptations, but they will certainly also be reflected physiologically in the pattern of the net photosynthetic response matrix. The development of these secondary physiological adaptations probably depends on the variation in the thallus operating environment occurring in a fairly predictable way on a seasonal basis. The stress responses, the net photosynthetic temperature optima and seasonal adaptations are discussed in Chapter 7.

2

The lichen environment: moisture

Walter (1931) has applied the term 'poikilohydric' to those groups of plants which do not have any active mechanism for enhancing water uptake or preventing rapid water loss. Poikilohydric plants such as lichens are dependent entirely on periods of rain, or dew, or high atmospheric humidity to achieve a satisfactory level of thallus hydration and a resultant vigorous rate of metabolic activity. This is in contrast to a homoiohydric plant which utilises active regulation of water uptake through its roots, transport through a vascular system, and active control of water loss by means of stomata. Accordingly lichens can be regarded as opportunistic, actively metabolising during periods of thallus hydration and maintaining very low levels of activity when 'air-dry'. It is thus not unexpected to find a close geographical correlation between the abundance of lichens and the distribution of rainfall. Similarly, habitats with strong continuous daily levels of solar radiation and hence high rates of evaporation tend to have a sparse lichen cover. The ecology of a lichen is controlled by a large number of parameters but certainly the water relations of a species are of central importance. The colloidal mechanism of water absorption and storage in a lichen thallus has been well documented by Blum (1973) and need not further concern us here. What are relevant are the pattern of liquid water availability, the rate of water uptake, the amount of water that can be retained in the thallus, the rates of evaporative losses and the specific levels of adapation in a lichen that lead to the optimisation of water usage in a particular environment.

2.1 Liquid water absorption

There is universal agreement that the absorption of liquid water by dry lichen thalli is a rapid process (Stocker 1927; Smyth 1934; Ried 1960*a*; Smith 1961, 1962; Blum 1973; Larson 1981). After the initial, rapid imbibition of liquid water, there is usually a more gradual absorption of water which may or may not continue for several hours. Unfortunately most of the experiments that have been used to study water gain (or loss) are, as Larson (1981) has pointed out, unrealistic ecologically. Most have relied on complete immersion to achieve thallus saturation, with the subsequent examination of evaporation limited to still-air conditions. This approach certainly obscures any interaction of thallus anatomy and morphology with wetting and drying rates. Such thallus modifications can play an extremely important role in the water relations of a lichen (see below). Further difficulties arise in defining exactly what constitutes 'full thallus saturation'. Blum (1973), in common with some goups of workers, removes the film of water on the lichen surface whereas others, including ourselves (eg. Kershaw 1977*b*,*c*; MacFarlane and Kershaw 1982), remove excess water only by shaking, so that a thin film of standing water is left covering the entire thallus surface. This latter approach may be more comparable with conditions in the field during and immediately after a period of rain, but there are very few field data to confirm this.

The results from immersion experiments show clearly that the majority of lichens achieve a high degree of thallus saturation within a 5 min period. However, some species require longer to hydrate fully and Blum (1973) reports that *Aspicilia esculenta* and *A. fruticulosa* require 17 and 26 min respectively. Similarly Larson (1981) demonstrates that *Stereocaulon saxatile* requires over 300 min. Blum (1973) observed after the initial rapid uptake of water that a few lichens can absorb considerable further quantities of water during more prolonged immersion periods, although he regards this as a state of over-saturation.

Larson (1981) adopted an alternative strategy which involved placing replicate lichen thalli in a wind tunnel with laminar wind flow and incorporating a nozzle which produced a stream of fine water droplets each approximately 10 μm in diameter. Full thallus saturation in most species examined was reached in 5-10 min (Fig. 20), somewhat longer than previously reported general values, and presumably reflecting the change in wetting technique. *Stereocaulon saxatile*, *Umbilicaria mammulata* and *U. vellea* required considerably longer to reach full hydration and Larson attributes the full range of values to the different surface area to weight ratios (*A*/*W*) observed in the species he examined. In addition to a number of lichen species, Larson examined several mosses and *Selaginella lepidophylla* (the 'resurrection plant'). The entire range of material shows a close correlation between the time required to achieve maximum thallus water content and the *A*/*W* ratio of each species replicate. This simple relationship suggests that for both mosses and lichens, the *A*/*W* ratio can be used to predict the amount of time needed for the plant to become saturated. In particular it shows that any plant exhibiting a very large *A*/*W* ratio, absorbs water very rapidly. Thus the filamentous lichens common in boreal and arctic regions

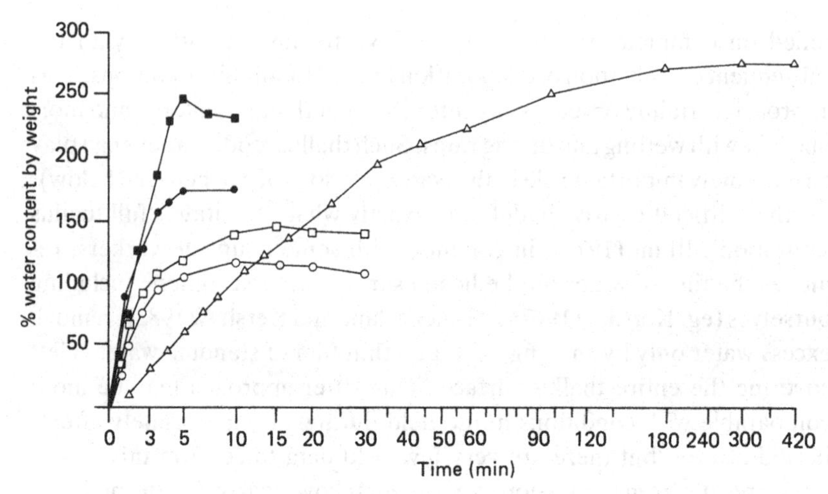

Fig. 20. Rate of uptake of water, during treatment with a mist of water droplets, in *Cladonia stellaris* (■–■), *Ramalina reticulata* (●–●), *Cladonia chlorophaea* (□–□), *Masonhalea richardsonii* (○–○) and *Stereocaulon saxatile* (△–△). (From Larson 1981.)

(e.g. *Usnea*, *Alectoria* and *Bryoria* species) which have been shown previously to dry out at an exceptionally rapid rate (Larson and Kershaw 1976; Larson 1979; and see Section 2.2 below) can also be expected to pick up water with equal speed. Conversely those plants which have a low *A*/*W* ratio reach saturation more slowly.

In addition to these more realistic experimental conditions, Larson (1981) included some imaginative morphological manipulations to examine the potential role, if any, of rhizinae in water uptake, by contrasting rates of uptake in replicate thalli before and after shaving the undersurface of the thallus. Similarly the differential rates of water uptake through the upper and lower surfaces of a thallus were examined before and after sealing the surfaces with a silicone preparation. In *Umbilicaria vellea* and *U. mammulata*, removal of the rhizinae from the lower surface not only markedly reduces the rate of water uptake but also results in a decreased quantity of water held at thallus saturation. In contrast, however, removal of the rhizinae from the lower surface of *Peltigera canina* did not significantly affect the rate of water uptake; this is contrary to the comments of Jahns (1973), who suggests a more active role for rhizinae during the absorption of water by this species. Silicone treatment of the upper and lower surfaces of *Umbilicaria vellea*, *U. mammulata* and *U. muhlenbergii* results in significantly different rates and amounts of water uptake, with the lower surface, and presumably the medullary hyphae, playing the dominant role. This is in general agreement with many earlier reports in the literature (see Blum 1973, for a full discussion). In *U. papulosa* and *U. deusta* the effects of silicone treatment are virtually identical for both surfaces, simply reducing the amount of water held at full thallus saturation without significantly affecting the rate of uptake. The actual change in water-holding capacity with silicone treatment is presumably a physical phenomenon, produced perhaps as a result of the penetration of the sealant into the medulla blocking some of the capillary spaces which are normally available for water storage. These changes may not therefore be significant.

The relationship between the rate of water uptake and the *A/W* ratio is of considerable interest and may be of fundamental significance, since it is also of central importance to the rate of evaporation of water from lichen thalli (see Section 2.2 below). It may be necessary to correct some *A/W* values when a more complete data set is available, since a hollow lichen will have a very different *A/W* value from another lichen of the same surface area and the true fundamental ratio may involve volume rather than weight alone. However, at present, this approach to the water relations of lichens represents a considerable step forward. It emphasises that although there is no active control of water uptake or loss, as one has in a higher plant, nevertheless a wide range of strategies for the control of effective water relations within a particular microenvironment has been evolved in many lichens. Just as the ecologist is familiar with the classical adaptations to low levels of water availability in xerophytic higher plants, so is it evident that the morphology of a lichen can represent a comparable level of adaptation to the xeric or mesic nature of the microhabitat. This is equally true for the rate of water uptake and the rate of water loss from lichens.

The measurement of the absolute thallus water capacity of a lichen presents severe problems in two main areas: that of definition of the term itself and that of selection of the unit of measurement to be used. Blum (1973) defines saturation in terms of the amount of water held internally, whereas in the field situation a variable amount of surface water is often present. This would cause considerably higher saturation values, a fact which should always be borne in mind when comparisons of data from different authors are made. The values given by Blum (1973) often correspond with the optimum level of thallus hydration generating the maximum rate of net photosynthesis (cf. Chapter 5). Expressions of absolute levels of thallus saturation are usually referenced to the oven (80 °C) dry weight of the thallus, but sometimes less stressful drying is required and a calcium chloride or even an air-dry reference point is acceptable. It is also some-

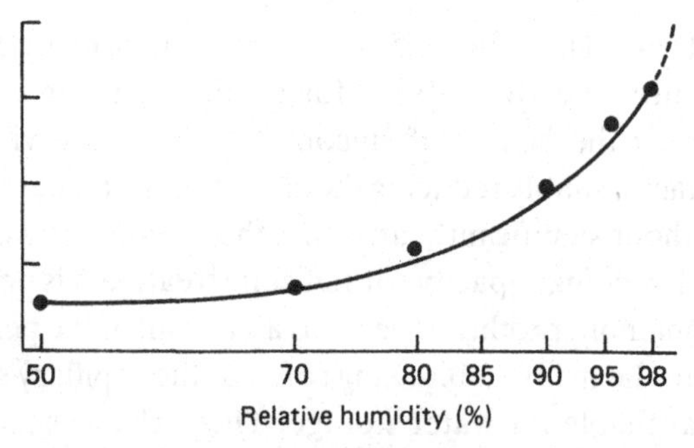

Fig. 21. The equilibrium between the level of thallus hydration and atmospheric humidity at 30 °C. (From Blum 1973.)

times useful to express levels of thallus hydration on a relative scale where 'full' thallus hydration is 100% relative saturation and oven-dry weight is 0%. Such an approach can usefully reduce a wide scatter of absolute values to single reference frame, but equally it can mask essential ecological information.

2.2 Water vapour absorption

There have been numerous studies on the rate of water uptake from both saturated and unsaturated atmospheres. For example Stocker (1927) found that a minimum of 6 days was required for an equilibrium to be reached between a dry lichen and a saturated atmosphere, in contrast to a maximum of 3 days reported by Butin (1954). Blum (1973) suggests this apparent discrepancy may be due to actual experimental levels of relative humidity being lower than the supposed 100%, since there is a close relationship between the amount of water vapour in the atmosphere and the final level of equilibrium water content of the lichen (Fig. 21). The higher the relative humidity the higher the equilibrium point between thallus water content and atmospheric moisture, but the longer it takes to reach the equilibrium point. However, there is universal agreement that lichens possess a remarkable ability to absorb water vapour from both saturated or partially saturated air. Since in the vapour phase there is no capillary-drawn liquid water involved, the final point of equilibrium is much lower than that attained by immersion of the thallus in water. Again, there is some discrepancy in the actual values given by different authors. Butin (1954), for example, suggests maximum values in the range 50-75% of the corresponding maximum level of hydration by water in the liquid phase. Ried (1960*b*), however, questions whether these high values might be due to condensation of liquid water on the thallus and suggests that a more realistic range is 30-50% of the liquid phase maximum level of thallus hydration. Blum (1973) gives values of 14% for *Collema flaccidum*, and 22% for *Aspicilia esculenta*, but shows that the majority of fruticose and foliose lichens fall within the 40-50% range. *Dermatocarpon miniatum* appears to be the most efficient at absorbing water vapour, achieving a value of thallus hydration which is 68% of the maximum level reached by immersion in water.

Unfortunately, however, all of the experimental data on both rates and quantities of water vapour uptake in lichens have been generated under very unrealistic environmental conditions and the findings cannot always be simply transferred to field conditions. Since the vapour pressure in air is strongly controlled by temperature (see below) it becomes extremely difficult to predict potential levels of thallus hydration and regrettably very few data involving direct measurement is available.

Heatwole (1966), working in open oak-pine-aspen forest in Michigan, examined the uptake of water vapour by a number of *Cladonia* species over a 2 day period (Fig. 22). The relative humidity of the air reaches 80-90% during the night and induces a level of thallus hydration of 35-40% in most of the species examined; the exception is *Cladonia uncialis*, which achieves a value of 50% water content by weight. This seems in good agreement with the laboratory-based values derived by Ried (1960*b*) and Blum (1973). The level of thallus hydration falls very steeply following sunrise (Fig. 22). In most habitats, under normal field conditions, the air is rarely fully saturated with water vapour for more than brief periods and it seems very unlikely that lichens in most environments would place any reliance on water in the vapour phase for their metabolic requirements. However, there are some environmental situations where high humidity and fog occur both regularly and frequently and furthermore represent the only available source of water. The ability of lichens to utilise this source of water effectively results in their abundance in areas which otherwise are extreme desert and largely devoid of vegetation. These deserts are all coastal with temperatures moderated by the proximity of the sea, but with extensive fogs induced by off-shore upwelling of cold water sometimes enhanced by adiabatic cooling. Large coastal deserts occur in three areas of the world: the Peruvian and Chilean Atacama Desert, the coastal Sonoran Desert in Baja California, and the Namib Desert in South-west Africa (Lange 1969; Lange, Schulze and Koch 1970*a*,*b*).

The formation of fog is a physical process and the effects of cooling or heating a parcel of partially saturated air are of considerable relevance to lichen water relations. When water evaporates into a fixed volume of air,

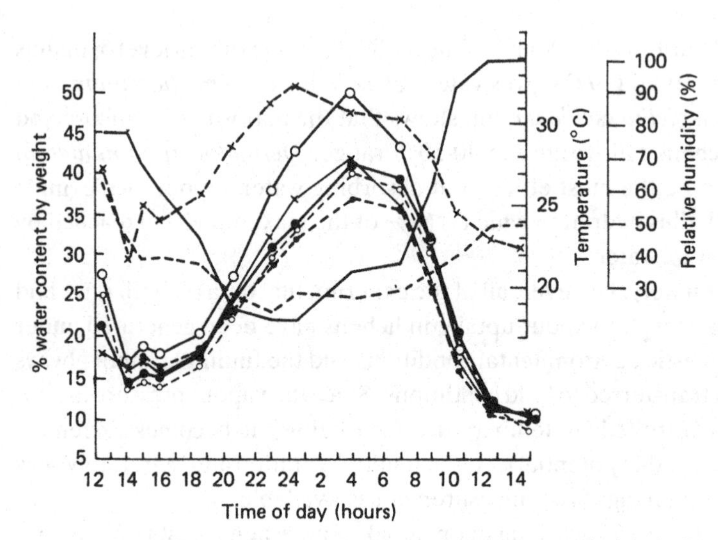

Fig. 22. Change in moisture content of *Cladonia uncialis* (○–○), *C. rangiferina* (○----○), *C. mitis* (●–●), *C. sylvatica* (●----●) and *C. stellaris* (×–×) in open oak-pine-aspen forest. Temperature at the lichen surface (—), temperature under the lichen mat (----) and % relative humidity (×----×) are also shown. (From Heatwole, 1966.)

the concentration of water vapour increases until an equilibrium is reached where the number of water molecules leaving the water surface is equal to the number of water molecules returning. This equilibrium position is dependent on the temperature of the air and is shown as the saturation vapour pressure curve in Fig. 23. The parcel of air represented at point *X* has a temperature T_1 (approximately 28°C) and a vapour pressure of 20 mbar. At this temperature it is unsaturated since the saturation vapour pressure at T_1 is *c*. 37 mbar. If the parcel of air is cooled below T_2 it can no longer hold the same amount of water vapour and the vapour will condense out as liquid water droplets. T_2 is the dew-point temperature. The vapour pressure of the parcel of air at T_1 can be represented either as an arithmetic difference from the saturation vapour pressure at that temperature (known as the saturation deficit) or as a percentage of the saturation vapour pressure (the relative humidity).

These terms give a measure of the relative drying powers of the parcel of air; this is important in Section 2.3 below. The formation of fog banks off the Namib Desert, for example, is simply due to the juxtaposition of the cold Benguela Current cooling the moisture-laden off-shore air below the dew-point temperature. Once a fog bank has formed, the top of the bank, and not the ground itself, becomes the active radiation surface, since water

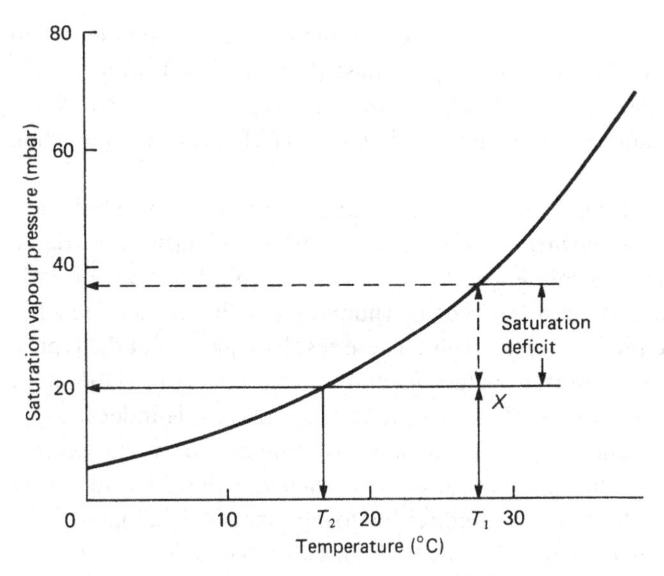

Fig. 23. The relationship between saturation vapour pressure and temperature. See text for details.

droplets very effectively emit long-wave radiation. This efficient energy loss further cools the fog bank and it can become progressively thicker, finally condensing as a downward moisture flux as the active surface continues to cool. Fog formation is aided by light winds which enhance the loss of sensible heat from the fog surface, but above a critical limit increased winds also develop turbulent mixing of the fog with drier air which results in the disappearance of the fog. The Namib fog zone is thus correlated with light winds. Meigs (1966) reports 285 days during the year with fog or dew in this area. With the annual rainfall only in the 2-5 cm range, surface condensation of this moisture largely reflects the only reliable source of water. As a result vascular plants are rare but windward sides of even small pebbles have well-developed lichen growth. This growth, dominated by *Caloplaca*, shows a number of morphological adaptations, with subfoliose forms directly intercepting fog water droplets and crustaceous forms at soil level utilising the runoff from fog condensation on the rock above (Vogel 1955; Rundel 1978*a*). Similar considerations apply to the coastal Sonoran Desert of Baja California as well as to the Atacama Deserts of South America (Rundel, Bowler and Mulroy 1972; Rundel 1978*a*; Nash *et al.* 1979).

The variation of morphology in the Namib Desert *Caloplaca* species is of considerable interest and raises the possibility that the rate of water vapour uptake in lichens may be controlled by the same surface area to weight ratio as in the uptake of water in the liquid phase (Larson, 1981). Blum (1973) similarly suggests that the absorption of water vapour depends to some extent on specific morphological and anatomical features of the lichen concerned.

Rundel (1974) has examined the water relations and morphological variation in *Ramalina menziesii* in California, where it is largely restricted to the coastal areas influenced by fogs. Under typical periods of early morning fog or under conditions of high humidity, thallus moisture reaches 30-35% above the oven-dry weight. These results suggest that the typical range of thallus saturation values given by Blum (1973) for laboratory experimental conditions of saturation vapour pressure is indeed duplicated in the field under equivalent conditions. Rundel also emphasises that the two main morphological variations of *R. menziesii* that have numerous perforations are particularly suited either to the zone of dense fogs or to the adjacent zone where there is only high humidity. However, the data of Larson (1983*c*) cast doubt on this simple interpretation. Rundel (1978*b*) similarly relates the range of morphological variation in the *Ramalina usnea* complex to a range of vapour pressure conditions which vary from upland habitats with frequent fogs to tropical coastal areas with consistently high humidities. The available evidence at the moment does suggest that the rate of water vapour uptake may indeed correlate with the thallus surface area to weight ratio. Furthermore, both interspecific and intraspecific morphological modifications may be ecologically advantageous in environments with contrasting humidities.

The effectiveness of water vapour as a source of moisture to support substantial rates of net photosynthesis in *Ramalina maciformis* has been well documented by Lange (1969) under laboratory conditions, and by Lange *et al.* (1970*b*). The net photosynthetic and respiratory response curves to changing thallus hydration in *R. maciformis* (Fig. 24) show a marked level of adaptation to a xeric habitat. Both net photosynthetic and respiratory rates increase very rapidly from thallus dryness up to *c.* 40% thallus moisture by weight, and above 60% thallus water content no further significant increases occur. At 30% water content by weight, *Ramalina* has achieved 50% of its full net photosynthetic capacity. Under laboratory conditions, *Ramalina* in equilibrium with an atmosphere of *c.* 100% relative humidity reaches 35% water content by weight and can achieve 90% of its maximal photosynthetic capacity achieved under conditions of full saturation with liquid water (Fig. 25). Under field conditions (Lange *et al.* 1970*a*) a very similar pattern is evident (Fig. 26). Before

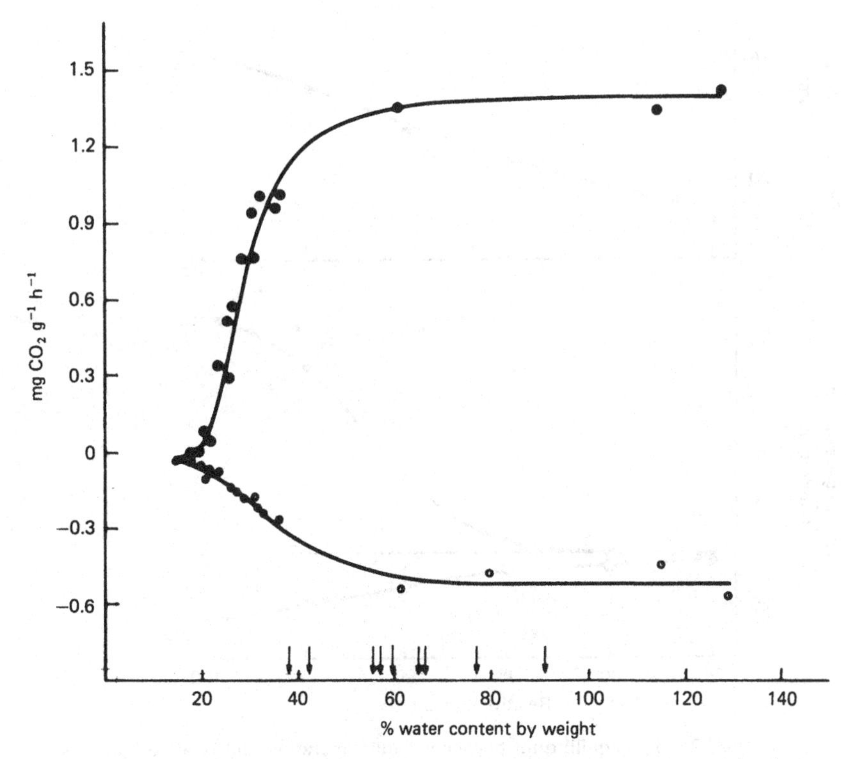

Fig. 24. Rates of net photosynthesis (•) and respiration (○) (mg CO_2 h^{-1} g^{-1}) in *Ramalina maciformis* in response to the degree of thallus hydration. Arrows indicate water contents recorded during periods of dew formation in the field. (From Lange 1969.)

sunrise the lichen thallus reaches 31% water content by weight under atmospheric conditions which preclude dew formation. By 7.00 a.m., with a thallus temperature of *c*. 23 °C and a light intensity of *c*. 30 klux (*c*. 300 μmol m^{-2} s^{-1}), *Ramalina* achieves a net photosynthetic rate of *c*. 0.3 mg CO_2 g^{-1} h^{-1}. By 8.30 a.m. thallus moisture has fallen to 13% and supports only a residual level of net photosynthetic activity. Furthermore this marked photosynthetic ability to utilise atmospheric humidity is present even after 16 weeks of drought stress, with the thallus maintained at 1% water content by weight (Fig. 27).

Under natural field conditions there is usually pronounced radiational cooling in the desert at night to below the atmospheric dew-point, resulting in the frequent formation of dew. Lange *et al.* (1970*a*) report the occurrence of dew in the Negev Desert on an average of 198 nights per year. The amount of dew usually corresponds to over one third of the annual precipitation in the form of rain. In 1962, a year of low rainfall, the amount of

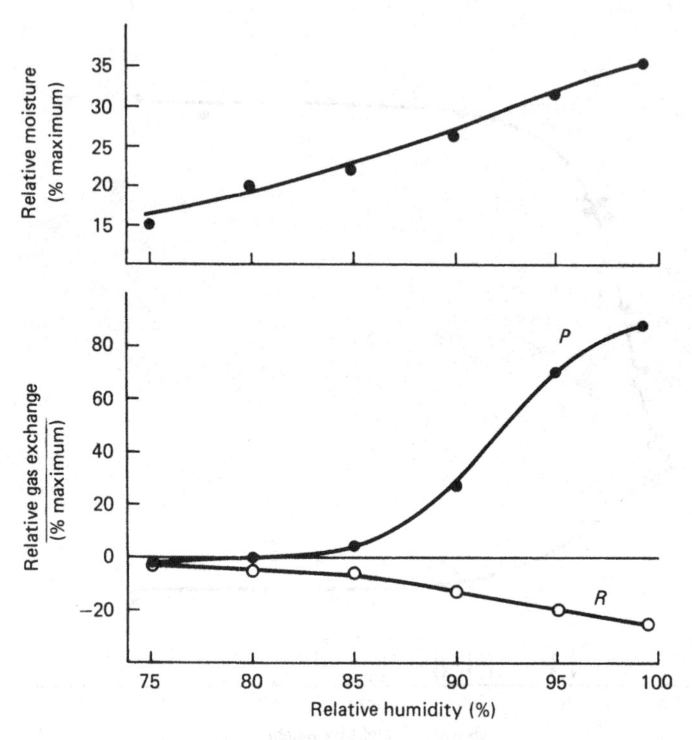

Fig. 25. The equilibrium between thallus hydration and relative humidity in *Ramalina maciformis* under experimental conditions, and the corresponding rates of relative gas exchange generated by these conditions. (From Lange *et al.* 1970*a*.)

dew actually exceeded the total rainfall. Under conditions of dew formation, thallus water contents of *R. maciformis* often reach 60%, with concurrent high levels of net photosynthesis for a short period in early morning (Fig. 28). On northerly exposures with a contrasting microclimate, the active net photosynthetic period is extended for much longer than on the easterly exposures (Fig. 29). Although both thallus temperature and light levels are lower, resulting in decreased net photosynthetic rates, evaporation rates are also much lower. As a result, the integrated yield from a northerly or easterly exposure is almost double that from a southerly exposure. Lange *et al.* (1970*a*) suggest that this interaction between the length of the photosynthetic active period and the attendant thallus temperature and illumination level probably accounts for the abundance of lichens on northerly aspects in the Negev Desert. This is an important finding and a concept which has a strong analogy with the ability of a lichen to maintain nitrogenase activity throughout the night when again thallus evaporation rates are extremely low (see Section 2.3 below).

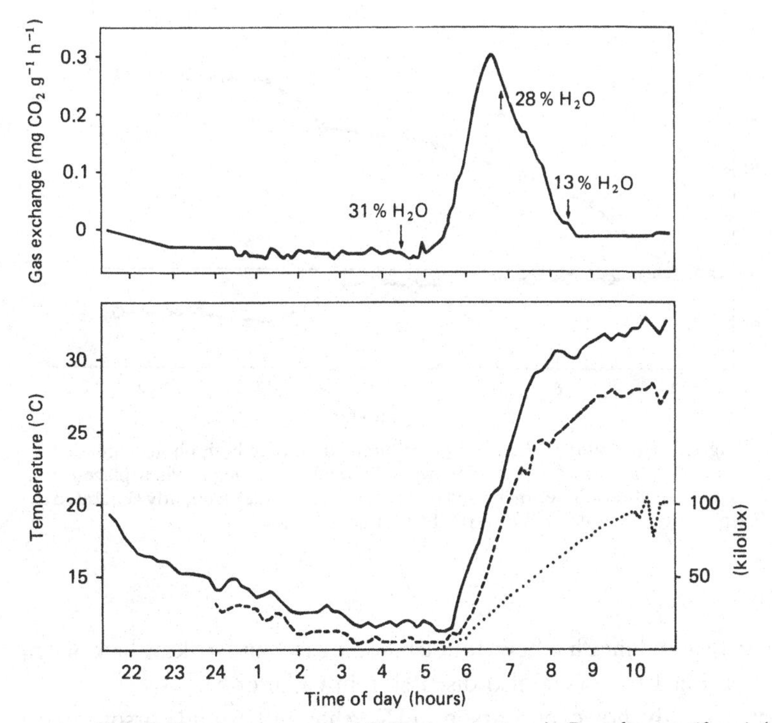

Fig. 26. The time sequence of thallus temperature in *Ramalina maciformis* (—), air temperature (----) and illumination (····), with the concurrent rates of gas exchange under field conditions. (From Lange *et al.* 1970*a*.)

2.3 Water loss: general comments

Because of the absence of stomata in lichens, there has been widespread acceptance of the idea that the evaporative loss of water from the thallus is a simple physical process and that the rate of evaporation is related in a simplistic way to the immediate physical environment. The general summaries by Smith (1962), Blum (1973) and Farrar (1973) all express this viewpoint. There has been no recognition that the adaptations of higher plants to a xerophytic habitat have any parallel whatsoever in lichen morphology and ecological distribution. However, it has also been widely recognised that although most lichens can tolerate extremely low levels of thallus hydration without damage, they do indeed have a wide range of specific ecological preferences. A number of attempts have been made to establish generalised groupings of species from 'xeric' or 'mesic' habitats, for example. Blum (1973), though, has questioned such approaches and pointed out discrepancies between absolute rates of evaporation within such arbitrary groups. Similarly the thickness of the cortex has also been related to the xeric or mesic nature of the habitat (Ertl 1951);

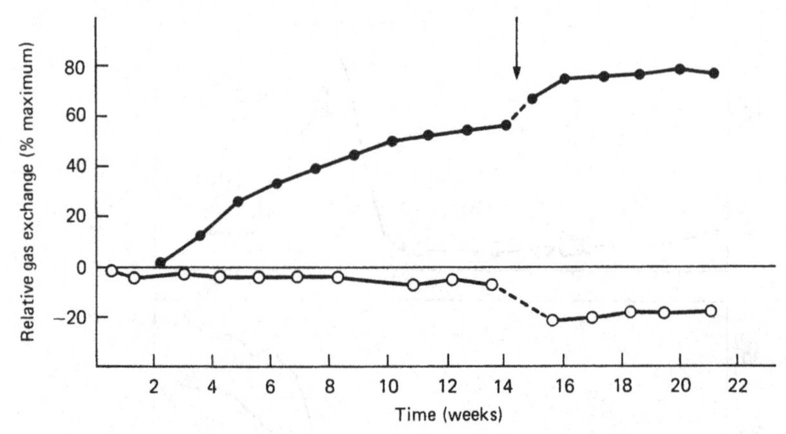

Fig. 27. The ability of *Ramalina maciformis* to initiate both photosynthetic (●) and respiratory (○) activity following 16 weeks of drought when placed in a saturated atmosphere. Increased metabolic activity is subsequently generated by spraying with water (arrowed). (From Lange 1969.)

close correspondence has been found in some cases and striking lack of it in others (Blum 1973) (see also discussion in Chapter 5).

More recently, however, Larson and Kershaw (1976) and Larson (1979) have shown, using controlled wind-tunnel experiments under diffuse and direct radiation conditions, that there are marked, although passive, morphological adaptations that are important in the water relations of lichens and are of considerable ecological significance. The experiments also show that the relationship of evaporation rate to the physical environment is anything but simplistic.

2.4 Physical aspects of evaporation

A detailed consideration of the physical process of evaporation is outside the scope of this book and attention is drawn to the particularly excellent treatments of this subject given by both Monteith (1964, 1973) and Oke (1978).

At an ecological level, the rate of evaporation from a wet lichen surface depends on four factors: an external source of energy, usually solar radiation; the ambient temperature and the temperature at the lichen surface; the atmospheric humidity both at the lichen surface and in the air; and finally the wind speed. The effects of these four environmental parameters are inextricably linked.

Monteith (1964) expresses evaporation (E) as:

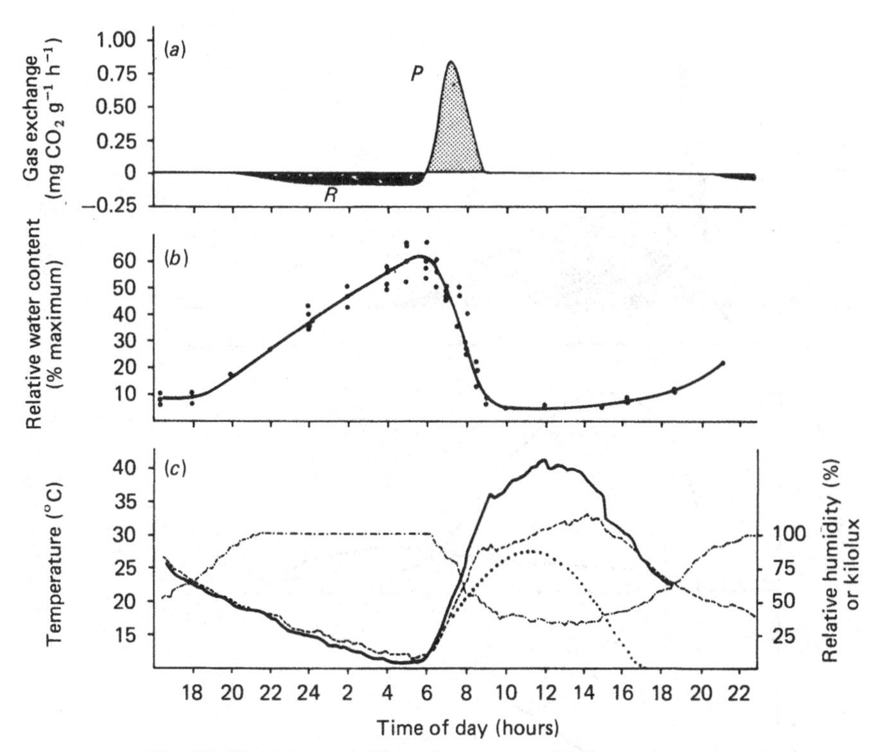

Fig. 28. The balance achieved between respiration and net photosynthesis in *Ramalina maciformis* after partial thallus hydration during the night by dew. (*a*) Gas exchange; (*b*) thallus relative water content; (*c*) concurrent thallus (—) and air temperatures (----), level of illumination (····) and atmospheric humidity (—·—). (From Lange *et al.* 1970*a*.)

$$E = \frac{\Delta H + \varrho c[e_s(T) - e]/r_a}{\lambda(\Delta + \gamma^*)} \qquad (4)$$

which is the basic equation derived originally by Penman (1963) expressing evaporation from a wet surface under normal conditions of solar radiation. H represents the effect of an input of energy (sensible heat) and is the term which largely drives and controls the rate of evaporation. In the absence of an input of energy, evaporation is considerably reduced and is then controlled by the remaining parameters, particularly $e_s(T)$, the saturation vapour pressure at the lichen surface which is at temperature T, and where e is the vapour pressure of the air. The difference is thus the vapour pressure gradient from the lichen surface to the air, and the greater the gradient the more rapidly water vapour will move from the lichen surface to the air. The movement of water vapour from the lichen is impeded by the

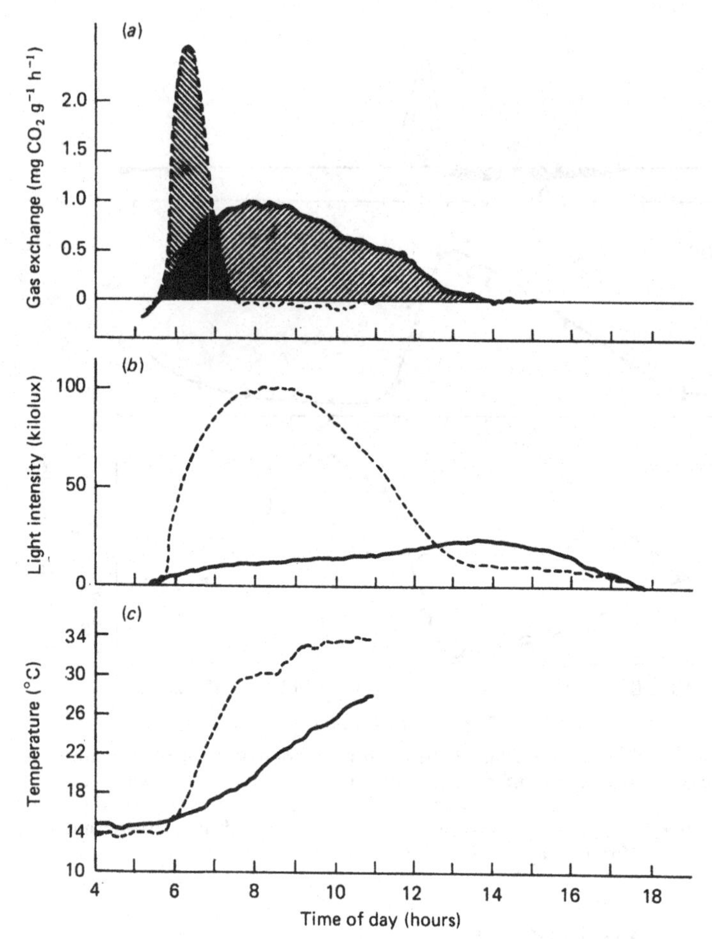

Fig. 29. (*a*) Gas exchange in *Ramalina maciformis* under field conditions, following thallus hydration by dew. Values for thalli growing on a northerly aspect (—) and on an easterly aspect (----) are shown. Assimilation yields are 4.63 and 2.42 mg CO_2 g^{-1} dry weight d^{-1} respectively. The corresponding levels of illumination and thallus temperature for northerly (—) and easterly (----) aspects are given in (*b*) and (*c*) respectively. (From Lange *et al.* 1970*a*.)

boundary layer (see Section 2.1 above) and by an internal resistance in the lichen thallus which will vary with the anatomy of different lichens. This combined resistance term (r_a) is further amplified in those species which grow as clumped podetia or as a complete mat, thus introducing a further resistance term, the canopy resistance. The boundary-layer resistance is directly proportional to its thickness which in turn is very dependent upon wind speed. Accordingly $1/r_a$ represents the control that wind speed exerts in the evaporative process by lowering the boundary-layer thickness

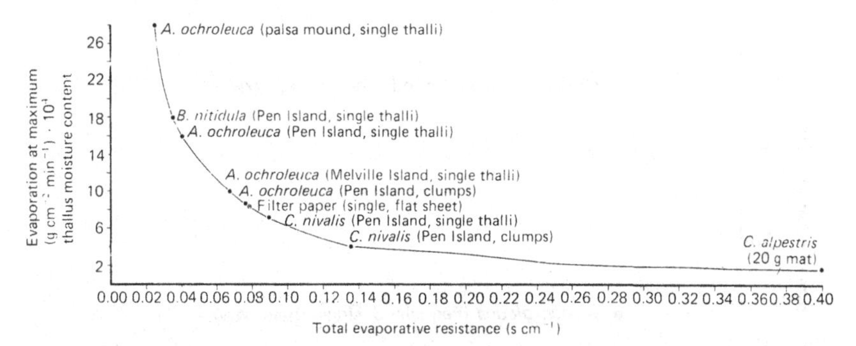

Fig. 30. The inter-relation between evaporation rate and total resistance to evaporation for a range of lichen morphologies, contrasted with the results for a single flat sheet of filter paper. (From Larson and Kershaw 1976.)

and thus increasing the rate of exchange of water vapour from the lichen surface to the air. Wind can only play a significant ecological role under natural field conditions, where there is usually a vapour pressure gradient. Under saturated conditions the continuation of evaporation is totally dependent on a further input of energy. Similarly a very large increase in wind speed reduces the boundary layer to such an extent that sensible heat loss (i.e. surface *cooling*) increases, and this can offset the advantage of rapid latent heat transfer. The remainder of the terms in Equation 4 can be regarded as weighting factors, determining the partitioning of radiant energy between evaporation and convection (Δ and γ^*), or physical terms which maintain the integrity of the units (λ = latent heat of vaporisation of water, c = specific heat of air, ϱ = density of air including water vapour).

The complexity of the evaporative process has largely been ignored in most of the work on the water relations of lichens, but clearly the morphology of a lichen can markedly affect r_a. Similarly lichen colour modifies thallus temperature (Section 1.5) and the resultant saturation vapour pressure ($e_s\,T$) at the lichen surface. In addition, all of the environmental parameters controlling evaporation vary enormously in the natural environment even at a microscale and offer a wide range of potential strategies for successful lichen growth.

2.5 Morphological control of evaporation

The control of evaporation from lichen thalli can be conveniently expressed in terms of the total evaporative resistance which the lichen offers to the transfer of latent heat. Larson and Kershaw (1976) document a number of evaporative resistances (Fig. 30) which are the result of several independent morphological features. Within a species, evaporative resist-

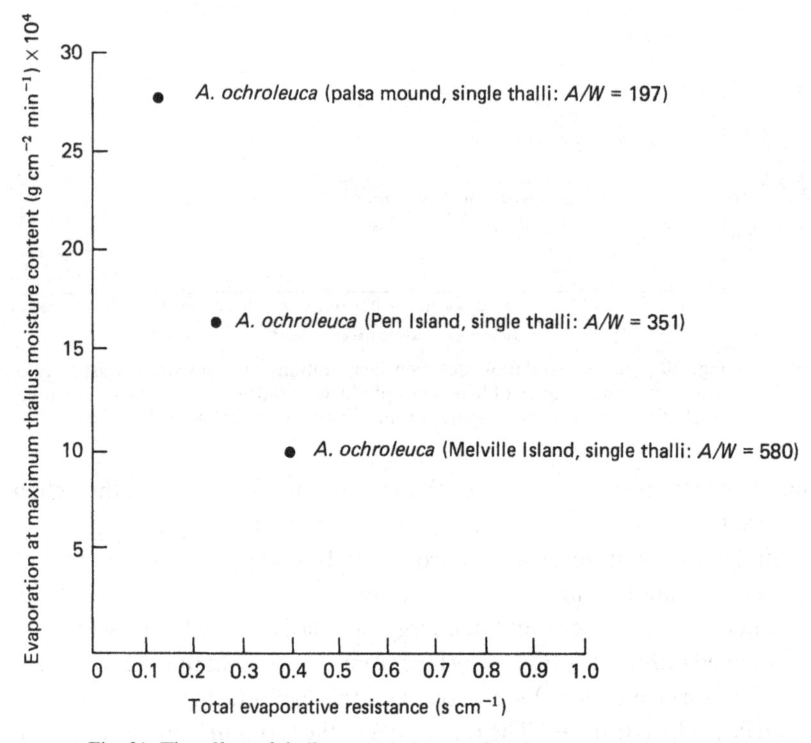

Fig. 31. The effect of thallus morphology and surface area to weight ratio (A/W) on evaporation of water from fully saturated thalli of *Alectoria ochroleuca*. (From Larson and Kershaw 1976.)

ance can be dramatically altered simply by changes in the surface area to weight ratio (A/W), which parallels the relationship for water uptake discussed above (Section 2.1). This is shown very clearly by *Alectoria ochroleuca* (Fig. 31) where three contrasting phenotypes have very different A/W ratios resulting in slight changes in r_a which produce large differences in evaporation rate. A more dramatic increase in evaporative resistance and hence a reduction of evaporative losses is also very evident in clumped rather than isolated podetia. Both *A. ochroleuca* from Pen Island and *Cetraria nivalis* (Fig. 30) increase their resistance term and reduce their evaporative losses when arranged in clumps. Effectively, there is an additional resistance term included in r_a – the canopy resistance – which in the case of mats of *Cladonia stellaris* is large, and results in extremely low rates of evaporation (Fig. 30).

The effect of the A/W ratio is also evident in filter paper controls (Larson and Kershaw 1976). There are differences in the time required to reach air-dryness by a flat single piece of filter paper, a crumpled sheet, and a flat

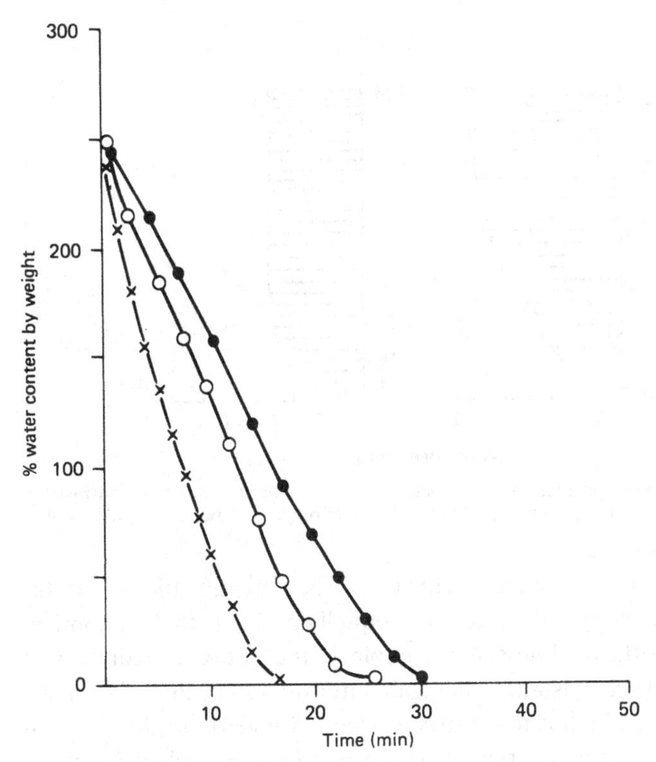

Fig. 32. Comparative drying curves for a single flat sheet of filter paper (×–×), a double sheet of filter paper (○–○) and a crumpled single sheet (●–●). (From Larson and Kershaw 1976.)

double sheet (Fig. 32). Doubling the thickness of the filter paper reduces the *A*/*W* ratio by a factor of 2 and virtually doubles the time to reach dryness. The double sheet holds effectively twice as much water which can only evaporate from an almost identical surface area and accordingly the process takes twice as long. More significantly perhaps, crumpling a single sheet also affects the time required to reach dryness even though the *A*/*W* ratio is unchanged. This is directly related to the differential exposures of the sheet to wind, resulting in lower evaporative rates from some segments of the paper. Such an adaptive strategy may well be seen in the convolutions and rugosities of some otherwise 'flat' lichen thalli.

It is evident from these results that there are at least three mechanisms by which a lichen may decrease or increase evaporative water loss from its surface: it may grow into a more finely branched or highly dichotomised structure to increase its *A*/*W* ratio; it may change its actual surface characteristics and water-holding capacity; or it may adopt a different growth

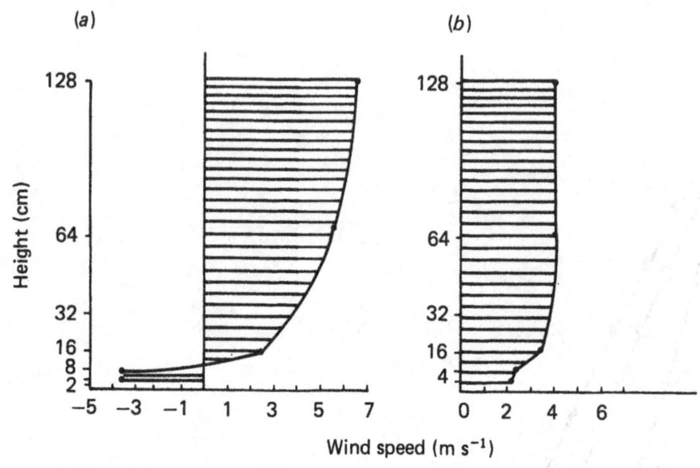

Fig. 33. Wind profiles over a raised beach in northern Ontario in contrasting ridge-top (*a*) and lower-slope (*b*) locations. (From Kershaw and Larson 1974.)

habit (clumped or mat-forming) in the field. These modifications all influence the air or canopy resistances. It is equally probable that anatomical modifications to the thickness, for example, of the cortex or medulla, will change the internal resistance and again alter the total value of r_a. Thus *Bryoria nitidula*, which often grows as isolated podetia in the field (in contrast to the palsa population of *Alectoria ochroleuca*, which is always densely clumped), exhibits by far the highest evaporative rate and the lowest evaporative resistance of any lichen examined so far. As a result it will rapidly lose water after periods of rain or mist, and will remain for extended periods of time at or near its optimal level of thallus hydration for maximum net photosynthetic activity (Fig. 30).

Larson (1979) presents a complex picture of the interaction between wind speed and radiation in the evaporative process from lichens. His data for *Umbilicaria* species show clearly that the increased evaporative flux induced by higher levels of radiation is substantially negated by higher wind speeds. This follows from the arguments presented by Monteith (1964) discussed above, where high wind speed reduces the boundary-layer thickness, enhances sensible heat exchange, lowers thallus temperature and thus *reduces* evaporation. However, the response to radiation under still-air conditions or at low wind speeds is extremely variable throughout the range of species examined. Larson concludes that species with low evaporative resistance dissipate increased radiant energy through increased latent heat transfer, whilst there is little increase in evaporation in species with a high evaporative resistance. Unfortunately a full energy

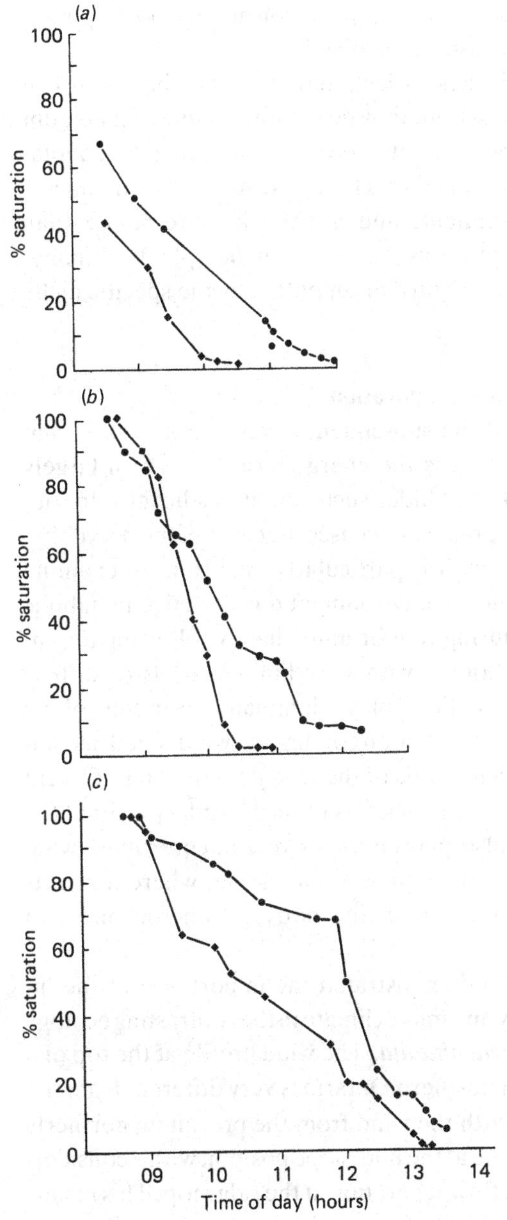

Fig. 34. Relative drying curves for *Cladonia stellaris* (●–●) and *Bryoria nitidula* (◆–◆) growing on the lower slope and ridge top respectively, of a raised beach in northern Ontario. (*a*) 30 July; (*b*) 31 July; (*c*) 4 August. (From Kershaw and Larson 1974.)

budget was not attempted and such an interpretation may be too simplistic. Thallus colour certainly may also be involved.

It is evident that although lichens lack active, metabolic control of evaporation through stomatal action, they do exhibit a significant amount of control through both intra- and interspecific morphological adaptations. These adaptations also correlate closely with the specific microclimate of the lichen environment, and are much more subtle than the obvious 'xerophytic adaptations' found in higher plants. Finally, morphological adaptations can be further amplified by the specific niche environment of a species.

2.6 Topographical control of evaporation

From Equation 4 and the subsequent discussion it is clear that under full radiation conditions it is the energy term (H) which largely controls the rate of evaporation. Under such conditions lichens dry out quickly and metabolic activity practically ceases. Because of the poikilohydric nature of lichens, however, it is particularly important to examine evaporation rates within the lichen environment during active metabolic periods. These periods are during rain or immediately following a rainstorm and under these conditions, with very limited levels of diffuse radiation, the other terms in Equation 4 play a dominant role in controlling evaporation rates. Under diffuse radiation conditions, wind speed becomes particularly important both in terms of the turbulent transfer of latent energy, and in the reduction of the thickness of the boundary layer at the lichen surface. Topography also plays a major role in controlling wind speed in the lichen environment, both at a microscale, where it affects individual thalli, and at a larger scale affecting the overall microclimate of a lichen association.

Kershaw and Larson (1974) demonstrated the importance of the interaction between topography and microclimate in the contrasting ecology of *Cladonia stellaris* and *Bryoria nitidula*. The wind profile at the top of a 3 m high raised beach ridge in northern Ontario is very different from the mid-slope profile (Fig. 33). With the wind from the prevailing northerly direction there is a 'still-air' zone at the mid-slope position with a considerable degree of turbulence and flow separation at the ridge top. This results in the development of contrasting surface microclimates at the mid-slope and ridge summit. In particular, there is a marked reduction of evaporation rate in the mid-slope 'still-air' zone. Under continuously cloudy skies on 31 July the relative thallus saturation measured by resistance grids (Harris 1969; Kershaw and Rouse 1971*a*) fell steeply, on the ridge top, to dryness by 13.30 hours; in contrast the mid-slope position dried much more slowly and by 15.00 hours ceased to dry any further (Fig. 34). A similar pattern is developed on 30 July with over a 2 h lag phase of drying in the mid-slope position. On 4 August under very low radiation conditions but with strong winds, a considerable differential of relative saturation level developed by 12.00 hours. At this point in the drying run the cloud cover dispersed and, good direct radiation conditions developed, which obscured the wind profile effect, and both sites dried very quickly and at almost the same rate (Fig. 34). Similar differences in drying patterns developed over quite minor topographic features; Kershaw and Larson (1974), for example, show dissimilar evaporation rates between a 9 cm high hummock dominated by *Dryas* and an adjacent hollow dominated by the lichen *Alectoria ochroleuca* (Fig. 35). The level of difference in evaporation rates in both the ridge summit/slope and hummock/hollow comparisons is quite small. Despite this, the direction of the differences is very consistent and when integrated continuously over time periods of several years, becomes appreciable. It clearly interacts, in the case of *Cladonia stellaris* and *Bryoria nitidula*, with their contrasting ecology.

There are comparable 'topographic' gradients of evaporation rates typically found in forest canopies and which develop in response to both the attenuation of energy and the wind speed, and with depth in the canopy. Denmead (1964) presents data on the marked reduction in latent heat flux corresponding with the attenuation of net radiation in a 10-year-old closed canopy stand of *Pinus radiata*, unthinned and unpruned. The generalised humidity and temperature relationship shows a great contrast in the environment below 3 m and that of the upper canopy. This is also reflected in the equivalent reduction in latent heat flux at the bottom of the canopy. It is apparent that the evaporative microclimate around the lower third of a tree trunk is quite different from that around the middle trunk and upper branches. The frequency of occurrence and abundance of several lichen species on the lower trunk may well reflect the mesic nature of this zone.

Harris (1969) showed that steep evaporative gradients occurred even in quite open groups of trees. Using resistance grids (Harris 1969; Kershaw and Rouse 1971*a*) placed in the top middle and lower trunk of a 12 m oak canopy, Harris found a 3-fold increase in evaporation rate between the lower trunk and the upper canopy (Fig. 36). He also points out that the aerodynamic resistance around a thick branch, for example, will be quite different from that of a thin branch and will lead to complex patterns of boundary-layer thickness at any one height in the canopy. Furthermore,

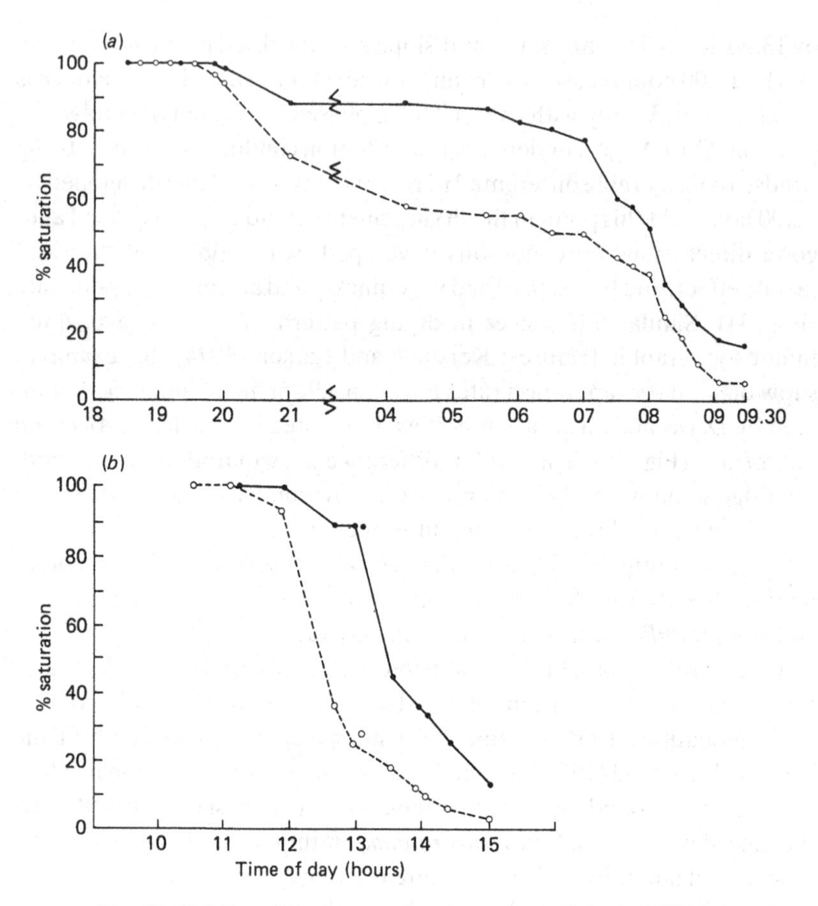

Fig. 35. Relative drying curves for *Dryas integrifolia* (●–●) and *Alectoria ochroleuca* (○–○) growing in hummock and hollow locations respectively on a lichen heath in northern Ontario. (*a*) Site 2, 16-17 August; (*b*) site 1, 19 August. (From Kershaw and Larson, 1974.)

evaporation from a lichen thallus will be profoundly influenced by the water-holding capacity of the underlying bark and its direct control of boundary-layer vapour pressure. Under full radiation conditions, aspect also strongly interacts with surface temperature (see Chapter 1), and although aspect may dominate the overall evaporative process, nevertheless there will always be a complex *seasonal* series of three-dimensional evaporative patterns established in tree canopies. The integration of these patterns is required for any complete understanding of the ecology of corticolous lichens.

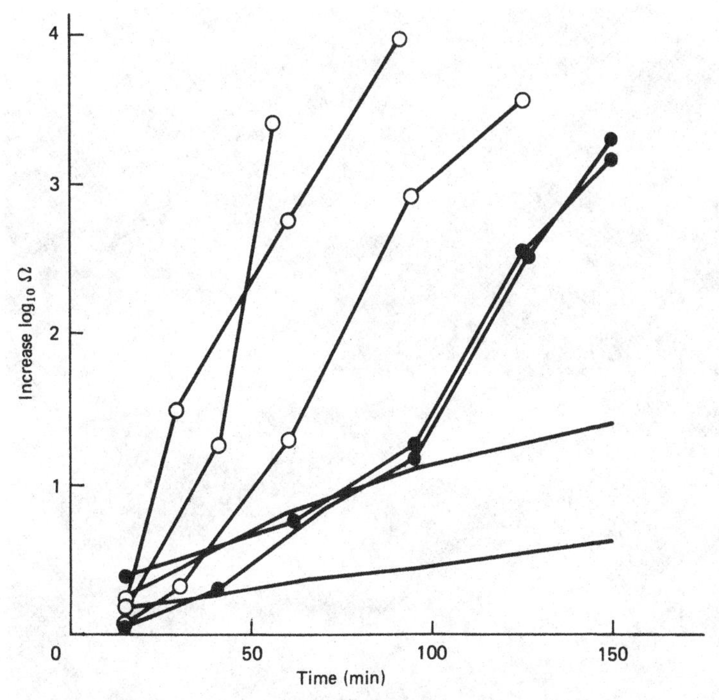

Fig. 36. Relative drying curves (measured as electrical resistance, Ω) for the upper canopy (○–○), middle canopy (●–●) and lower trunk zone) (—) in an open oak woodland in south-west England. (From Harris 1969.)

2.7 Lichen canopy microclimate

Just as a tree or grass canopy generates its own specific microclimate, lichens which develop a clumped growth habit or form an extensive mat can also markedly influence the soil surface microclimate, producing very often in its place an intensely stratified canopy microclimate. A particularly notable example of this is the extensive carpets of lichens found in the northern boreal, circumpolar spruce-lichen woodlands. Kershaw (1977*a*) has discussed the lichen woodlands of the northern boreal forest region of Canada and concludes that they represent a successional stage in the recovery sequence following forest fire which leads to closed-canopy woodland. With the extremely fire-susceptible nature of the lichen woodland, however, only rarely does the tree density reach a sufficient level to suppress lichen growth; a reburn and the subsequent recovery cycle occurs first (see also Hustich, 1951; Ahti, 1959; Fraser, 1956;). Mature *Cladonia stellaris* woodland is characterised by a uniform surface lichen mat of closely packed *C. stellaris* podetia, up to 10 cm and sometimes

Plate 2. *Cladonia* woodland, a characteristic vegetational type of circumpolar dry eskers, drumlins and ridges.

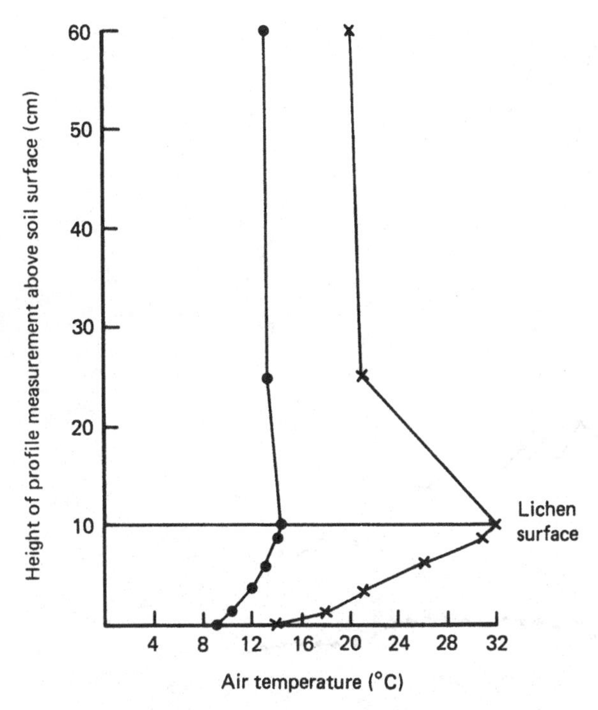

Fig. 37. The temperature profile above and within a *Cladonia stellaris* mat in spruce-lichen woodland in northern Quebec in the morning (●–●) and at solar noon (×–×). (From Kershaw and Field 1975.)

even 15 cm in depth (Plate 2). The highly compact nature of the podetial branching in *C. stellaris* produces a canopy unusual in both form and microclimate. Kershaw and Field (1975) have documented the extreme temperature and humidity profiles which develop on a daily basis. In August, temperatures at the surface of a 10 cm *C. stellaris* mat reach *c.* 32 °C at midday; corresponding soil surface temperatures are only *c.* 14 °C (Fig. 37) and there is a gradient of almost 20 deg C through the 10 cm canopy. In September very similar gradients develop even under moderate radiation conditions with intermittent sunshine through a broken cloud cover (Fig. 38). At the same time there is an extreme gradient of relative humidity, with values of almost 100% in the lower 4 cm of the lichen canopy contrasting with *c.* 30% at the lichen surface and an ambient value of *c.* 60% (Fig. 39).

The effect of the *Cladonia* canopy is to moderate profoundly the soil temperature profile as well as the evaporation of moisture from both the soil and lower levels of the lichen mat. This is also evident in *Stereocaulon* mats. There is as a result an extreme gradient of drying within the lichen

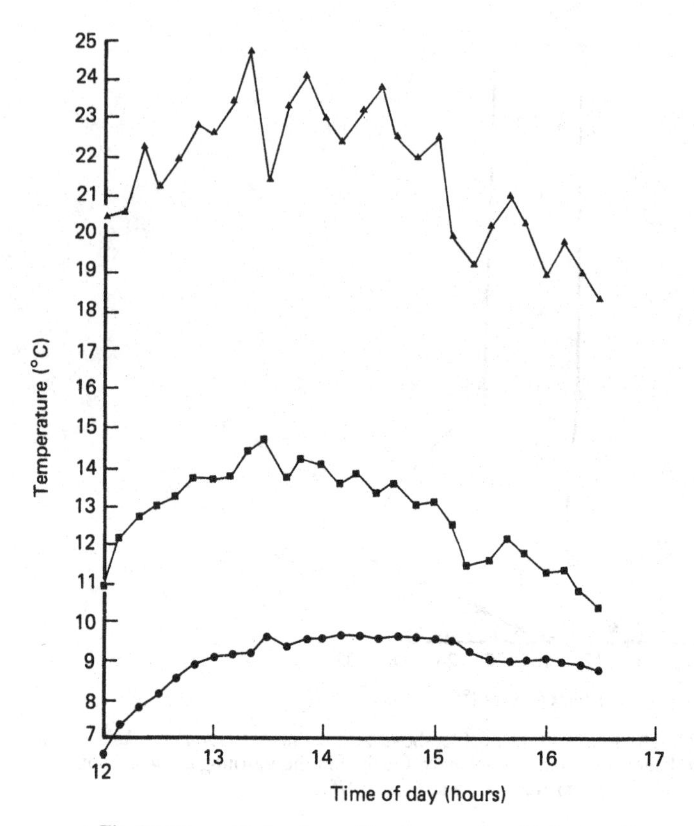

Fig. 38. The vertical pattern of temperature in a *Cladonia stellaris* mat under good radiation conditions in early September. Height above soil surface: ●–●, 0 cm; ■–■, 3.75 cm; ▲–▲, 8.75 cm. (From Kershaw and Field 1975.)

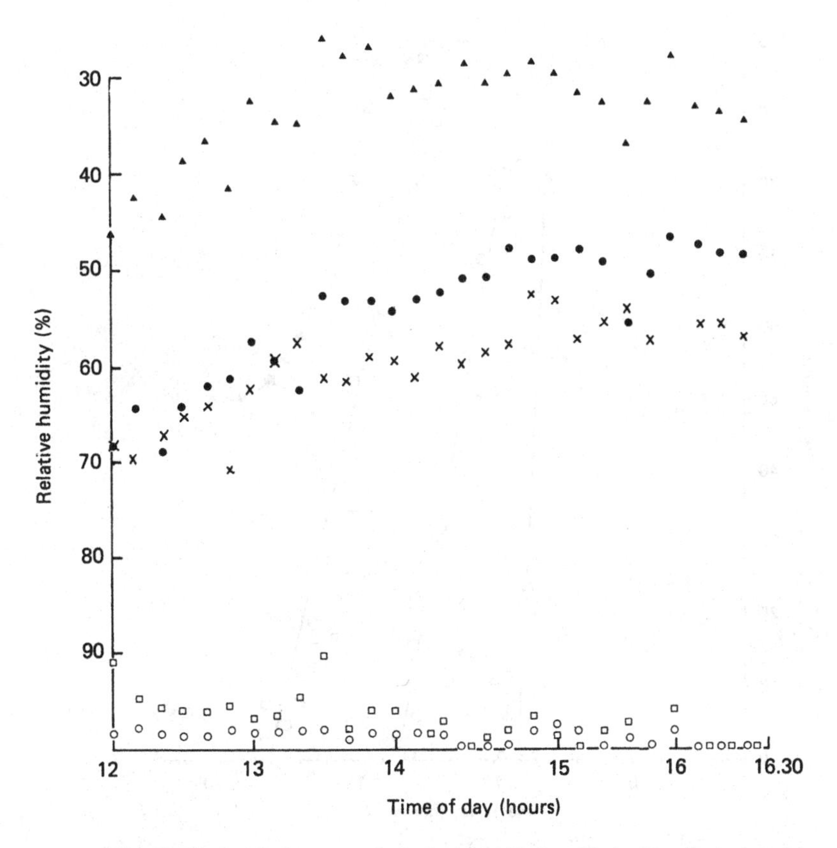

Fig. 39. The vertical pattern of relative humidity in a *Cladonia stellaris* mat under good radiation conditions in early September. Height above soil surface: ○, 0 cm; □, 3.75 cm; ×, 8.75 cm; ●, 10 cm; ▲, 60 cm. (From Kershaw and Field 1975.)

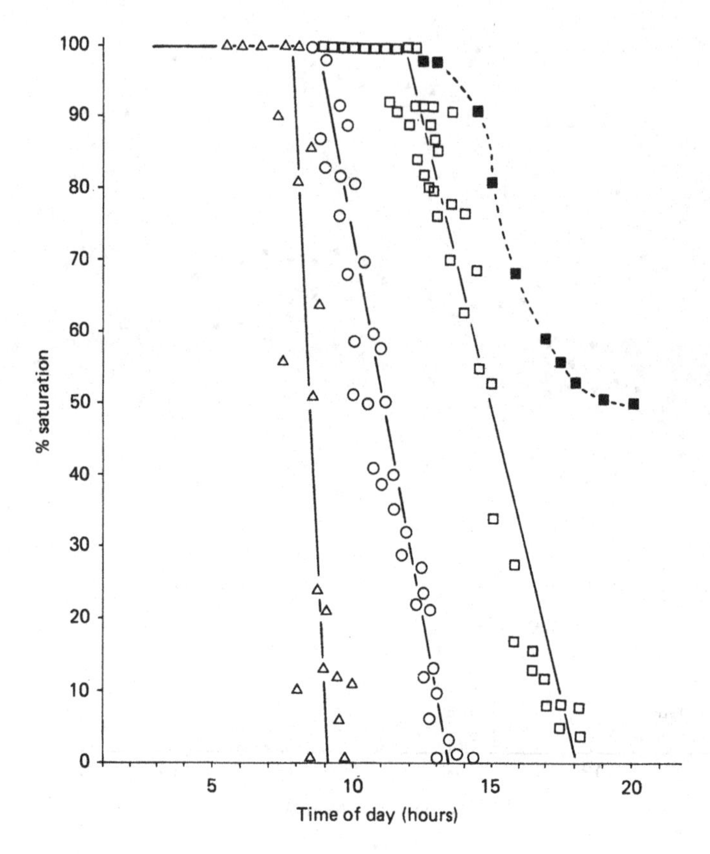

Fig. 40. The vertical pattern of drying in a *Cladonia stellaris* mat in spruce-lichen woodland in northern Ontario, under good radiation conditions in mid-summer. Probe depths: △, surface; ○, 5 cm; □, 7 cm; ■, 10 cm. (From Kershaw and Rouse 1971*a*.)

mat itself (Fig. 40), the complete mat only drying out during mid-summer and then only under full radiation conditions extending over a period of days. The lichen surface acts as an extremely efficient mulch and very effectively controls soil moisture throughout the entire summer (Kershaw and Rouse, 1971*a*). Despite the distribution of *Cladonia* woodland largely on very well drained, sandy soils laid down during the last ice age, these soils remain at field capacity throughout the entire summer period (Fig. 41). Removal of the lichen surface by fire results in full evaporative losses from the soil, producing extremely xeric summer conditions. The close interaction between soil microclimate and the lichen surface not only plays an important role in the entire successional sequence following fire, but

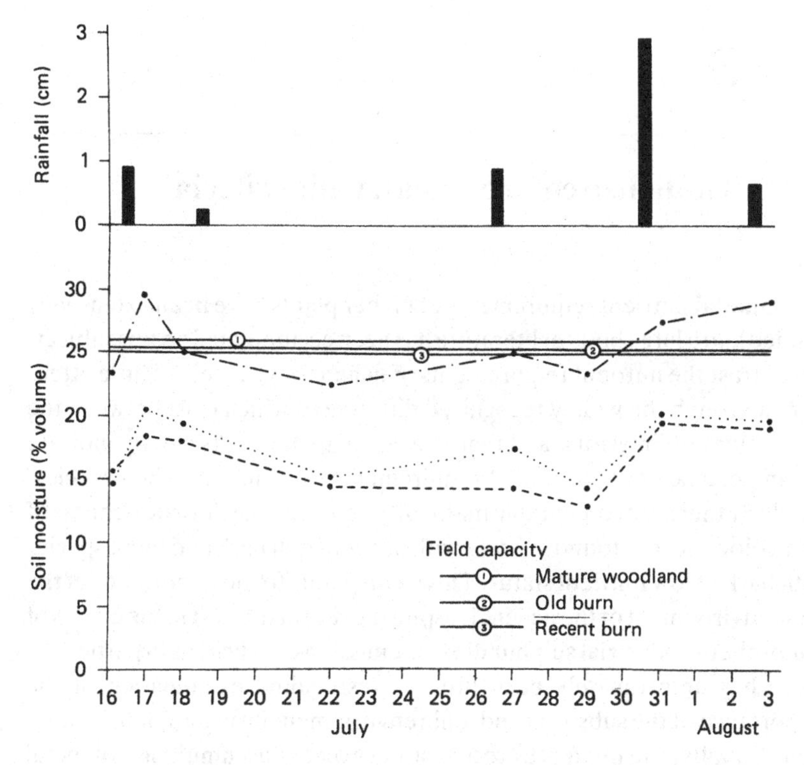

Fig. 41. The summer pattern of rainfall, soil field capacity (at 100 mbar suction) and average soil moisture levels in mature lichen woodland (—·—), lichen woodland burnt recently (····) and 16 years previously (----). (From Kershaw and Rouse 1971*a*.)

also moderates the soil temperature and moisture to such an extent in the mature woodland that tree growth and seedling establishment are both considerably restricted.

3

The lichen environment: ionic criteria

The mineral nutrient requirements of higher plants have been extensively studied and a large body of literature has been accumulated on this subject. In contrast the nutrient requirements of lichens have received little attention, although the widely recognised differences which exist between the lichen flora of limestone and that of acidic igneous rocks clearly indicate the importance of the mineral requirements of different lichen species. Similarly there are comparable major differences in the floristic richness of corticolous lichens found on trees with nutrient rich bark and those species with bark of low nutrient status. This again points to the importance of the ionic environment of the lichen. Despite the recent interest in the effects of atmospheric industrial sulphur dioxide emissions, on lichens in particular, there has unfortunately been little corresponding examination of the importance of the substrate and ionic environment in an unpolluted situation. Equally, the numerous recent studies on the accumulation of metal pollutants in lichens have also fallen short of a realistic treatment at an ecological level and have not examined, for example, the equivalent levels of accumulation and uptake of calcium (Ca^{2+}), aluminum (Al^{3+}) and potassium (K^{+}). Although the former studies have done much to focus attention in detail on the actual mechanisms of ion uptake in lichens and have extended our knowledge in this area to a considerable extent, the basic information on uptake of essential ions such as Ca^{2+} and K^{+}, for example, in undisturbed habitats is still missing.

There are thus three specific areas where there is a considerable amount of information available: on the patterns of species distribution in relation to contrasting substrates which have quite different ionic environments; on metal ion uptake; and on sulphur dioxide/substrate interactions. It is, however, not possible at the present time to link these separate data sets to form a coherent picture of the importance of the ionic environment in the ecology of lichens. Unfortunately there are still considerable gaps in our knowledge and any attempt to bridge these gaps, as here, necessitates extrapolation of data from higher plant nutritional ecology, which may or may not be valid.

3.1 The elemental content in lichens and its location and form

Nieboer, Richardson and Tomassini (1978) and Nieboer and Richardson (1981) present a comprehensive summary from the current literature of the range of elemental levels found in lichens. The data they present show tremendous ranges correlating with different or contrasting environments and in particular 10-fold increases in metal content near ore smelters. The seashore environment yields similarly high values of sodium, magnesium and calcium content in lichens. However, it is important to recognise that the exact cellular location of different ions in a lichen may be quite specific and, furthermore, that the ions may also be present in quite different forms. The origin of the elements found in lichen thalli is twofold: particulate atmospheric 'fall-out' and ionic solutions. The latter may be delivered to the thallus either directly as rainfall or indirectly as surface runoff. The form and hence the availability of the metallic elements, in particular, from these contrasting sources can be quite different.

Nieboer *et al.* (1978, 1982) present very convincing evidence of the widespread occurrence of trapped particulates in lichens, not only those adjacent to industrial development but also those in very remote areas. The data are derived from a number of studies of two geologically diverse areas of Canada that are remote from industrial activity (Mackenzie Valley, Northwest Territories; and southern New Brunswick). The geology of these areas includes sandstones, limestones, shales and granites. The iron/titanium ratios for a wide range of rock types are remarkably constant, presumably because their respective oxides originally crystallised together. As a result, although the absolute magnitudes of the two elements vary for the Canadian shield rock types the iron/titanium quotient has an average value of 7.2. A plot of the iron/titanium ratios for a range of species of the genus *Cladonia* collected from non-industrial areas similarly has a slope of 7.0 ± 0.2 (Fig. 42), and clearly points to the presence of these two elements largely in the form of particulates in many lichens. The exhaustive treatment (requiring hydrogen fluoride digestion and pyrosulphate fusion) necessary to release titanium from ashed or acid-digested lichens, also infers its presence in the form of oxides, its normal elemental form in particulates. Similar considerations apply to the corresponding theoretical titanium ratios with chromium, vanadium and nickel, which also agree closely with the observed levels of these elements found in lichen thalli from rural areas of Canada (Nieboer *et al.* 1978). The intercept on the ordinate in Fig. 42 of 25 ± 6 ppm iron thus appears to be that level of iron sufficient to provide the nutritional requirement of lichens. Consequently elevation of non-essential elements such as chromium, vanadium, nickel

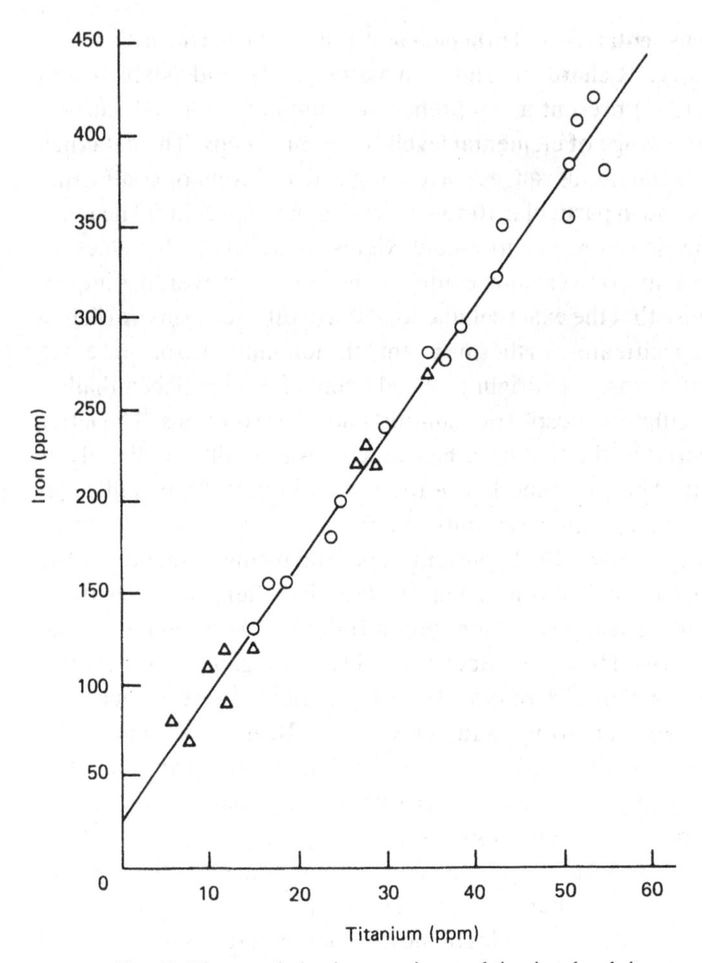

Fig. 42. The correlation between iron and titanium levels in a range of *Cladonia* species collected near St John's, New Brunswick (○) and in the Canadian arctic (△). (From Nieboer *et al.* 1978.)

and lead above the theoretical level expected from the equivalent titanium content gives a useful index of contamination by these elements from extraneous sources.

Detailed confirmation of particulate metallic fall-out is presented by Garty, Galun and Kessel (1979). Their study demonstrates by means of scanning and transmission electron microscopy the extensive extracellular deposition of particulates, mainly in the medulla of the lichen *Caloplaca aurantia*. The elemental composition of particles integrated in the lichen tissue was also compared with that of dust particles collected from the surface of the lichen colonies using energy dispersive x-ray analyses and a close correspondence was established.

It is evident that a distinction has to be made between 'uptake' and 'accumulation' of all elements and in particular those metals that are known to be potentially toxic to living organisms. Substantial amounts of metals may accumulate in a particulate, non-available form in the large intercellular spaces of the medulla. Simple elemental analysis of the thallus might suggest a remarkable degree of tolerance to such high concentrations of metals (Lounamaa 1956; Maquinay *et al.* 1961; Lawrey and Rudolph 1975; Nash 1975; etc.) which would only be true if actual metal ions were involved. Even then the available evidence to date suggests that many metal ions are only bound extracellularly and again are not metabolically available. The evidence for this extracellular location of metal ions is derived from the classic paper by Tuominen (1967) which has been extensively amplified and corroborated by Brown (1976), Nieboer *et al.* (1976*a*,*b*) and Goyal and Seaward (1981, 1982*a*,*b*). (See also Nieboer *et al.* (1978, 1982) for an excellent summary.)

3.2 Mechanisms of uptake and release of cations

Two mechanisms can be distinguished in the cation uptake by plants. Active uptake of a cation is closely linked to the metabolism of an organism and usually a close correlation is observed between ion uptake and respiration rate. There is also a strong reduction or complete elimination of cation uptake by metabolic poisons or inhibitors. The mechanism of active uptake is complex and of less significance here than passive uptake of cations, which is a purely physico-chemical process resulting in extensive extracellular ion exchange in lichens. Passive uptake has considerable importance in studies of the apparent metal tolerance of some lichens and accordingly will be examined in detail.

3.2.1 *Ion exchange theory and extracellular metal exchange in lichens*

Ions may be incorporated into an ion exchange resin either by weak or strong field exchange, or by electrolyte sorption. The identity of the particular functional group repeated along the polymer chains of the resin determines whether the exchange is of the weak- or strong-field category. If the field of the functional group is weaker than that provided by polar water molecules, no direct association between the cation and the fixed ionic charge of the resin occurs. The hydrated metal ion is instead weakly held by hydrogen bonding to the functional group. Conversely if the functional group is a weak-acid anion, such as $-COO^-$, the exchange mechanism is said to be of the strong-field type. At low pH, such functional groups are protonated and the ion exchange capacity becomes a function of pH. These groups, when not protonated however, provide a strong electrostatic field which readily overcomes the weak field of the water dipoles from the coordination sphere of the hydrated metal cation and a coordinate metal complex forms with the functional group. If more than one functional group binds to the same metal ion, chelation or ring-formation occurs. Different metal ions have different tendencies to form coordinate bonds and those with a high charge and small ionic radius are preferred. These preferences yield a specific metal ion selectivity pattern which differs from the equivalent pattern of selectivity seen in weak-field exchange (Helfferich 1962; Rieman and Walton 1970; Puckett *et al.* 1973).

The functional groups for cation exchange are usually weak-acid carboxyl groups which are largely non-ionic at low pH and dissociate at a specific pH. The dissociation constant, K, is usually expressed as

$$pK_a = -\log K_a. \tag{5}$$

When the pH drops below the pK value, the group is non-ionic and effectively does not contribute to the ion exchange ability of the resins. Naturally occurring cellular functional groups can be found as several forms (see below).

The alternative method of uptake, electrolyte sorption, is fundamentally different from ion exchange in that a particular ion is taken up at a rate which is inversely proportional to the magnitude of the Donnan potential that develops across the resin and solution phases. No new ions appear in solution and, furthermore, desorption is readily achieved by elution with distilled water. As Nieboer *et al.* (1978) have indicated, it is important to distinguish between desorption of electrolytes by elution with distilled water and hydrolysis. For example Lawrey and Rudolph (1975) reported that significant amounts of iron and aluminium could be leached from *Cladonia cristatella* and this could well be in the form of hydroxo complexes of Fe^{3+} and Al^{3+} which can be formed even when these ions are complexed to weak-acid functional groups.

Tuominen (1967), in an outstanding contribution, applied the basic concepts of ion exchange in large resin molecules to strontium uptake in *Cladonia stellaris* and established a number of basic criteria. When the thallus of *C. stellaris* is submerged in a dilute solution of strontium chloride, the concentration of the solution decreases rapidly to an equilibrium value whilst concurrently the pH of the solution also decreases and new cations,

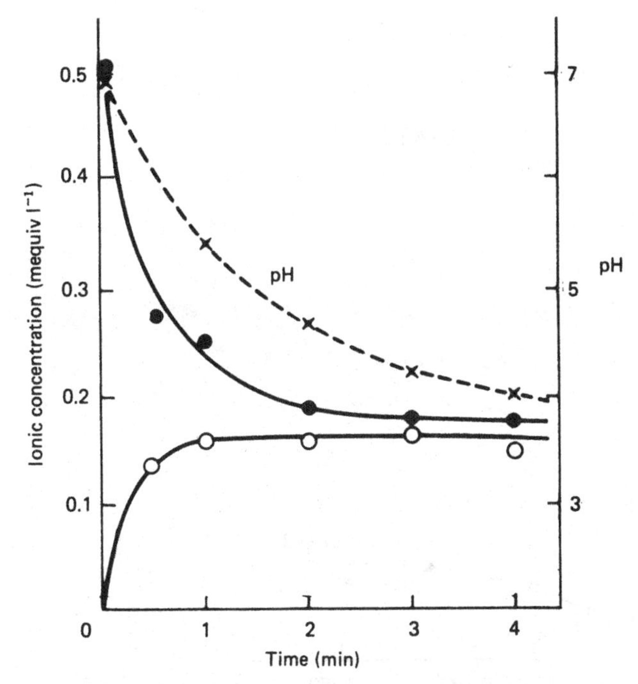

Fig. 43. The exchange of strontium (●) and calcium (○) by a *Cladonia stellaris* thallus placed in a solution of strontium chloride. (From Tuominen 1967.)

in this case Ca^{2+}, appear in the solution (Fig. 43). In other replicates from different locations, potassium, sodium, calcium and magnesium all appear in different proportions. After strontium uptake, Sr^{2+} ions cannot be eluted from the lichen thallus by repeated washing in distilled water. The uptake of Sr^{2+} is strongly dependent on pH (Fig. 44) as well as the concentration of Sr^{2+} in the external solution. These findings clearly point to an ion exchange mechanism being involved rather than simply electrolyte sorption. Thus the failure to elute Sr^{2+} with distilled water eliminates sorption as a major component of ion uptake, and the marked pH dependency also suggests strong field interactions with weak-acid carboxyl groups, which are predominantly non-ionic below about pH 3. Tuominen (1967) concludes that cation uptake in *C. stellaris* is not necessarily pure cation exchange but may be complicated by coordination compounds of Sr^{2+} and/or hydrogen bonding between carboxyls.

These findings were confirmed by a number of additional studies. Handley and Overstreet (1968) were able to remove Ca^{2+} ions from *Ramalina reticulata* using sodium chloride solutions which, as a pretreatment, in-

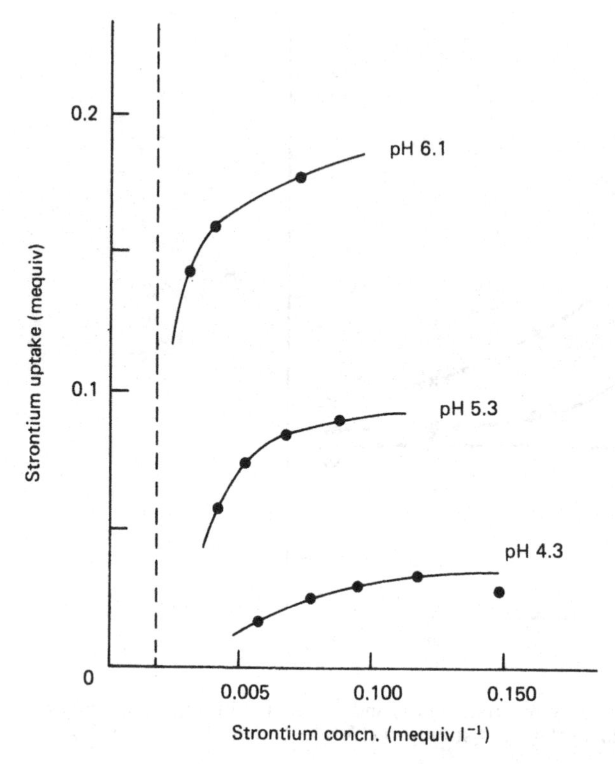

Fig. 44. Uptake of strontium by *Cladonia stellaris* as a function of pH and external concentraton of strontium in the solution. The dashed line indicates the analytical limit of detection. (From Tuominen 1967.)

creased exchanged lead and potassium from *Cladonia rangiformis* with hydrochloric acid solutions or nickel solutions. These early studies are well summarised and extended by Puckett *et al.* (1973) who established the selectivity sequence iron ⩾ lead ⩾ copper ⩾ nickel, zinc > cobalt and confirmed that the uptake mechanism was indeed ion exchange modified by metal-complex formation (see also Brown 1976). Nieboer *et al.* (1976*b*, 1978) have formulated a model representative of the whole process (Fig. 45). The hydrated cell wall, A, contains the functional groups which bind cations and perhaps anions, onto the hyphal (or algal) cell wall surface. Zone B is a reaction zone in which there are free, unreacted or uncomplexed electrolytes; this is identical under equilibrium conditions with the bulk solution, zone D. The diffusion from the external ionic environment D is across zone C, a film which limits the rate of ion uptake or release and is not affected by convection or stirring.

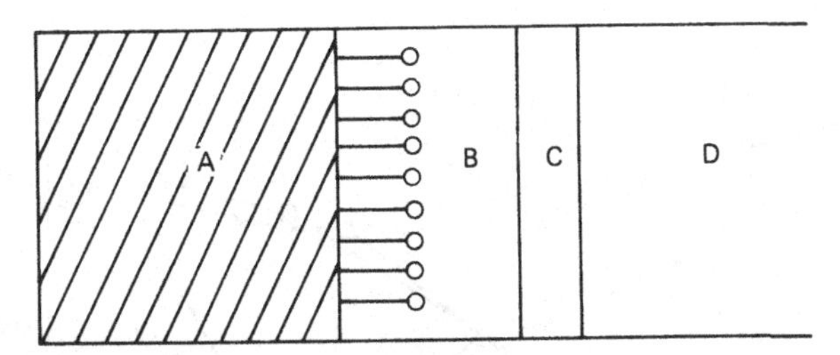

Fig. 45. Diagrammatic representation of the exposed surface of a wetted or submerged lichen thallus. Zone A, ion exchange surface (e.g. hyphal cell wall); zone B, reaction zone in which the functional groups or binding sites are 'dissolved'; zone C, undisturbed diffusion layer or 'film'; zone D, bulk solution zone. (From Nieboer *et al.* 1976*b*.)

The process of cation exchange can be represented in the following way:

$$M^{2+} + 2HA \rightleftharpoons MA^{+} + A^{-} + 2H^{+}. \tag{6}$$

The metal ion, M^{2+}, diffuses from the bulk solution (zone D) into the reaction zone B. For example the double-charged M^{2+} could displace two hydrogen ions ($2H^{+}$) in interacting with two fixed and protonated anionic functional groups (2HA) localised in zone B on the lichen cell walls. The displaced hydrogen ions diffuse out into the bulk solution leaving in zone B a charged metal complex, MA^{+} , with a negatively charged functional group A^{-}. Electrical neutrality is maintained in zone B and the charge equivalents of the ions leaving and entering are conserved. Thus the exchange ratio for the exchange of Ni^{2+} for Sr^{2+} is 1 : 1 whilst Sr^{2+} for H^{+} is 1 : 2. The anionic functional groups involved in cation exchange have p*K* values of 2-6 log units (Tuominen 1967; Nieboer *et al.* 1976*a*; Richardson and Nieboer 1981). The evidence to date indicates that carboxylic or hydroxycarboxylic acids are involved, but as yet the exact identification of the polymer has not been achieved. Tuominen suggests uronic acid polymerates may be involved and of considerable importance particularly to epiphytic species (see Section 3.2.2).

3.2.2 *The evidence for intracellular ion uptake*

Brown and Slingsby (1972), Puckett (1976), Brown (1976) and Buck and Brown (1979) have all provided firm evidence for the intracellular location of K^{+} in lichens. The evidence comes from a variety of experimental sources, which showed, for example, either that K^{+} is not exchanged with the molar ratio that would be expected for maintenance of the charge equivalents when active metabolism was not involved, or that

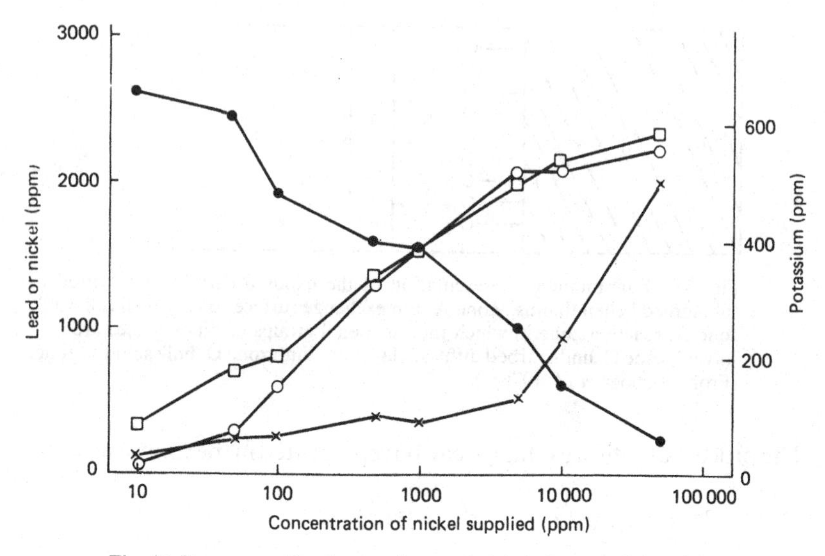

Fig. 46. Recovery of lead, potassium and nickel after supplying nickel at various concentrations to living *Cladonia rangiformis*. ×, potassium displaced by nickel; ○, lead displaced by nickel; ●, lead retained by lichen; □, nickel absorbed by lichen. (From Brown and Slingsby 1972.)

treatment of living lichen thalli in a manner which caused death and loss of integrity of cell membranes resulted in the subsequent massive efflux of potassium into the medium. Brown and Slingsby (1972), for example, exchanged the lead bound to the functional groups on the walls of *Cladonia rangiformis* using Ni^{2+} ions (Fig. 46). With increasing nickel concentration the amount of lead retained in the lichen decreased whilst the amount recovered in solution steadily rose, closely approaching complete replacement of lead by nickel. In contrast the potassium recovered in solution increased only slightly with increasing concentration of nickel, representing the small amount of K^+ bound to the functional exchange groups. At high nickel concentrations (*c.* 7000 ppm) there is an abrupt efflux of potassium, possibly as a result of nickel replacing calcium in the cell membranes and allowing the intracellular potassium to flood out. Brown and Slingsby (1972) also showed that increasing normalities of hydrochloric acid finally resulted in loss of membrane integrity again accompanied with massive efflux of potassium. At low normalities of hydrochloric acid there was some K^+/H^+ exchange pointing to the presence of small amounts of potassium linked to the functional groups of the cell walls.

Puckett (1976) confirmed and amplified these findings and showed, for example, that heat-killed discs of *Umbilicaria muhlenbergii* (120 °C for 24

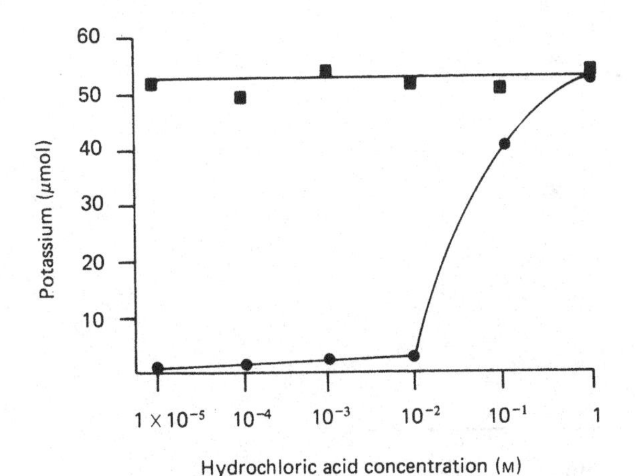

Fig. 47. Elution of potassium from live (●) and heat-treated (■) *Umbilicaria muhlenbergii* with varying molarities of hydrochloric acid. Discs (50) of *U. muhlenbergii* were shaken in 50 ml of hydrochloric acid varying in molarity from 1 $\times 10^{-5}$ to 1 M for 3 h. Similar heat-treated discs (120 °C for 24 h) were incubated in the same manner. The acid solutions were then analysed for potassium. (From Puckett 1976.)

h) immediately released K^+ (Fig. 47). It is now evident that thallium (Nieboer *et al.* 1976*b*; Burton, Le Sueur and Puckett 1981) and zinc (Wainwright and Beckett 1975; Brown 1976), as well as copper, mercury and silver (Cu^{2+}, Hg^{2+} and Ag^+) are all taken up intracellularly, at least in part (Puckett 1976). These ions are of particular interest since only concentrated solutions of Ni^{2+}, Pb^{2+}, Co^{2+} and Cd^{2+} induce release of cellular K^+ whereas even very dilute solutions of Cu^{2+}, Hg^{2+} and Ag^+ are very effective in inducing a rapid loss of potassium (Fig. 48). Hg^{2+} and Ag^+ are particularly damaging to lichen metabolism, eliminating photosynthetic activity completely after even short-term exposure to a dilute solution (Fig. 49).

Nieboer and Richardson (1980, 1981) have suggested that metal ions may be classified into three groups: Class A (e.g. K^+, Ca^{2+} and Sr^{2+}), a borderline or intermediate group (e.g. Zn^{2+}, Ni^{2+}, Cu^{2+} and Pb^{2+}), and class B (e.g. Ag^+, Hg^{2+} and Cu^+). Class A ions are characterised by a strong preference for oxygen-containing binding sites (carboxylate and phosphate anions for example) while class B metal ions show a strong affinity for nitrogen- and sulphur-containing binding sites (imidazole moiety of histidine, and sulphydryl (–SH) groups). Class B ions and those borderline ions with Class B affinities are strongly detrimental to lichens either by themselves and/or in combination with sulphur dioxide (see

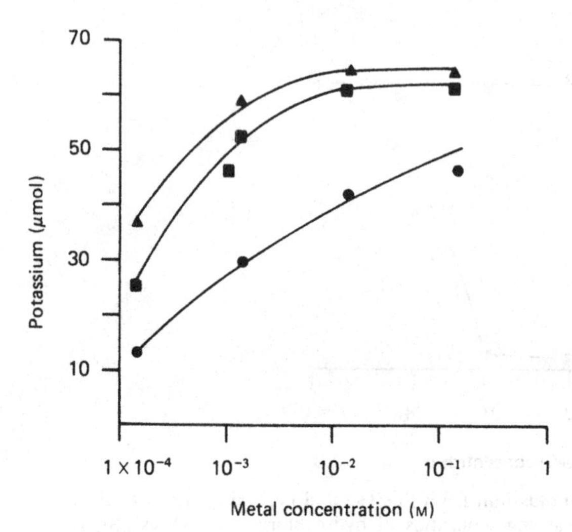

Fig. 48. The loss of potassium from *Umbilicaria muhlenbergii* as a function of the copper (●), mercury (▲), and silver (■) concentration. Discs (120) of *U. muhlenbergii* were incubated in solutions of copper chloride, mercuric chloride or silver nitrate which varied in concentration from 2×10^{-4} to 2×10^{-1} M for 3 h. The amount of potassium lost into the medium was then determined. (From Puckett, 1976.)

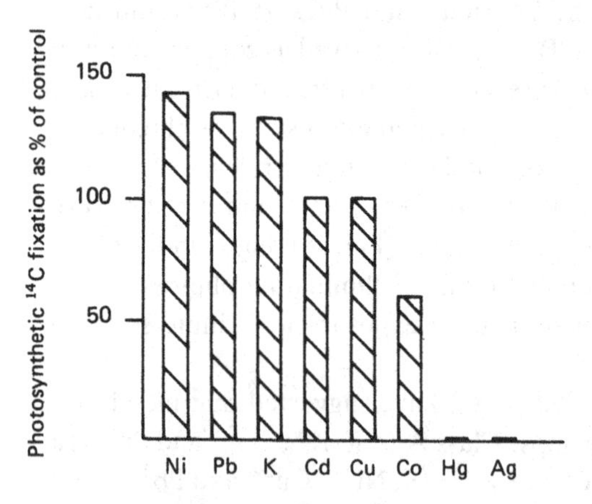

Fig. 49. The effect of short-term exposures to heavy metals on photosynthesis by *Umbilicaria muhlenbergii*. Discs (10) of *U. muhlenbergii* were incubated in 4 ml of each metal-ion solution (1×10^{-2} M as the chloride salt, with the exception of lead and silver) for a period of 1 h. The discs were then washed thoroughly and allowed to photosynthesise in 4 ml of distilled water containing 10 Ci $NaH^{14}CO_3$ for a further 1 h period. (From Puckett 1976.)

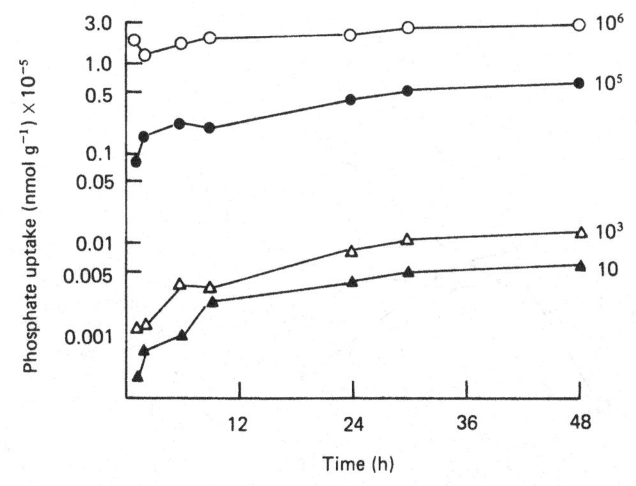

Fig. 50. Uptake of inorganic phosphate by *Hypogymnia physodes* at different concentrations of phosphate. Approximately 0.1 g lichen was placed in 200 cm^3 phosphate solution at the given concentrations (mol^{-1} dm^{-3}), labelled with 0.001 μCi $^{32}P_i$. At the times shown 1 cm^3 of medium was withdrawn for counting. (From Farrar 1976*b*.)

below) (Nieboer *et al.* 1979; Richardson *et al.* 1979). Their detrimental effects reflect that, in part at least, they are taken up intracellularly.

Nieboer and Richardson (1980) give an extremely useful and detailed survey of the potential mechanisms of toxicity as well as the toxicity sequences that have been established for a range of organisms, and Ochiai (1977) offers an in-depth treatment. In brief, toxicity mechanisms can be examined under three categories: (1) blocking of the essential biological functional groups of active biomolecules; (2) displacement of the essential metal ions found in certain classes of biomolecules; and (3) modifying the active conformation of biomolecules. The high toxicity of class B ions is readily explained by their ability to participate in all three of Ochiai's toxicity mechanisms. For example, Hg can bind to –SH groups at catalytically active centres such as enzymes, or can displace endogenous borderline ions such as Zn^{2+} from metallo-enzymes, resulting in unfavourable conformational changes and rendering them inactive. Similar considerations apply to the toxicity of borderline ions but a detailed discussion is not essential here and the reader is referred to Ochiai (1977) and Nieboer and Richardson (1980).

3.3 Anion uptake in lichens

With the exception of phosphate and the special case of sulphite uptake (see below) remarkably little is known about anion accumulation

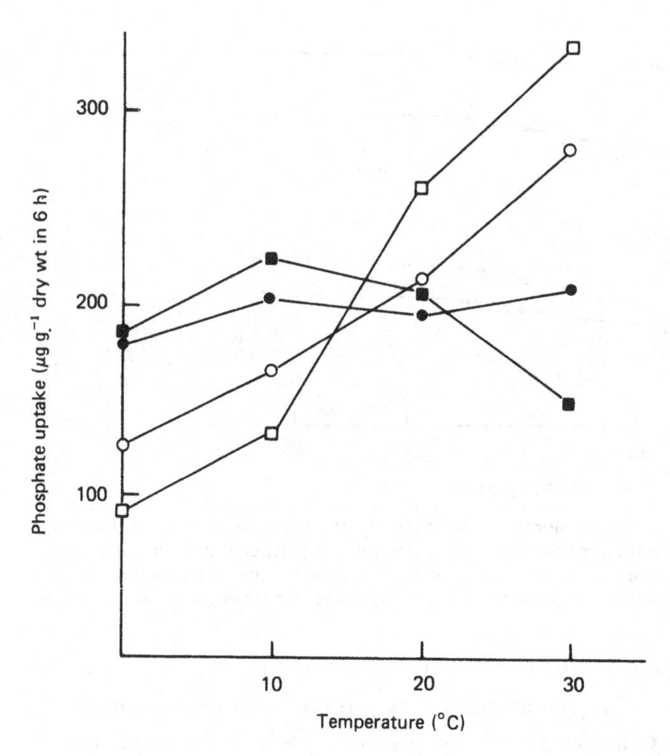

Fig. 51. The uptake of phosphate by *Xanthoria parietina* (circles) and *Parmelia saxatilis* (squares) at different temperatures in the light (3500 lux) and dark. Phosphate was measured by the molybdate method. Filled symbols, dark; open symbols, light. (From Fletcher 1976.)

by lichens. Smith (1960) showed that discs of lichen absorb quite large amounts of phosphate from potassium phosphate solutions and suggests (unpublished data) that the process is active since it is inhibited by anaerobic conditions and dinitrophenol. Subsequently (Farrar 1976*b*) using *Hypogymnia physodes* showed that the amount of inorganic phosphorus taken up was very dependent on the concentration in the medium (Fig. 50) and was indeed rapidly inhibited by dinitrophenol. Farrar also showed that uptake from dilute solutions was non-exchangeable whereas some of the phosphate absorbed from solutions of high concentration could be exchanged. Two uptake mechanisms could be involved, both leading to the accumulation of substantial quantities of polyphosphate. Fletcher (1976) has also shown that uptake of phosphate is very dependent on temperature, but only in the light (Fig. 51), and there seems little doubt in the face of the overall evidence that phosphate uptake is indeed an active process.

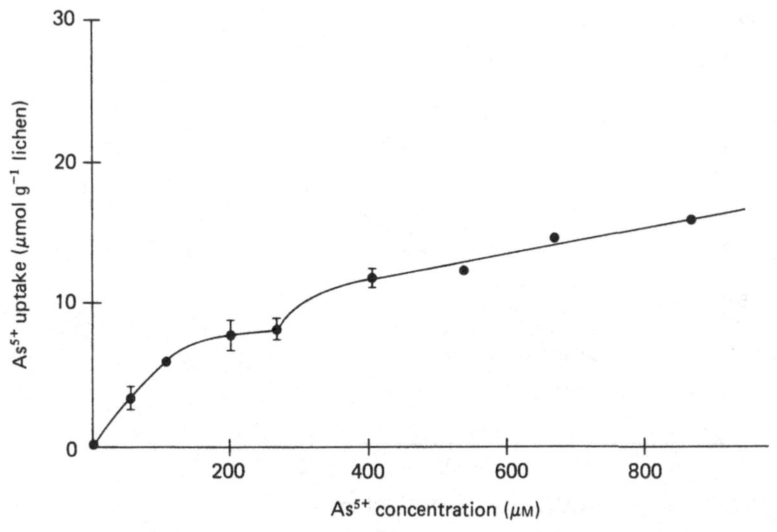

Fig. 52. The dependence of As^{5+} uptake in *Umbilicaria muhlenbergii* on the concentration of the external solution. $V_{max} = 0.37 \pm 0.03$ μmol g^{-1} min^{-1}, K_m 360 ± 60 μmol. The vertical bars show the S.E.M. (From Richardson *et al.* 1984.)

Recently Richardson *et al.* (1984) have substituted arsenate as an experimental anion for uptake studies since it appears that it is taken up by the same system as phosphate (Beever and Burns 1980) but can be measured with much greater ease and accuracy. As^{5+} uptake seems to be biphasic (Fig. 52) with K_m values of *c.* 100 and 700 mol respectively in contrast to typical K_m values either in the range 1-10 μmol or 100-1000 μmol for phosphate uptake in free-living fungi (Beever and Burns 1980). Current views of fungal anion uptake, particularly phosphate, suggest a carrier-mediated mechanism. The energy for the process is provided by an accompanying movement of protons down a proton gradient, with H^+ returned from the cell to maintain the gradient. A further proton is ejected per molecule of phosphate taken up, to prevent an increase in acidity of the cell. At the same time K^+ ions are taken up to maintain the overall charge balance (Beever and Burns 1980). Richardson *et al.* (1984) show that during arsenate uptake there is a release of K^+ with 3 μmol of K^+ ions being released for every 3 mol of arsenate ions taken up (Fig. 57). They suggest that K^+ efflux may replace at least partially the corresponding H^+ efflux in fungi. Furthermore, the interactions between arsenate uptake and the uptake of other anions are quite complex (Fig. 54; E.H. Nieboer, unpublished data). There is rapid decline of arsenate uptake with increasing concentration of phosphate in the solution, which further substantiates the

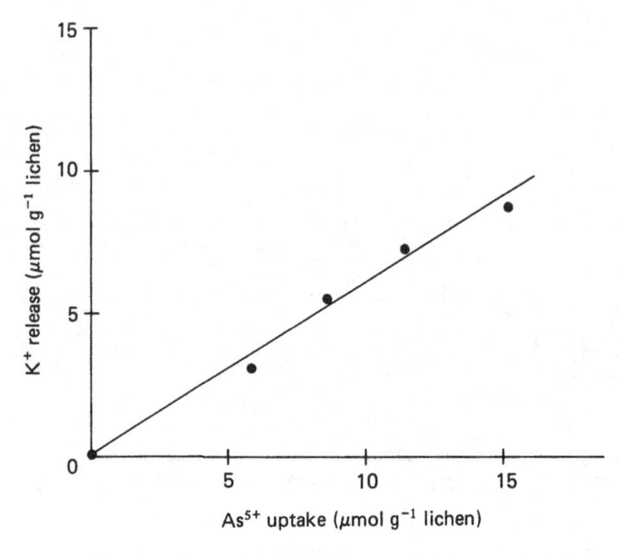

Fig. 53. The concurrent release of K^+ during As^{5+} uptake by *Umbilicaria muhlenbergii*. (From Richardson *et al.* 1983.)

likelihood of a common carrier. Similarly there is an equivalent reduction in level of arsenate uptake with increasing sulphite (HSO_3^-) ion concentration, but a marked biphasic enhancement of arsenate uptake with the increased concentration of sulphate ions. Thus sulphate ions are 'piggy-backed' along with arsenate using a common carrier and this results in a significant increase in the level of photosynthetic inhibition in the presence of both ions.

Clearly the uptake of anions by lichens is often an active process with certainly first-order anionic interactions involved but potentially significant secondary and tertiary interactions as well. It is equally evident that the data we have available so far are still very limited and urgently require extending.

3.4 The special case of sulphite uptake in relation to atmospheric pollution

There is an extensive body of literature now available on the lichen deserts which occur around major urban areas and areas of industrial activity. These deserts have developed largely as a result of sulphur dioxide emissions from the burning of fossil fuels. Gilbert (1973) gives a particularly good early summary of the literature which has subsequently been amplified by a number of publications and has been covered in depth by

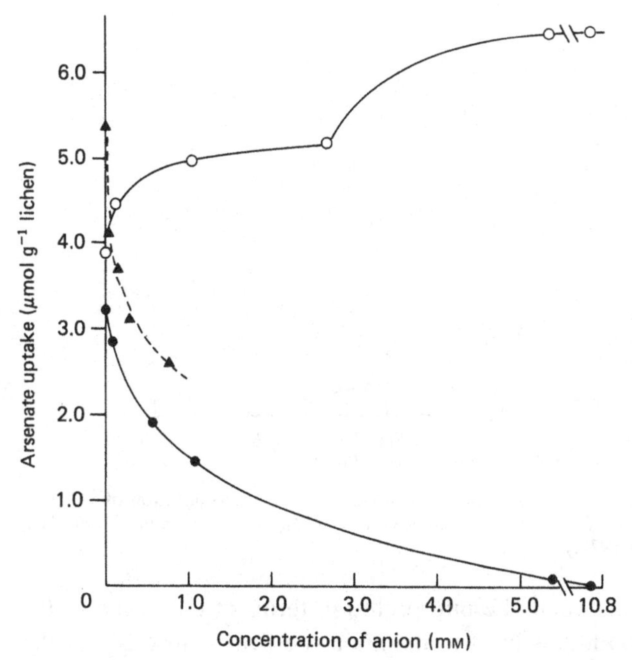

Fig. 54. The interaction of arsenate uptake with the concentration of other anions in solution: ○, sulphate; ▲, sulphite; ●, dihydrogen phosphate (all at pH 4.6). (E.H. Nieboer, unpublished data.)

Ferry, Baddeley and Hawksworth (1973). However, no treatment of the physiological ecology of lichens is complete without at least an adequate coverage of the subject, since the effects of sulphur dioxide emissions are now so widespread and have resulted in some very characteristic and stable distribution patterns in a number of lichen species.

3.4.1 *The effects of sulphur dioxide on lichen metabolism*

The pioneer studies by Pearson and Skye (1965) and Rao and LeBlanc (1966) both demonstrated that fumigation with sulphur dioxide in the gaseous form resulted in chlorophyll degradation but unfortunately both used unrealistically high concentrations. This difficulty in working with sulphur dioxide in the gaseous phase has caused sulphite solutions to be used in most subsequent work on its toxic effects. Gilbert (1968) avoided the difficulties of making experimental fumigant gases by using 'city air' with *c.* 0.01 ppm sulphur dioxide. His results showed that over a period of 40 days, respiration rates and chlorophyll content in *Ramalina farinacea* both decreased whereas in *Lecanora conizaeoides*, which is resistant to atmospheric pollution, they remained constant. However,

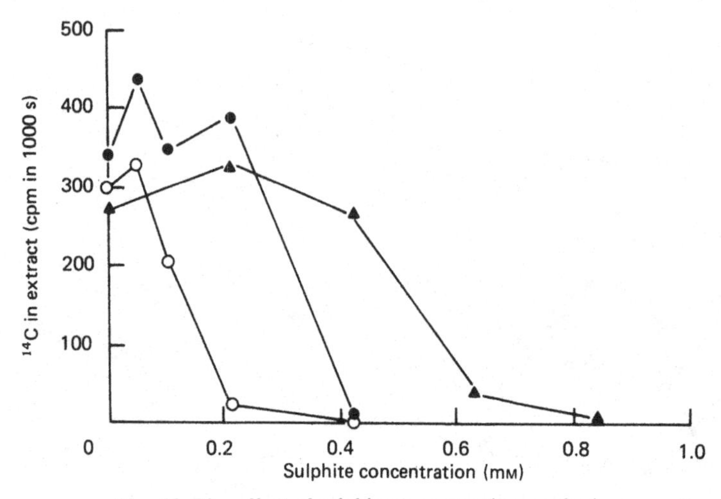

Fig. 55. The effect of sulphite concentration on the incorporation of $^{14}CO_2$ in *Usnea subfloridana* (○), *Parmelia physodes* (●) and *Lecanora conizaeoides* (▲). (From Hill 1971.)

these results could equally be interpreted in terms of the uncontrolled environmental conditions used to maintain the lichens throughout the experimental period (Hill 1971).

Hill (1971), using sulphite solution, showed conclusively that photosynthetic incorporation of $^{14}CO_2$ was very strongly affected in *Usnea subfloridana* by low concentrations (0.2 mM) of sulphite solution. This concentration had little effect on *Parmelia physodes* or *Lecanora conizaeoides*. Similarly *Parmelia* was markedly affected by 0.4 mM sulphite solution which, in turn, had only a slightly deleterious effect on the photosynthetic rate of *Lecanora* (Fig. 55). When sulphate was used instead of sulphite, it had little effect on the rate of $^{14}CO_2$ incorporation in *Usnea* over the same concentration range used for the sulphite treatment (Fig. 56). Furthermore Hill showed that the sensitivity of *U. subfloridana* was strongly pH-dependent. At pH 5-7.5 incorporation of $^{14}CO_2$ was not affected by sulphite solution but below pH 5 photosynthesis was greatly reduced and was completely eliminated below pH 4 after 3 h of sulphite treatment (Fig. 57).

Although Hill tentatively suggested that the ionic form of the sulphur dioxide molecule in solution was perhaps important, it was left to Puckett *et al.* (1973) to emphasise strongly this aspect of the overall problem. Thus, in all four species examined at pH 3.2 photosynthesis was completely eliminated by incubation in 75 ppm aqueous sulphur dioxide, whereas at pH 4.4 only in *Cladonia stellaris*, the most sensitive of the four species, was

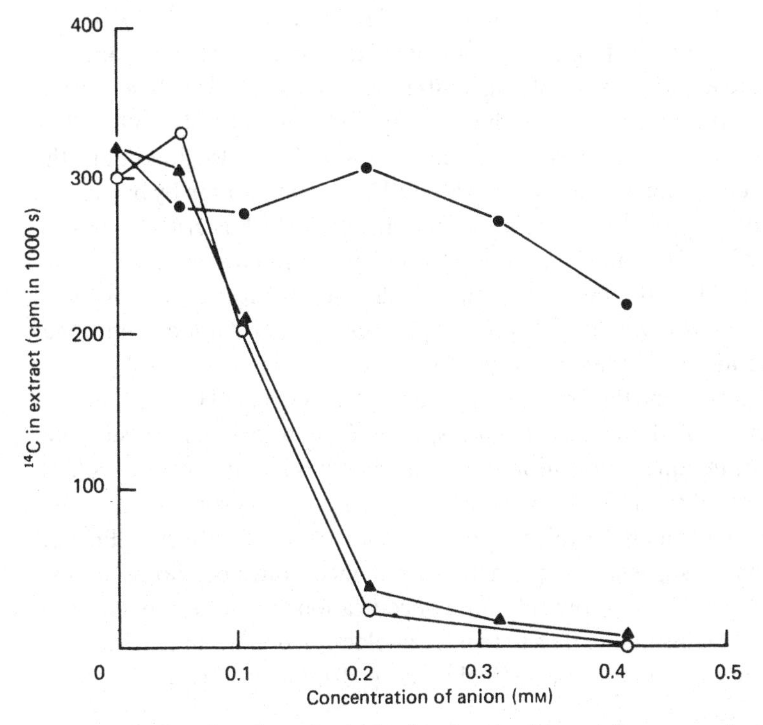

Fig. 56. The effect of sulphite (▲,○) and sulphate (●) concentration on the incorporation of $^{14}CO_2$ in *Usnea subfloridana*. (From Hill 1971.)

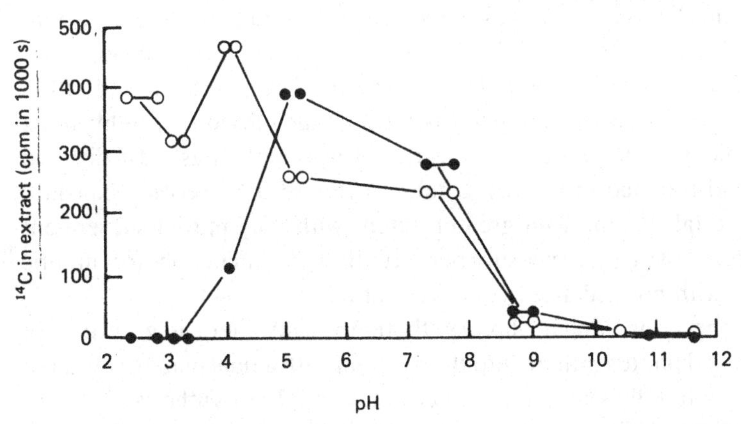

Fig. 57. The effect of pH (○) and pH of sulphite solution (●) on the incorporation of $^{14}CO_2$ in *Usnea subfloridana*. (From Hill 1971.)

photosynthesis completely eliminated (Fig. 58). Puckett *et al.* (1973) concluded that indeed pH and the ionic state of the solution are important, and that there is a wide range of sensitivities between different lichen species. Further, the mechanism of the toxicity includes not only chlorophyll breakdown, which is markedly evident at low pH, but also potentially the interference with electron flow to $NADP^+$ and competition by bisulphate ions with phosphate for the same binding site, which is critical in phosphorylation. Türk and Wirth (1975) also demonstrated a marked dependence on pH of the degree of sulphur dioxide damage to *Hypogymnia (Parmelia) physodes* and *Xanthoria parietina*. Again *Parmelia physodes* was the more tolerant of the two species.

Richardson and Puckett (1973) have emphasised that pH determines the proportion of the different ionic species (sulphurous acid, bisulphite, sulphite) in aqueous solution, and that each has specific attributes. Sulphurous acid results in the irreversible oxidation of chlorophyll at low pH, whereas at higher pH levels electron transport chains are affected. Puckett *et al.* (1973) suggest that the actually enhanced rates of photosynthesis which are evident at low sulphite concentrations relate to the ability of sulphite and bisulphite ions to behave as electron donors and this appears to be a general phenomenon (Puckett *et al.* 1973; Hill 1971; cf. Fig. 55 and Fig. 58).

Subsequent work has shown that some lichen species can recover from a single exposure to sulphur dioxide whilst others cannot (Hill 1974); that repeated exposures are cumulative; and that in addition to the range of toxicity mechanisms previously reported, membrane damage is also involved (Puckett *et al.* 1974, 1977; Nieboer *et al.* 1979; Richardson *et al.* 1979). Thus, $^{14}CO_2$ fixation by *Umbilicaria muhlenbergii* is reduced within 15 minutes of exposure to sulphur dioxide but the subsequent ability to fix carbon dioxide is largely recovered after 6 h by samples washed in distilled water and retained in a moist condition (Fig. 59). A second exposure reduces total fixation to a greater extent with only partial subsequent recovery, whilst a third exposure period entirely eliminates photosynthetic activity, with no evidence of recovery at all.

With the reduction in photosynthetic capacity there is an efflux of ^{14}C-labelled photosynthate into the medium with a final massive efflux of carbohydrate following the actual cessation of photosynthesis (Fig. 60). There is similarly a large efflux of potassium following exposure to aqueous sulphur dioxide which points to an intracellular source of K^+ in addition to the exchangeable K^+ available extracellularly (Puckett *et al.* 1977). This sulphur-dioxide-induced damage to membranes can be modified by class

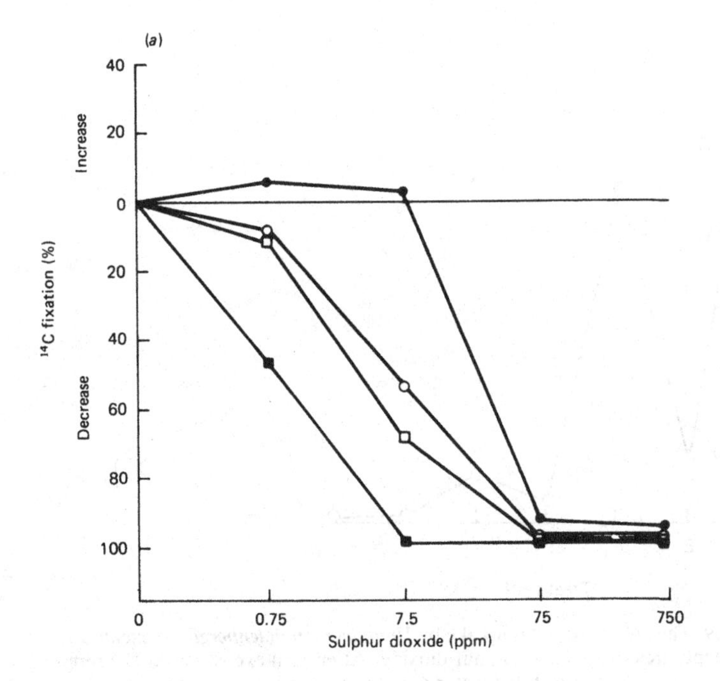

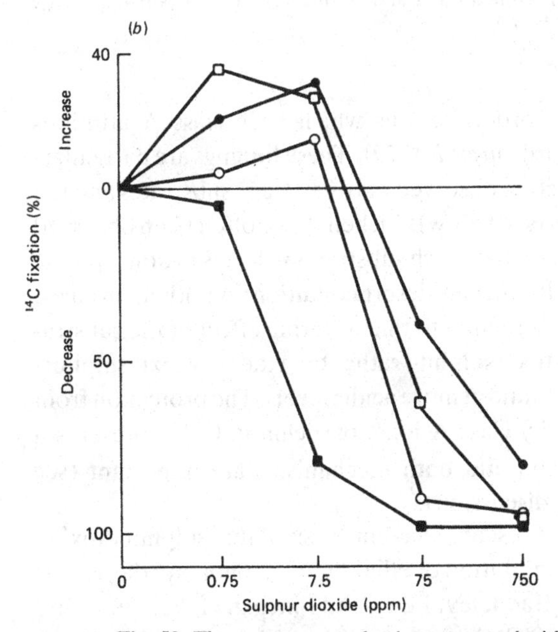

Fig. 58. The percentage reduction compared with controls, of $^{14}CO_2$ uptake in lichens preincubated in solutions of sulphur dioxide buffered to pH 3.2 (*a*) and pH 4.4 (*b*). ■, *Cladonia stellaris*; □, *C. deformis*; ○, *Umbilicaria muhlenbergii*; ●, *Stereocaulon saxatile*. (From Puckett *et al.* 1973.)

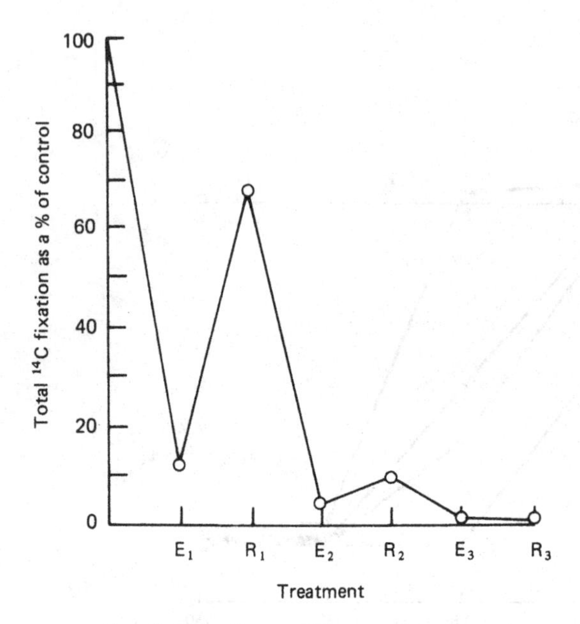

Fig. 59. The effect on $^{14}CO_2$ uptake by *Umbilicaria muhlenbergii* of repeated 15 min exposures to aqueous sulphur dioxide. After each exposure (E_1-E_3) replicates were washed in distilled water and given a 6 h recovery period (R_1-R_3). (From Puckett *et al.* 1977.)

A metal ions and those borderline ions which have class A affinities (Nieboer *et al.* 1979; Richardson *et al.* 1979). These findings are particularly significant to the interaction observed between Ca^{2+} substrates and the ecology of suburban areas (see below). Lichens have often been observed growing quite successfully on base-rich substrates in levels of atmospheric pollution which produce full lichen-desert conditions on other surfaces. This phenomenon has been related to the buffering effect of the substratum, which produces non-toxic sulphite rather than the very toxic sulphurous acid or bisulphite ions found at more acidic levels. The protection from sulphur dioxide afforded by class A ions, particularly Ca^{2+}, presents a complementary explanation and both mechanisms are important (see p.88 for a more detailed discussion).

Although the majority of research has emphasised that sulphur dioxide toxicity is linked to direct inhibition or elimination of photosynthesis, it is evident from the work of Baddeley, Ferry and Finegan (1971, 1972) and Richardson and Nieboer (1983) that other aspects of lichen metabolism are all potentially involved and that, in particular, both respiration and nitrogenase activity are also markedly affected. Baddeley *et al.* (1971) report

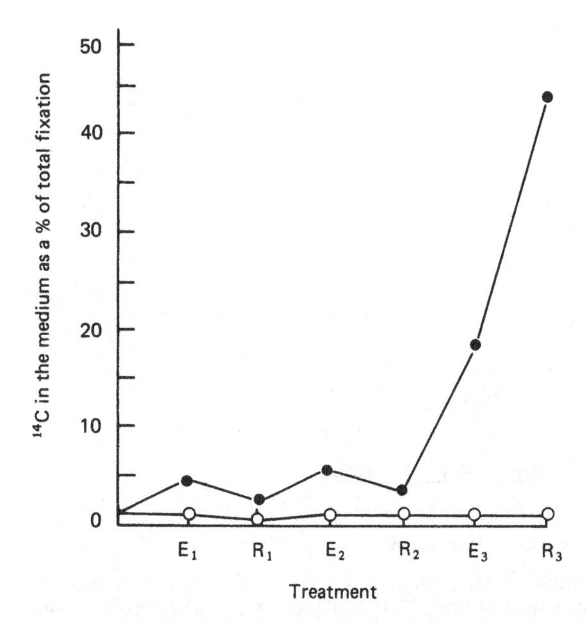

Fig. 60. The effect of repeated 15 min exposures to aqueous sulphur dioxide on the proportion of ^{14}C found in the photosynthetic medium of *Umbilicaria muhlenbergii*. After each exposure (E_1-E_3), replicates were washed in distilled water and given a 6 h recovery period (R_1-R_3). (From Puckett *et al.* 1977.)

an analogous interaction between pH, sulphur dioxide fumigation and the elimination of respiration in *Cladonia impexa* (Fig. 61) which points yet again to the extreme toxicity of sulphur dioxide in the sulphurous acid ionic form at low pH. Hällgren and Huss (1975) examined the effect of sodium bisulphite ($NaHSO_3$) solution on nitrogenase activity in *Stereocaulon paschale* and showed there was almost complete inhibition at 5×10^{-4} M bisulphite at pH 5.8 (Fig. 62). At pH 6.5, however, nitrogenase activity is reduced but not eliminated and photosynthesis at both pH levels is not significantly affected until 5×10^{-3} M bisulphite. Again there is a marked stimulation of photosynthesis at low bisulphite concentrations and the same interaction between toxicity and pH (Fig. 62).

3.4.2 *The ecological effects of sulphur dioxide*

Although the sensitivity of lichens to 'smoke' pollution from major areas of industrial activity has been suspected since the mid-nineteenth century, it was not until ambient concentrations of sulphur dioxide could be monitored in the field that a full realisation of the magnitude of the problem became evident. There is now a very extensive literature available on the subject and no attempt will be made here to

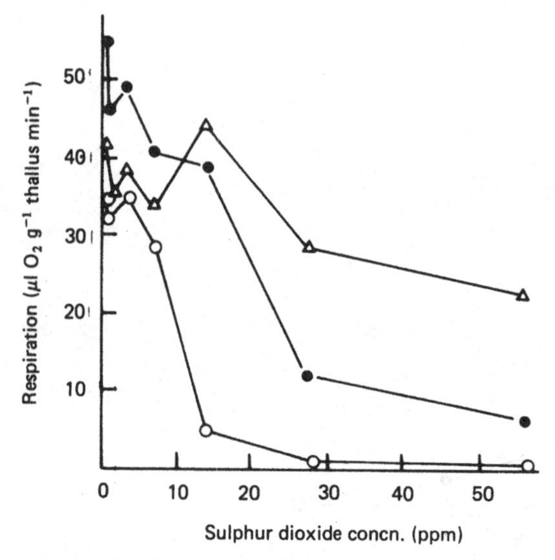

Fig. 61. The inhibition of respiratory activity in *Cladonia impexa* by sulphur dioxide in solution buffered to pH 3.2 (○), pH 4.2 (●) or pH 6.2 (△). (From Baddeley *et al.* 1971.)

cover it comprehensively. Rather, the basic patterns imposed by sulphur dioxide fumigation will be discussed at an ecological level and the general concepts will be established from a small number of selected and well-documented examples. For a more comprehensive treatment the reader is referred to Gilbert (1973), Ferry *et al.* (1973), Hawksworth (1973), Hawksworth and Rose (1976), De Wit (1976) and Richardson and Nieboer (1981, 1983).

The majority of the earlier published work is an extensive documentation from a wide range of geographical locations of the lichen deserts associated with either large urban and industrial areas, or point sources of sulphur dioxide emission. The results are generally expressed as maps and zones and much of the work is repetitive, adding a broader data base rather than new concepts. By 1960-5 correlations were formally established between equivalent zonations of ambient sulphur dioxide concentration and zonations of lichen diversity, as more and more species were eliminated by increasing levels of atmospheric pollutants. The causal nature of the correlation was simply assumed until the experimental work of the late 1960s and early 1970s elaborated the mechanisms of toxicity under controlled laboratory conditions.

The marked zonation of species sensitivity correlating with zones of

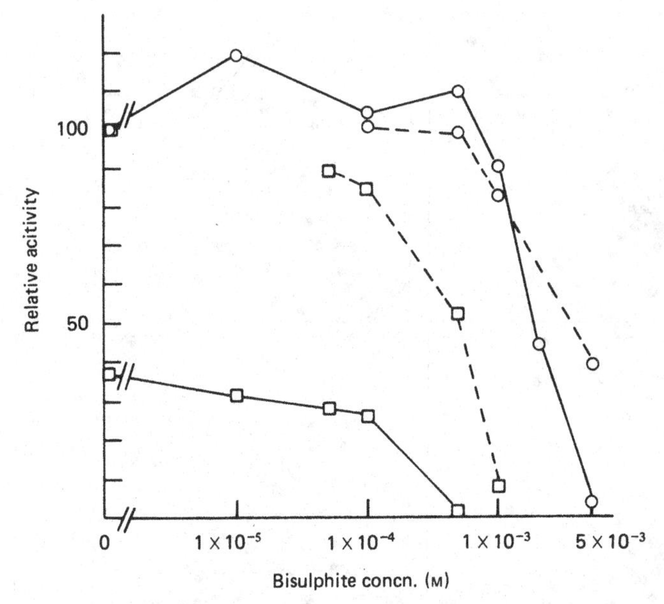

Fig. 62. Relative activities of photosynthesis (○) and nitrogenase (□) in *Stereocaulon paschale* treated with bisulphite solution buffered to pH 5.8 (—) or pH 6.5 (— —). (From Hällgren and Huss 1975.)

sulphur dioxide emission is well illustrated by the data of Morgan-Huws and Haynes (1973) on the distribution of epiphytic lichens around an oil refinery in Hampshire, England. The bulk of sulphur dioxide emission is from the furnace stacks, which thus constitute a point source in an area which is otherwise rural agricultural land. The prevailing wind is from the south-west and the resultant pollution plume also affects what was largely a rural area to the south-east of Southampton but is now an area with increasing levels of ribbon-urbanisation. The lichen survey was restricted to macrolichens on well developed trees (diameter at breast height > 0.5 m) which were free-standing. There is a very clear-cut order of species tolerance, with at one extreme *Lecanora conizaeoides*, which occurred abundantly at all sites examined, and at the other, *Parmelia perlata*, which was completely absent from areas immediately adjacent to the refinery and to a considerable distance downwind of the emission source (Fig. 63). Morgan-Huws and Haynes recognise four zones of lichens which correlate well with the observed zonation of sulphur dioxide concentration (Fig. 64).

Rao and LeBlanc (1967) and Gordon and Gorham (1963) present an even more spectacular example of extreme levels of sulphur dioxide pollution on boreal forest in North America and its epiphytic lichen flora. Blown

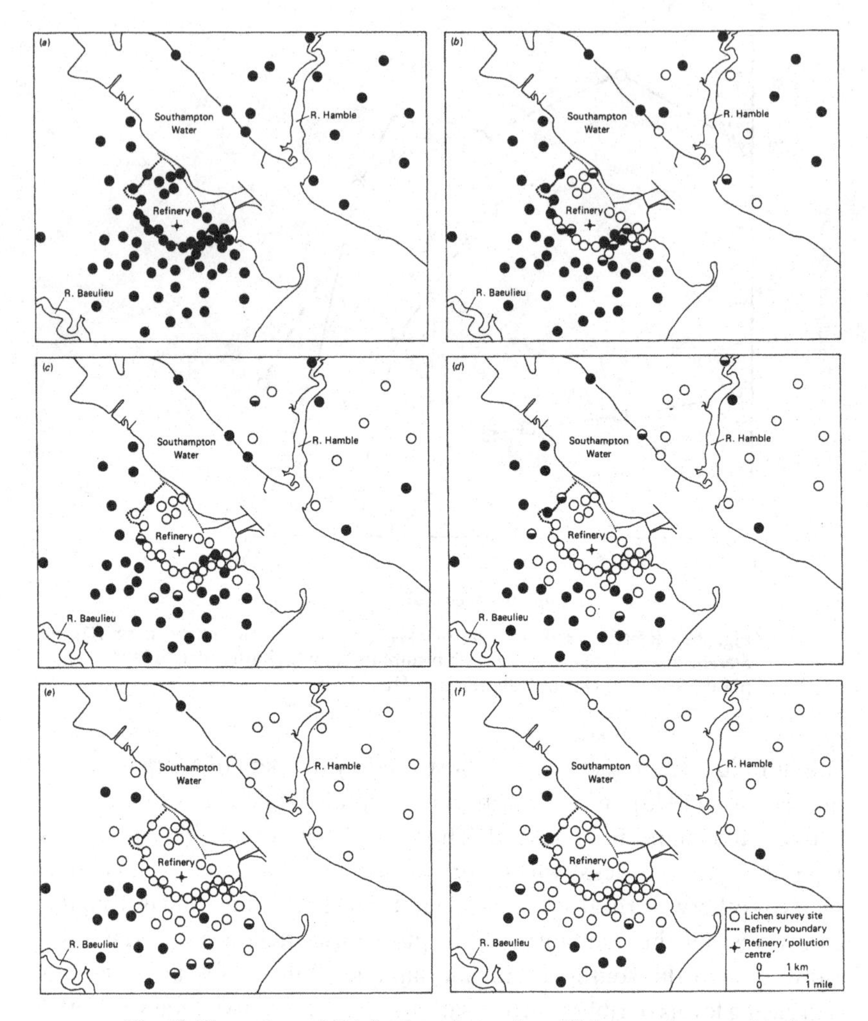

Fig. 63. The pattern of distribution of a number of lichen species on mature trees adjacent to an oil refinery in south-west England. (*a*) *Lecanora conizaeoides*, (*b*) *Hypogymnia physodes*, (*c*) *Evernia prunastri*, (*d*) *Parmelia caperata*, (*e*) *Usnea* spp. and (*f*) *Parmelia perlata* agg. ○, absent; ●, present; ◒, present but only at the base of the trunk. (From Morgan-Huws and Haynes 1973.)

by strong prevailing winds consistently from the south-west, the sulphur dioxide plume from an iron-sintering plant at Wawa in northern Ontario has produced visible damage to higher plants at least 20 miles (32 km) north-east of Wawa. Gordon and Gorham (1963) distinguish four categories of damage ranging from very severe damage with virtually complete elimination of all vegetation to moderate damage where only the tips of the overstorey trees have been killed leaving a normal complement of under-

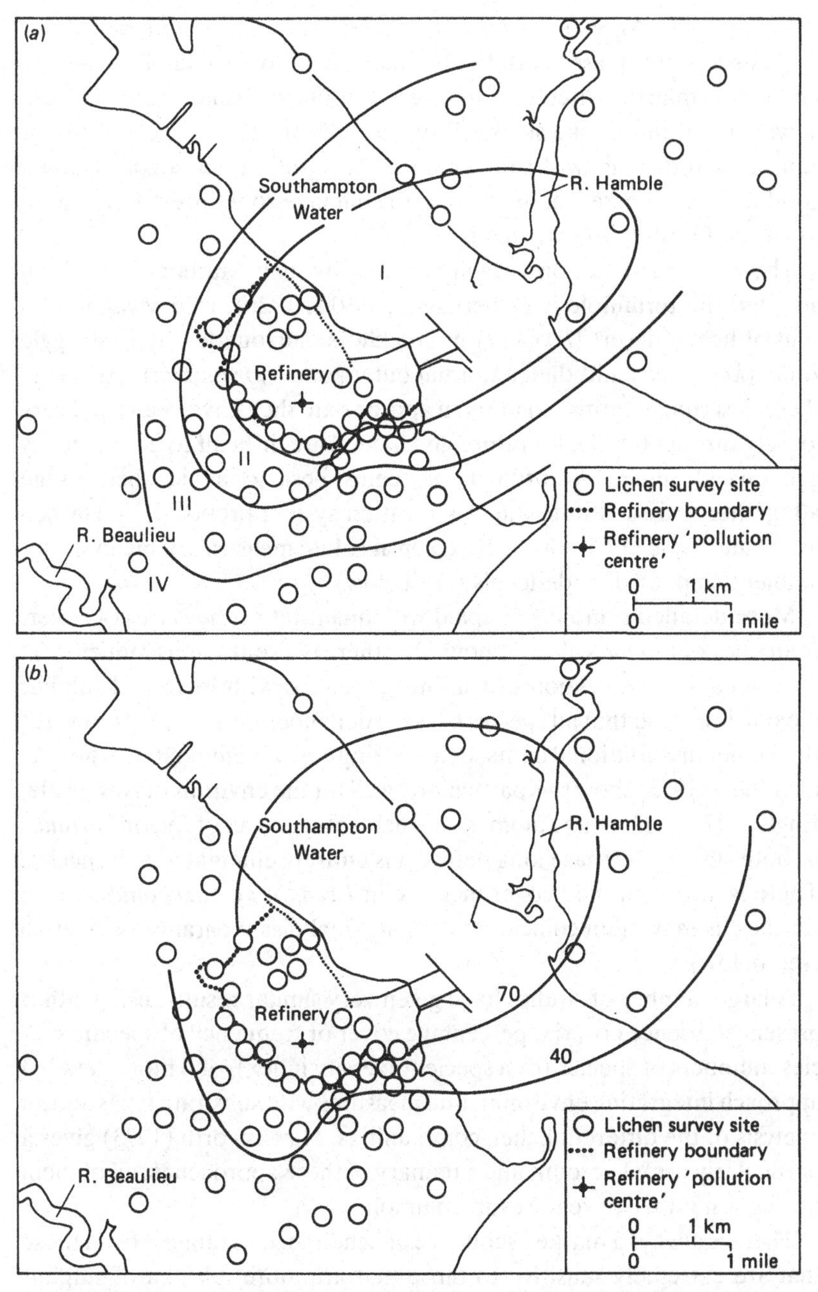

Fig. 64. (*a*) The major zonation of lichens and (*b*) the corresponding zonation of ambient sulphur dioxide levels (μg SO_2 m^{-3}) around the same oil refinery as in Fig. 67. (From Morgan-Huws and Haynes 1973.)

storey species. An examination of the distribution of epiphytic species suggests that they are considerably more sensitive to sulphur dioxide fumigation than the vascular plants, resulting in visible damage up to 48 km downwind of the smoke stacks. Rao and LeBlanc (1967) correlate their equivalent four zones with soil sulphate concentration. These zones show a good degree of correspondence with the damage zones previously established by Gordon and Gorham.

The zone classifications which are used by many authors have often adopted the terminology of Sernander (1912, 1926), who recognised a central lichen desert (*lavöken*) in Stockholm surrounded by a 'struggle zone' (*kampzon*) and then a normal outer zone (*normalzon*). Although these descriptive terms might seem appropriate they have been used very loosely throughout the literature and indeed are difficult to define in any generalised sense. Accordingly it seems best to avoid their usage altogether and substitute the classification system proposed by Hawksworth and Rose (1970) that is based on absolute measures of mean winter ambient sulphur dioxide levels (see below).

More detailed sampling coupled with quantitative measures of cover, frequency or biomass always shows that there is a continuous reduction in biomass and a depression of fruiting associated with increased sulphur dioxide levels, so that all species become depauperate and sterile towards their inner distributional limits, before being entirely eliminated. The data of Gilbert (1973) show this particularly well for the environs of Newcastle, England (Fig. 65), where biomass and percentage cover of *Evernia prunastri* both start to decrease long before it is entirely eliminated as a species. There is also a marked cover increase in *Lecanora conizaeoides*, which reaches its maximum coincident with the final disappearance of Evernia (see below).

A large number of studies have given very similar results, using either presence/absence criteria, percentage cover or frequency of specific species, numbers of species (or a species diversity index), or a more detailed approach integrating environmental measures with substrate types and an analysis of the different lichen communities. Hawksworth (1973) gives a particularly useful account and summary of these approaches, all of them having a number of features in common:

There is always a marked sequence of lichen species ranging from those that are extremely sensitive to those that are more tolerant of sulphur dioxide pollution. *Lecanora conizaeoides*, for example, a widespread crustaceous tree epiphyte in Europe, has long been recognised as pollution-tolerant and indeed its tolerance has been confirmed experimentally

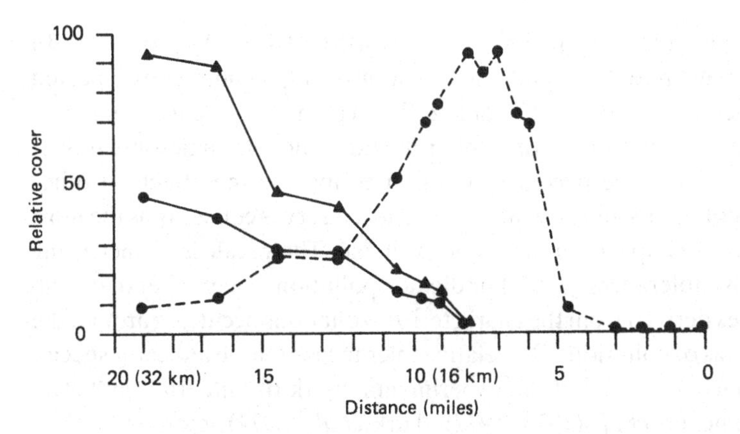

Fig. 65. Relative cover and biomass (oven dry weight) of two lichen species on ash trees along an east-west transect to the west of Newcastle, England. *Evernia prunastri* biomass (▲–▲) and cover (●–●); *Lecanora conizaeoides* cover (●----●). (From Gilbert 1969.)

(see Fig. 55). This species was first recorded in the nineteenth century. Laundon, 1967, indicates 1881-82 as the date when Crombie revisited the Walthamstow area of London and found *L. conizaeoides* 'frequent on pales'. Gilbert (1973) suggests that it is the most widespread epiphytic lichen throughout most of Europe in areas of extreme atmospheric pollution. It is replaced by *Bacidia chlorococca* in North America but Ahti (1965) reports the appearance of *L. conizaeoides* in Newfoundland and it is interesting to speculate that it eventually may also become widespread in North America. *Stereocaulon pileatum* is another example of a lichen which can apparently tolerate high ambient levels of sulphur dioxide; and taking advantage of the lack of competition from other lichen species in urban areas, it has extended its ecological range considerably (Kershaw, 1963; Laundon, 1973). Before 1955, when it was first collected near London, it was restricted to siliceous rocks in the upland areas of Great Britain, but by 1963 it was established in abundance on suburban brick walls in several districts of London. It is significant that Gilbert (1973) records *S. pileatum* as the first lichen to colonise exposed rocks adjacent to a smelter at Fort William in Scotland.

In addition to the handful of species which can tolerate fairly high levels of pollution, *Parmelia physodes* is probably the most widespread lichen on a circumpolar basis found close to areas of extreme sulphur dioxide pollution. It is, however, of a secondary level of tolerance compared with *L. conizaeoides* (cf. Fig. 55). At the opposite end of the tolerance scale are a large number of species, particularly epiphytes, which are extremely sensitive to sulphur dioxide pollution. Hawksworth and Rose (1976) proposed a 10-point scale based on epiphytic lichens growing on both eutrophicated and non-eutrophicated bark (Table 2). Each point on the scale is related to an ambient concentration of sulphur dioxide, and although some refinement may finally be necessary in order to include the effects of other gaseous pollutants and possible synergisms, it serves currently as the most objective yet simple reference scale available. Their scale is defined using the relative tolerances to sulphur dioxide pollution of a number of indicator species derived from the characteristic zonations occurring around the major areas of pollution. The relative tolerances of these indicator species are similarly reflected in the experimental work of Gilbert (1968), Hill (1971), Puckett *et al.* (1973, 1977), Türk *et al.* (1974), etc.

The second fundamental aspect which has emerged from the plethora of descriptive papers on the sensitivity of lichens to air pollution, is the very marked interaction between substrate and the apparent sensitivity of a lichen to sulphur dioxide fumigation. Again, the data of Gilbert (1965, 1969, 1973) show this very clearly. Although the number of species found in all habitats decreases with increasing pollution, the most sensitive species are eliminated first from trees, secondly from sandstone walls and lastly from asbestos roofs (Fig. 66 and Fig. 67). In other urban areas concrete, mortar or other basic substrata in addition to asbestos, produce a very characteristic and abundant lichen cover composed of a small number of species which, although intolerant of sulphur dioxide fumigation, nevertheless extend their range considerably inwards towards a city centre on these base-rich substrata. It is now evident from the work of Richardson and Puckett (1973), Türk and Wirth (1975), Nieboer *et al.* (1979), and Richardson *et al.* (1979), that the explanation of this phenomenon has a number of facets. The basic nature of limestone, mortar and asbestos substrates will buffer dissolved sulphur dioxide to a considerable extent, so that it is in the form of sulphite rather than the more toxic ionic forms present at medium and low pH (Puckett *et al.* 1973). In addition, however, a high concentration of Ca^{2+}, Mg^{2+} or other class A ions imparts directly a degree of protection against sulphur dioxide toxicity (Richardson *et al.* 1979). A basic substratum thus has a dual role.

Gilbert (1976) describes a spectacular example of these effects brought about by the alkaline dust emanating from the crushing machinery in limestone quarries in Derbyshire, England. In North Derbyshire the ambient average winter sulphur dioxide levels are 65 $\mu g\ m^{-3}$, which results normally in the ubiquitous appearance of the corticolous Conizaeoidion association. However, as a direct consequence of quarrying activities in the

Table 2. *Qualitative scale for the estimation of sulphur dioxide air pollution in England and Wales using epiphytic lichens*

Zone	Non-eutrophicated bark	Eutrophicated bark	SO_2 (μg m^{-3})
0	Epiphytes absent	Epiphytes absent	?
1	*Pleurococcus viridis s.l.* present but confined to the base	*Pleurococcus viridis s.l.* extends up the trunk	> 170
2	*Pleurococcus viridis s.l* extends up the trunk; *Lecanora conizaeoides* present but confined to the base	*Lecanora conizaeoides* abundant; *L. expallens* occurs occasionally on the base	~ 150
3	*Lecanora conizaeoides* extends up the trunk; *Lepraria incana* becomes frequent on the base	*Lecanora expallens* and *Buellia punctata* abundant; *B. canescens* appears	~ 125
4	*Hypogymnia physodes* and/or *Parmelia saxatilis* or *P. sulcata* appear on the base but do not extend up the trunk; *Lecidea scalaris*, *Lecanora expallens* and *Chaemotheca ferruginea* often present	*Buellia canescens* common; *Physcia adscendens* and *Xanthoria parietina* appear on the base; *Physcia tribacia* appears in S	~ 70
5	*Hypogymnia physodes* or *Parmelia saxatilis* extend up the trunk to 2.5 m or more; *P. glabratula*, *P. subrudecta*, *Parmeliopsis ambigua and Lecanora chlarotera* appear; *Calcium viride*, *Lepraria candelaris*, *Pertusaria amara* may occur; *Ramalina farinacea* and *Evernia prunastri* if present largely confined to the base; *Platismatia glauca* may be present on horizontal branches	*Physconia grisea*, *P. farrea*, *Buellia alboatra*, *Physcia orbicularis*, *P. tenella*, *Ramalina farinacea*, *Haematomma coccineum* var. *porphyrium*, *Schismatomma decolorans*, *Xanthoria candelaria*, *Opegrapha varia* and *O. vulgata* appear; *Buellia canescens* and *X. parietina* common; *Parmelia acetabulum* appears in E	~ 60
6	*Parmelia caperata* present at least on the base; rich in species of *Pertusaria* (e.g. *P. albescens*, *P. hymenea*) and *Parmelia* (e.g., *P. revoluta* (except in NE), *P. tiliacea*, *P. exasperatula* (in N)); *Graphis elegans* appearing; *Pseudevernia furfuracea* and *Alectoria fuscescens* present in upland areas	*Pertusaria albescens*, *Physconia pulverulenta*, *Physciopis adglutinata*, *Arthopyrenia alba*, *Caloplaca luteoalba*, *Xanthoria polycarpa* and *Lecania cyrtella* appear; *Physconia grisea*, *Physcia orbicularis*, *Opegrapha varia* and *O. vulgata* become abundant	~ 50
7	*Parmelia caperata*, *P. revoluta* (except in NE), *P. tiliacea*, *P. exasperatula* (in N) extend up the trunk; *Usnea subfloridana*, *Pertusaria hemisphaerica* *Rinodina roboris* (in S) and *Arthonia impolita* (in E) appear	*Physcia aipolia*, *Anaptychia ciliaris*, *Bacidia rubella*. *Ramalina fastigiata*, *Candelaria concolor* and *Arthopyrenia biformis* appear	~ 40
8	*Usnea ceratina*, *Parmelia perlata* or *P. reticulata* (S and W) appear: *Rinodina roboris* extends up the trunk (in S) and *Arthonia impolita* (in E) appears	*Physcia aipolia* abundant; *Anaptychia* ciliaris occurs in fruit; *Parmelia perlata*, *P. reticulata* (in S and W), *Gyalecta flotowii*, *Ramalina obtusata*, *R. pollinaria* and *Desmaziera everniodes* appear	
9	*Lobaria pulmonaria*, *L. amplissima*, *Pachyphiale cornea*, *Dimerella lutea* or *Usnea florida* present; if these absent crustose flora well developed with often more than 25 species on larger well lit trees	*Ramalina calicaris*, *R. fraxinea*, *R. subfarinacea*, *Physcia leptalea*, *Caloplaca aurantiaca* and *C. cerina* appear	> 30
10	*Lobaria amplissima*, *L. scrobiculata*, *Sticta limbata*, *Pannaria* spp. *Usnea articulata*, *U. filipendula* or *Teloschistes flavicans* present to locally abundant	As 9	'Pure'

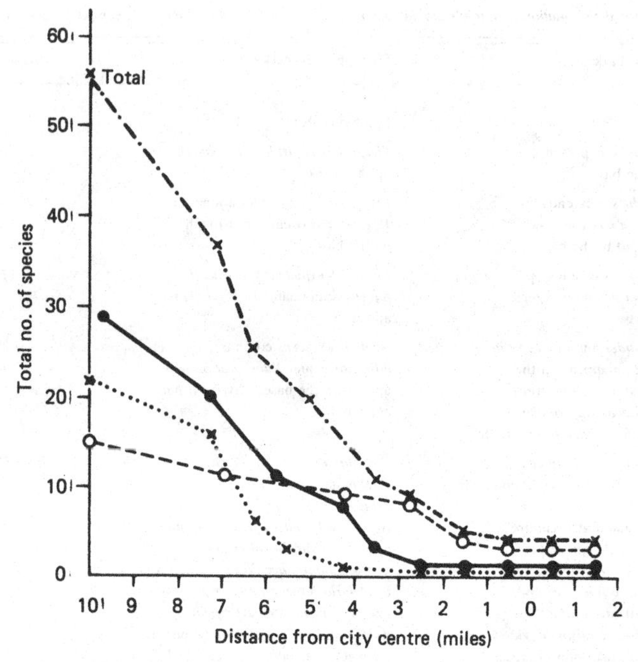

Fig. 66. Transect showing that the number of lichens growing on the tops of sandstone walls (●–●), on asbestos roofs (○----○) and on 'standard' ash trees (×····×) decline as Newcastle is approached from the west. (From Gilbert 1965.)

area, the pH of all tree bark is elevated above pH 6.5 on trees adjacent to the dust source, resulting in the replacement of the Conizaeoidion association by the Xanthorion association containing up to 30 species per tree. As a result trees growing near to heavy dust-pollution sources in the area carry one of the richest epiphytic floras in Great Britain (Gilbert, 1976).

Laundon (1967) also emphasised the importance of substrate on the type and richness of the lichen flora in the suburbs of London; the greatest species diversity occurs on limestone memorials. Laundon recognised too that the fact that *Caloplaca heppiana* formed well-developed associations in old graveyards but was absent from nearby cemeteries of more recent date, pointed to its relict nature. Presumably the propagule establishment phase of some lichens may be considerably more sensitive to sulphur dioxide fumigation than the mature thallus.

3.5 The mineral ionic environment in the field

It is now evident from the discussions presented in Sections 3.1 and 3.2 above, that many of the simple elemental analyses of lichen thalli that have appeared in the literature may be of little relevance to an

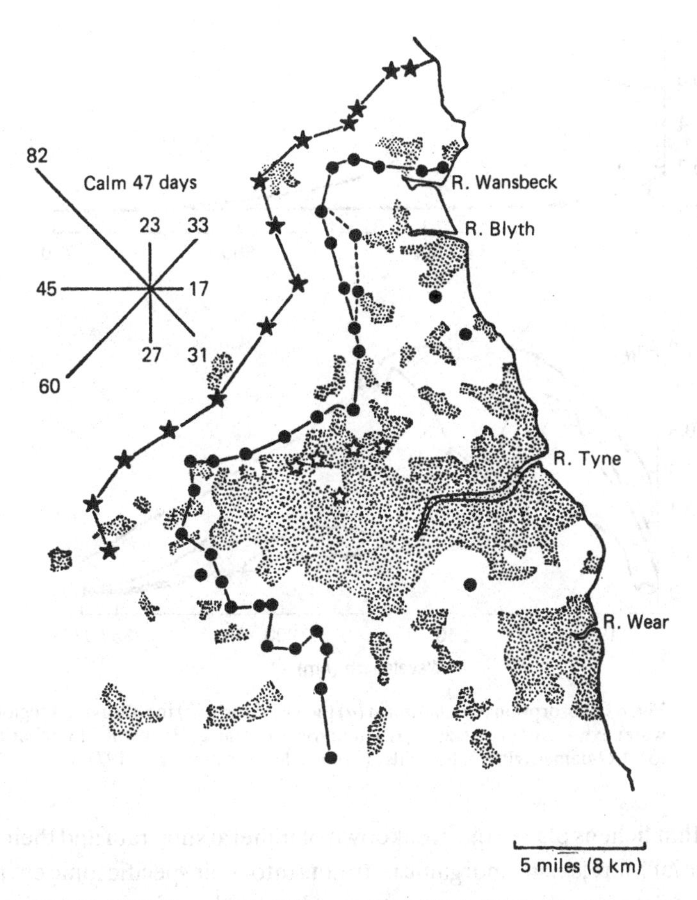

Fig. 67. Map of lower Tyne Valley and adjacent areas showing the continuously built-up area (stippled) surrounded by concentric lichen deserts: on asbestos (☆), on sandstone (●) and on ash trees (★). Wind star for Newcastle 1960-64. (From Gilbert 1965.)

understanding of the ecology of the lichens. The binding of cations to extracellular uptake sites as well as the potential widespread occurrence of particulate material within the lichen thallus presents a formidable problem which necessitates a detailed and exact analysis of the form, location, source, concentration, season of availability and potential physiological importance of both cations and anions. The ionic composition of rain water, soil water, stem flow and runoff from rock surfaces can be obtained fairly readily, but the active nutrient requirements of lichens are completely unknown and with our present inability to grow them successfully under controlled conditions, nutritional growth experiments are impossible.

There has also been a considerable degree of controversy over the role, if

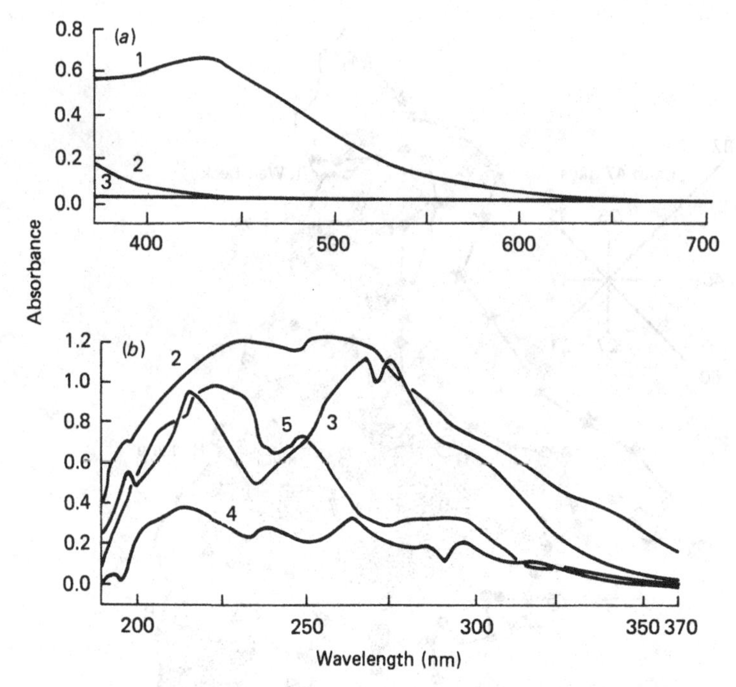

Fig. 68. Absorption of radiation in (*a*) the visible and (*b*) the ultraviolet region, by water extracts of (1) salazinic, (2) fumarprotocetraric, (3) evernic, (4) lobaric and (5) 4-*O*-demethylbarbatic acids. (From Iskander and Syers 1971.)

any, that lichens play in the breakdown of mineral substrata and their own potential for releasing inorganic nutrients into their specific ionic environment. Classically lichens were suggested to be the primary colonisers of rock surfaces and the initial agents of rock weathering and soil formation. More recently this simple lithosere concept has been challenged. However, several recent studies have now shown that some lichen acids have at least a limited solubility in water and furthermore that when in solution can indeed interact with a number of mineral types.

3.5.1 *Solubility of lichen acids and their chelation properties*

It has usually been considered that lichen compounds are insoluble in water (A.L. Smith 1921; D.C. Smith 1962; Haynes 1964) and as a result their active role in mineral weathering has been ignored or discounted. Iskander and Syers (1971) have examined a number of lichen compounds of depside, depsidone configuration (Asahina and Shibata 1954; Culberson 1969) for their solubility by shaking in distilled water and examining the supernatants spectrophotometrically in both the visible and

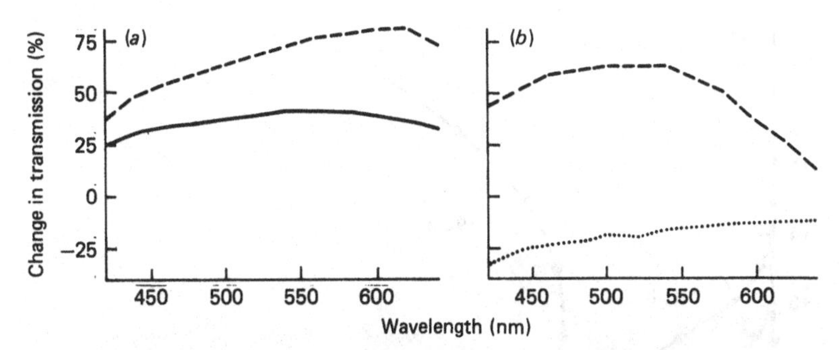

Fig. 69. Percentage change in light transmission following treatment of granite (----), marl (····) and mica (—) suspensions with (*a*) *Parmelia conspersa* and (*b*) *P. furfuracea* homogenised thalli. (From Schatz 1963.)

ultraviolet regions of the spectrum. Their results show clearly that although only salazinic acid gave a coloured supernatant with clear absorption in the visible region, all of the lichen compounds examined have strong absorbence in the ultraviolet region (Fig. 68). Lack of a coloured supernatant is not an adequate criterion of the insolubility of a lichen compound. Since the structure of many of these molecules is consistent with the presence of groups which can function as electron donors, their solubility in water suggests strongly that they are probably also involved in metal complexing reactions under field conditions. Indeed, Schatz (1963) has shown a number of marked interactions by a number of lichen species with mica, granite or marl. Lichen thalli as well as purified physodic and lobaric acid induced coloured complex formation which was examined spectrophotometrically in the visible range. *Parmelia conspersa* was particularly reactive with granite and mica, the supernatant solutions developing a reddish colour within 3-4 h (Fig. 69). Of equal interest is the reactivity of a Colorado specimen of *P. furfuracea* with granite and mica.

More recently Iskander and Syers (1971) have quantified the amounts of iron, aluminium, calcium and magnesium ions released from silicate materials by solutions of lichen acids. The results clearly show the release of these cations, presumably as complexes. Thus the occurrence of any two of the electron donor groups –OH, –CHO, and –COOH in ortho positions (adjacent positions) can result in chelation or the formation of a ring structure with an otherwise unavailable cation. Both Jackson and Keller (1970) and Ascaso, Galvan and Ortega (1976) have demonstrated chemical and morphological changes in rocks by *Stereocaulon vulcani* and *Parmelia conspersa* and there seems little doubt that many lichens can actively contribute to their own ionic environment. Such activity will not be necessarily restricted to saxicolous or terricolous species since the probably

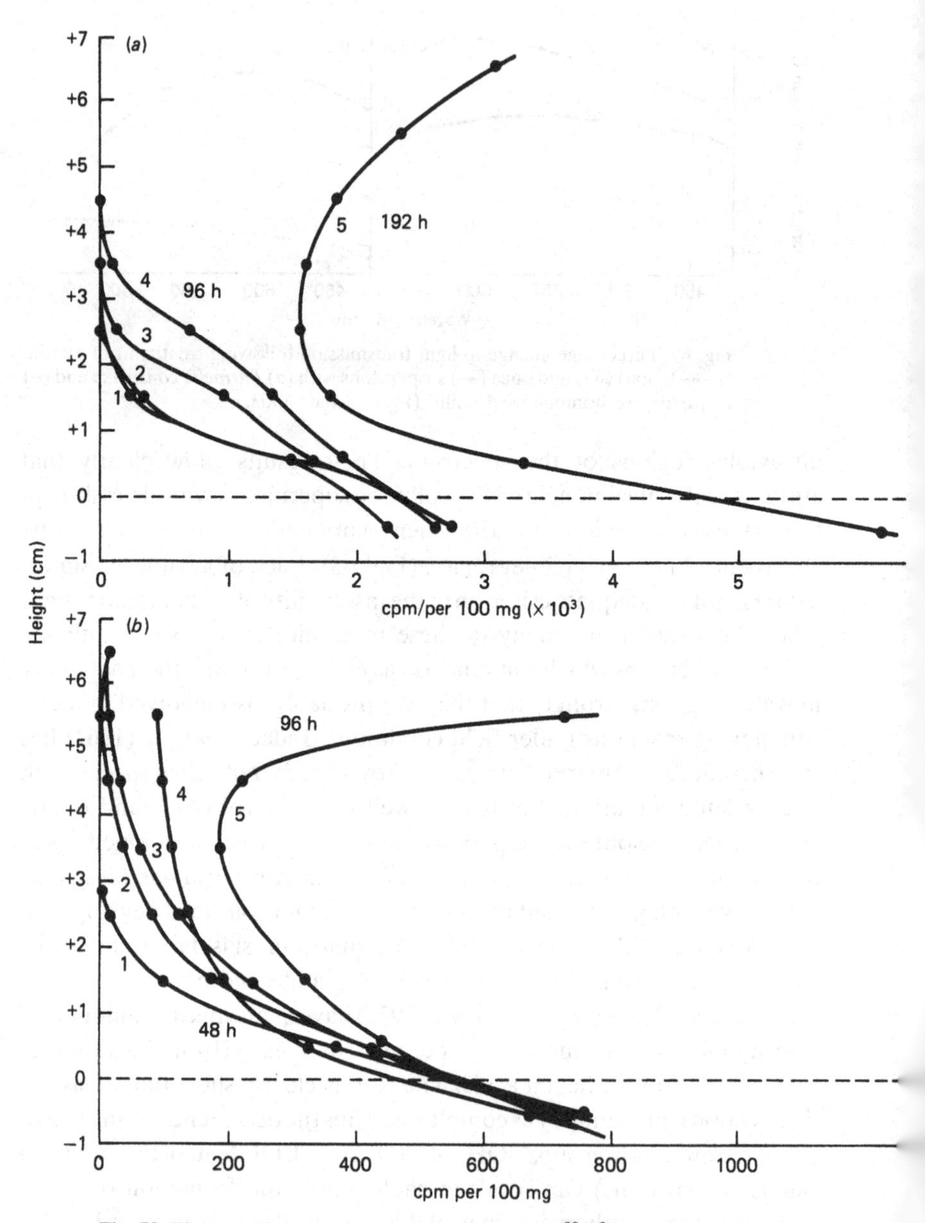

Fig. 70. The rate of movement of (*a*) 50 μCi per 7 ml $^{90}Sr^{2+}$ and (*b*) 1.1 μCi per 7 ml $^{137}Cs^{+}$ in fully hydrated podetia of *Cladonia stellaris* under a range of drying gradients induced by varying quantities of silica gel above the experimental podetium. In (*a*): 1, 0.00 g silica gel; 2, 1.20 g; 3, 2.39 g; 4, 3.97 g; 5, 5.90 g. In (*b*): 1 and 2, 0.00 g silica gel; 3, 0.58 g; 4, 1.65 g; 5, 3.59 g. (From Tuominen 1968.)

widespread existence of particulate material within the thalli of corticolous lichens will similarly provide, either by chelation or as an interaction with the acidic nature of the thallus, an important inorganic source of cations (Nieboer *et al.* 1978), and this is implicit in the data of Gilbert (1976) discussed above (Section 3.4.2). During periods of thallus hydration, the chelation of any particulates trapped in the medulla will provide inorganic ions which can then be translocated throughout the thallus. Again translocation is dependent on periods of thallus hydration, and Tuominen (1968) has shown that the movement of cations within the thallus is consistent with a diffusion model, modified by ion exchange.

The bottom 1 cm portions of 7 cm long, fully hydrated but dead podetia of *Cladonia stellaris* were immersed either in ^{137}Cs or ^{90}Sr solutions and left in closed tubes with a silica gel 'moisture sink' at the top of each tube to generate a moisture gradient. Varying quantities of silica gel were used to alter the gradients over different time periods. The results are clear cut (Fig. 70) with bivalent $^{90}Sr^{2+}$ travelling much more slowly than $^{137}Cs^{+}$ from the base to the apex of the podetia. For example, after 4 days in a moisture gradient induced by 0.58 g of silica gel, ^{137}Cs is detectable at the ultimate tips of the podetia whereas under a more extreme gradient $^{90}Sr^{2+}$ is only detectable halfway up the podetia. It is also evident that the more extreme moisture gradients markedly affect the rates of translocation and, as Nieboer *et al.* (1978) point out, the marked moisture and temperature gradients which also develop in lichen mats will result in the very effective uptake of mineral nutrients from the substratum with accumulation in the actively growing tips. Tuominen (1968) shows that the data are consistent with a simple diffusion model which is modified by ion exchange criteria. The very efficient uptake of $^{90}Sr^{2+}$, for example (Fig. 70), particularly reflects this ion exchange aspect of the whole process.

Corticolous lichens have an additional nutrient source from the canopy throughfall as well as from stem flow. The amounts of organic and inorganic matter derived from a sessile oak canopy, for example, have been well documented by Carlisle, Brown and White (1966, 1967) and represent a significant increase over the limited nutrient content of rainfall. The nutrient content of the canopy and particularly the bark of different tree species does vary considerably and probably affects at least to some extent the range of species which are specific to certain trees. This has been reviewed exhaustively by Barkman (1958), but due to the limitations imposed by the absence of more critical pH and nutrient data, and a lack of detailed microclimate information, it will not be included here.

Lange, Reiners and Heier (1976) and Pike (1978) present data which show that there is a considerable potential for an actual loss of inorganic nutrients from the epiphytic lichen thalli themselves, at least immediately following rehydration. It is not clear from their experimental data whether the demonstrated flux of cations was partially or totally from inter- or extracellular sources, and it is currently impossible to predict the effect throughout the wetting and drying cycles of an input of canopy bark nutrients as throughfall and stem flow on the overall ionic balance of a lichen. There is an additional complication in that Tuominen (1967) suggests that epiphytic lichens have much higher levels of uronic acid, one of the cation uptake sites, and may thus rapidly and efficiently replace their initial nutrient losses which have occurred immediately following hydration.

3.5.2 *Substrate pH*

It is evident from many accounts that the pH of the natural substratum of a lichen is of considerable importance (Hale, 1955; Culberson, 1955; Barkman 1958; Brodo, 1968, 1973; James, Hawksworth and Rose 1977; etc.) and the markedly different speciation of limestone rocks in particular has long been recognised. However, no information on the possible mechanisms of a direct interaction between pH and lichen distribution is available. The evidence from higher plant ecology accumulated over the past 50 years suggests that it is very unlikely that there is indeed any direct link between hydrogen or hydroxyl ion concentration and the ecology of a species. It is much more probable that the variation with pH of the solubility of a number of cations is responsible for the marked calcicole-calcifuge distribution of many lichen species. This has been well documented for higher plants and in particular the solubilities and ionic forms of iron, manganese and aluminium which change markedly with pH and which are all potentially toxic have been implicated in distribution patterns. In addition phosphorus absorption may be impeded as the ionic form $H_2PO_4^-$ at the acidic end of the pH range shifts to HPO_4^{2-} at neutrality.

Clymo (1962) investigated the ecology of *Carex lepidocarpa* and *C. demissa*, characteristic species of wet 'calcicole' and wet 'calcifuge' habitats respectively. His data show clearly that the calcicolous *C. lepidocarpa* would be excluded from a habitat with Al^{3+} concentrations above *c.* 1 ppm and Ca^{2+} concentrations below *c.* 30 ppm. Clarkson (1969) similarly defines three major attributes of calcifuge species: (1) an ability to deal with a low calcium supply; (2) a low phosphorus supply due to its precipitation at high Al^{3+} concentrations; and (3) tolerance of high Al^{3+} ion concentrations.

The toxic effects of Al^{3+} in higher plants include extensive membrane damage particularly in the choroplast. Thus, Hampp and Schnabl (1975) have shown that $^{14}CO_2$ fixation in spinach is significantly inhibited by 10 mol l^{-1} Al^{3+} and at 500 mol l^{-1} Al^{3+} the inhibition is 90% compared with untreated controls. The activity of both ribulose-5-phosphate kinase, and RuBP carboxylase is barely affected by Al^{3+} concentrations below 500 mol l^{-1}. It seems probable that there is a similar level of sensitivity to Al^{3+} in calcicole lichen species, with a corresponding ability in a calcifuge species to tolerate the higher levels of Al^{3+} associated with low pH. Confirmation of this suggested relationship of siliceous and calcicolous species is one of the more outstanding gaps in our knowledge of lichen ecology. It is also evident that corticolous species will be equally exposed to a range of ionic environments induced by the contrasting pH of the bark of the host tree, although the impact of the ionic criteria will often be strongly modified by the moisture, thermal and light regimes of the environment. The seashore environment presents an even more complex and confusing situation, as do surfaces exposed to bird excrement. There is no general agreement as to the definitive features these apparently contrasting environments have in common but it seems likely that ionic criteria are centrally important (Brodo, 1973; Fletcher, 1976). Again there is an outstanding gap in our understanding of the maritime lichen environment.

3.5.3 *Allelopathy*

The potential activity of lichen aqueous extracts as allelopathic agents was investigated several times in the 1960s during the period of general ecological interest in direct inhibition of one species by another. Follmann and Nakagara (1963) applied thallus homogenates and aqueous extracts of *Sticta weigelii* and demonstrated both a delay of seed germination in *Phaseolus* and an inhibition of growth. Similarly Miller *et al.* (1963) and Miller and Schaefers (1964) using *Umbilicara papulosa* extracts demonstrated its allelopathic properties on several species of fungi and higher plants. Ethanol extracts of macerated lichens also inhibited root growth in germinating cucumber seeds. However, these approaches are unrealistic in a field situation and open to the same criticisms that were made of much of the earlier allelopathic work. Pyatt (1967) used both an aqueous extract of homogenised *Peltigera canina* and aqueous washings of sterile grit upon which the lichen had been 'grown' and kept moist for several days, which Pyatt suggests is closely similar to natural field conditions. His data for root initiation and growth in seeds of *Festuca ovina* and *F. rubra* (both of which compete with *Peltigera* in its natural habitat), show strong inhibition by both extracts, although leaf growth was not significantly affected. More recently Lawrey (1977*a*) suggested, from the evidence of moss spore inhibition by acetone extracts of a number of *Cladonia* species, that lichen acids may be the active allelopathic components. He has subsequently demonstrated that evernic acid and squamatic acid were indeed both capable of inhibiting moss spore germination (Lawrey, 1977*b*).

Although the majority of the evidence available on allelopathy is taken from laboratory experimental situations and is of questionable validity in the field, there are very few field data to complement it. Brown and Mikola (1974) attempted to show the allelopathic effects of *Cladonia stellaris* under field conditions but unfortunately their data lack environmental control and can be interpreted in a number of ways. For example, the extensive and deep cover of *C. stellaris* in northern boreal regions interacts to a considerable extent with the growth rate of the associated spruce trees. It can act as a physical barrier preventing seed germination and establishment, and it can induce conditions at the soil surface which strongly inhibit seed germination and establishment. Both mechanisms are probable. Furthermore the lichen mat strongly affects the thermal root environment of mature trees and thus can potentially affect the growth rate of its major light competitor (Kershaw and Field, 1975; Rouse, 1976; Kershaw, 1977*a*). Brown and Mikola (1974) showed that aqueous extracts of *C. stellaris* in particular, actively inhibited a number of cultured fungi found commonly as ectomycorrhizae. *Paxillus involutus*, for example, in pure culture was not only inhibited by all of the lichen extracts but would not form the usual mycorrhizal association with *Pinus silvestris* at all when in the presence of *C. stellaris* extract. As a result there was a marked reduction of ^{32}P uptake by the *Pinus* seedlings.

This work has subsequently been extended to the field situation by mulching tree seedlings with mats of *C. stellaris*. However, no controls were established such as mulching with straw or plastic, and the apparent inhibition of seedling growth can be readily explained by changes in soil microclimate, particularly soil temperature, which will also interact with phosphorus uptake. The problem has recently been re-examined by Cowles (1981) who investigated a number of combined treatments on the growth of mature spruce trees. The treatment plots were established to examine four possible mechanisms by which the *C. stellaris* mat could influence tree growth; (1) soil temperature, (2) soil moisture, (3) nutrient

Table 3. *Summary of the different treatments used in studying growth of mature spruce trees to vary soil temperature and moisture, presence of allelopathic agents, and nutrient cycling*

Treatment	Symbol	Description
Control	CTRL	Control plot, no perturbations
Control, bare (two plots)	CBAR	Lichen cleared to serve as control for other cleared areas
Clear polyethylene (four plots)	EXPT	Cleared and covered with clear polyethylene to cause rise in soil temperature and fall in soil moisture
Clear polyethylene with fertiliser (two plots)	EXPH	Cleared and covered with clear polyethylene after fertilisation with 112 kg P ha^{-1}
Bare, fertilised: insulated	BFI	Covered with opaque polyethylene after lichen cleared and area fertilised with 112 kg P ha^{-1}
Bare, fertilised: no. 1	BF1	Fertilised with 112 kg P ha^{-1} after removing lichen
Bare, fertilised: no. 2	BF2	Fertilised with 112 kg P ha^{-1} and 168 kg N ha^{-1} after removing lichen
Lichen extract	LE	Covered with opaque polyethylene after lichen cleared and aqueous extract of lichen applied.
Lichen, fertilised: no. 1	LF1	Woodland area fertilised with 112 kg P ha^{-1}
Lichen, fertilised: no. 2	LF2	Woodland area fertilised with 112 kg P ha^{-1} and 168 kg N ha^{-1}
Lichen, fertilised: no. 3	LF3	Woodland area fertilised with 168 kg N ha^{-1}

interactions, and (4) allelopathy. Plots were cleared of all lichen and clear plastic was used to modify soil temperature and moisture, or was combined with the application of an aqueous extract of the lichen mat. The amount of extract added was equivalent to the average yearly rainfall. The combination of treatments and their results on the growth of spruce are given in Table 3. Although all the desirable treatment combinations could not be included, it is evident that removal of the lichen mat – which resulted in increased soil temperatures, lower soil moisture, and possibly a loss of nutrient input from the decomposing base of the mat – induced a significant reduction in branch elongation in the spruce (CBAR, Table 4). Lichen extracts (LE) also markedly reduced branch growth and there is little doubt that the effect of lichen leachate is a real phenomenon in lichen woodlands. It is not a dominant factor, however, since tree growth with the intact mat (CTRL) was significantly better than without it (CBAR). Application of fertiliser significantly increased growth but not when accompanied by removal of the lichen mat (BFI and BFI). The loss of soil moisture in all treatments appears to be profoundly important and Cowles (1982) concludes that the lichen mat enhances tree productivity because of

Table 4. *Growth response in 1980 expressed as % change of bole diameter or branch elongation measurements*

Plot symbol	*n*	Branch elongation (% change)
CTRL	222	NA
BFI	81	−9.61
BF1	80	0.2[a]
BF2	81	28.9*
CBAR	162	−15.6*
EXPH	161	−7.74*[b]
EXPT	260	−5.18
LE	81	−53.0*
LF1	81	9.39*
LF2	82	35.3*
LF3	80	46.3*

NA, not applicable.
[a] Mean was significantly less than control mean in 1979.
[b] Mean was significantly greater than control mean in 1978 and 1979.
* $P < 0.05$.

the maintenance of soil moisture and despite the allelopathy coupled with low soil temperatures.

In conclusion it appears that in some instances allelopathy is indeed of some importance in allowing certain lichen species to avoid undue competition from higher plants, mosses or perhaps other lichens.

4

Nitrogen fixation in lichens

Nitrogen fixation involves primarily the conversion of atmospheric nitrogen to ammonia and is mediated by the enzyme nitrogenase. It constitutes the first step which leads to amino acid and hence protein synthesis in a variety of organisms. In lichens, it is restricted to those species which contain heterocystous blue-green algal (cyanobacterial) symbionts. The process is one of chemical reduction and is markedly inhibited by oxygen in cell-free extracts. In heterocystous blue-green algae the enzyme is excluded from the deleterious effect of ambient oxygen tension by a thick heterocyst wall. The nitrogenase enzyme requires a source of reductant and a supply of energy in the form of ATP. The studies so far that have examined the environmental control of nitrogenase activity in lichens have focused almost exclusively on carbohydrate availability in the dark or photosynthetic activity in the light, both of which control the supply of energy in the form of ATP.

In fact these considerations, and that of moisture availability, dominate the control of the field level of nitrogenase activity on a continuous basis. The requirement for water reflects, of course, the poikilohydric nature of all lichens and their lack of metabolic activity in the absence of at least some degree of hydration.

Nitrogen fixation is normally measured by use of the ^{15}N isotope of nitrogen. Although this yields exact data, it is a cumbersome and lengthy procedure. More recently the reducing abilities of the nitrogenase enzyme have been exploited by substituting nitrogen by acetylene in excess. Acetylene is efficiently reduced by the enzyme to ethylene, which can be readily and quickly measured by gas chromatography, providing an alternative and easier measure of nitrogenase activity. As a result the majority of the work on nitrogenase activity in lichens has utilised this approach and the fixation rates of nitrogen are assumed (often incorrectly) to follow a comparable pattern.

All lichens with heterocystous blue-green algal phycobionts so far examined contain active nitrogenase (Tables 5 and 6). *Nostoc* is by far the most important phycobiont involved, either as the constituent algal component of the thallus, or as a secondary algal component contained in superficial or internal cephalodia. Millbank and Kershaw (1973) and Millbank (1976) give a useful general survey of the lichen genera and their respective algal counterparts. However, the more recent work now available indicates that there is little doubt that all lichens containing heterocystous phycobionts will show nitrogenase activity. The evidence also suggests that the rates of activity are comparable with those of free-living blue-green algae (Millbank, 1974*a*). Considerable variation in levels of nitrogenase activity has been found in replicate thalli of a given species and

Table 5. *Nitrogen content, nitrogenase activity, and heterocyst frequency in lichens with one blue-green phycobiont*

Lichen species	Thallus N content %	nmol C_2H_4/mg^{-1} thallus N h^{-1}	nmol C_2H_4/mg^{-1} thallus dry wt h^{-1}	Heterocyst frequency (% total algal cells)	Nos. of thallus specimens examined
Collema auriculatum	3.60	15.10	0.54	2.7	1
C. crustatum	4.39	n.d.	n.d.	2.1	1
C. fluviatile	4.37	3.52	0.15	n.d.	
C. furfuraceum	4.09	22.67	0.93	4.0	2
C. subfervum	3.85	7.34	0.28	2.6	6
Ephebe lanata	5.80	7.60	0.44	n.d.	
Leptogium burgessii	5.51	38.12	2.10	4.5	1
L. cyanescens	n.d.	n.d.	n.d.	2.7	1
L. lichenoides	3.87	22.1	0.86	2.6	5
L. sinuatum	4.31	14.81	0.64	2.4	4
L. teretriusculum	4.50	2.22	0.10	n.d.	
L. tremelloides	n.d.	n.d.	n.d.	3.5	1
Lichina confinis	3.73	4.17	0.16	2.4	1
L. pygmaea	6.49	2.93	0.19	4.6	1
Lobaria scrobiculata	2.77	10.63	0.29	3.9	3
Massalongia carnosa	4.27	19.20	0.82	5.5	1
Nephroma laevigatum	4.22	23.89	1.01	4.1	3
N. parile	n.d.	n.d.	n.d.	4.9	3
Pannaria microphylla	5.14	n.d.	n.d.	4.5	1
P. pezizoides	6.37	1.66	0.11	3.9	1
P. rubiginosa	2.25	15.67	0.35	n.d.	
Parmeliella atlantica	3.90	58.31	2.27	7.3	3
P. plumbea	4.76	18.85	0.90	5.9	1
Peltigera canina	3.28	118.0	3.87	4.87	1
P. evansiana	n.d.	n.d.	n.d.	4.7	6
P. polydactyla	3.5	102.41	3.58	5.8	5
P. praetextata	n.d.	n.d.	n.d.	4.4	10
P. venosa	n.d.	n.d.	n.d.	7.8	3
Placynthium nigrum	2.73	6.48	0.18	2.0	1
P. pannariellum	4.61	15.33	0.71	n.d.	
Polychidium muscicola	3.19	24.85	0.79	2.6	1
Pseudocyphellaria thouarsii	3.90	5.89	0.23	6.7	1
Sticta fuliginosa	4.19	16.34	0.68	6.0	3
S. limbata	3.93	34.46	1.35	4.9	7

n.d., not determined.

even the most rigorous control of pretreatment and experimental variables does not eliminate this problem. Kershaw (1974) commented on the large degree of variability that exists even between replicates taken from the same thallus: these can show two-fold differences in their level of nitrogenase activity (cf. Fig. 74 below). Kershaw (1974) attempted to explain these differences by expressing rates of nitrogenase activity in *Peltigera* in terms of algal numbers, but met with a conspicuous lack of success. With the more recent evidence provided by Hitch and Millbank (1975*a*,*b*) an alternative and more likely explanation may be found in the variation of heterocyst frequency in different replicate thalli or even in different parts of a single thallus. Whatever the reasons for the variability, it is essential to use an adequate number of replicates for each combination set of environmental parameters, throughout the full range of seasonal, temperature and light variation. As a result, the examination of the full range of the response of nitrogenase activity to moisture, temperature, light and time of year is a considerable undertaking.

4.1 The interaction between nitrogenase activity and thallus hydration

The response pattern of nitrogenase activity following thallus

Table 6. *The heterocyst in the blue-green phycobiont filament of cephalodiate lichens*

Lichen species	Heterocyst frequency (% of total algal cells)	No. of estimates
Dendrisocaulon umhausense	1.9-15.2	6
Lobaria amplissima	21.6	2
L. laetevirens	30.4	5
L. pulmonaria	35.6	2
Nephroma arcticum	14.1	1
Peltigera aphthosa	21.1	13
Placopsis gelida	15.0	7
Psoroma hypnorum	10.1	2
Solorina crocea	17.8	6
S. saccata	14.7	3
S. spongiosa	5.8-54.7	12
Stereocaulon vesuvianum	20.5	2

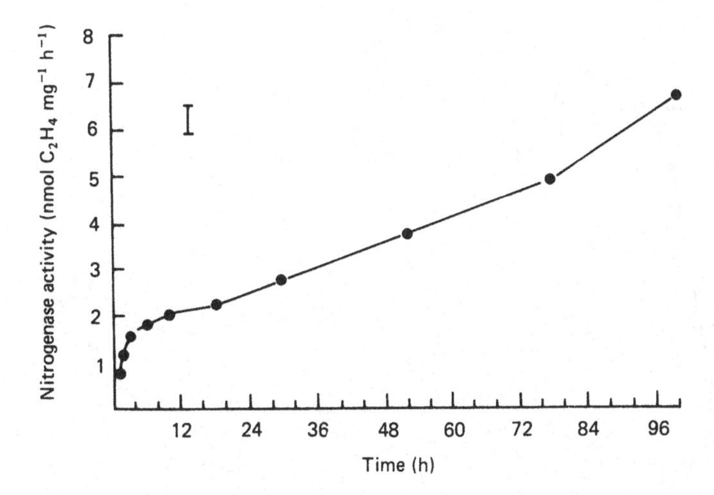

Fig. 71. The effect on nitrogenase activity in *Peltigera polydactyla* of incubation under a continuous atmosphere of 10% acetylene/90% air mixture. The error bar shows the maximum standard error. (From MacFarlane *et al.* 1976.)

rehydration has been examined by MacFarlane *et al.* (1976) in combination with the effects of long-term exposure to acetylene. As any time course experiment using ethylene production as a measure of nitrogenase activity concurrently involves long-term exposure of the material to acetylene, it is essential to separate these two effects.

The stimulating effect of acetylene on nitrogenase activity is usually quite clear. For example in *Peltigera polydactyla* after 100 h of exposure to acetylene, levels of fixation are 6 times greater than in the thalli exposed to air between incubations (Fig. 71). This result is in agreement with the findings of Nielson, Rippka and Kunisaura (1971) and Bone (1971*a*,*b*) and emphasises the need for careful control of experiments involving long-term exposure to acetylene. Stewart (1974) interprets this stimulation of nitrogenase activity as being due to the resulting nitrogen depletion causing a 'derepression' of nitrogenase synthesis. Over and above the long term effects of exposure to acetylene there is a gradual increase in nitrogenase activity following the rehydration of the thallus. The rehydration response time for *Peltigera* appears generally to be from 3 to 5 h. Thus, in *P. praetextata* the nitrogenase activity increases sharply for the first 2 h after rehydration even though full thallus saturation is achieved within the first hour (Fig. 72). The rates continue to increase for a further hour and thereafter remain more or less constant over a short-term experimental period. The same hydration response is also evident in the first 5 h following rehydration of *P. polydactyla*, with a rapid increase in activity over the

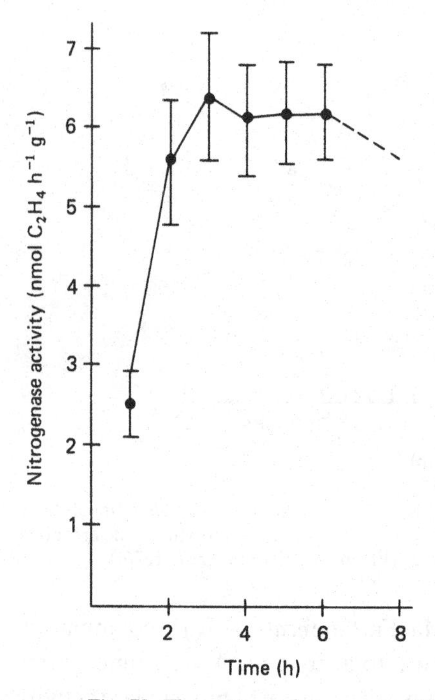

Fig. 72. The pattern of recovery of nitrogenase activity in *Peltigera praetextata* following rehydration of the thallus. (From MacFarlane *et al.* 1976.)

initial 3 h period but with smaller rate increases continuing up to 5 h.

In other species the recovery pattern may be somewhat different. Henriksson and Simu (1971) demonstrated the remarkable ability of the lichens *Collema tuniforme* and *Peltigera rufescens* to recover their nitrogenase activities immediately when moistened after varying periods of desiccation. However, they did not examine the time course of recovery. Hitch (1971) similarly examined the recovery of nitrogenase activity using *Lichina confinis*, *Peltigera canina* and *Collema crispum* as test material. Nitrogenase activity began after lag periods of 60, 20 and 35 min, respectively, as thallus moisture contents rose from the original level of 20% of oven-dry weight to 200% within 5 min. Again, however, these responses to wetting were investigated for 2 h only. The re-establishment of maximum rates of activity in response to moisture pretreatment could be expected to vary from species to species and accordingly must be examined and controlled in each lichen species under investigation. Some of the absolute rates of nitrogenase activity reported in the literature and the subsequent estimates of total nitrogen fixed per year are thus of dubious worth, since rarely has the hydration response been taken into account.

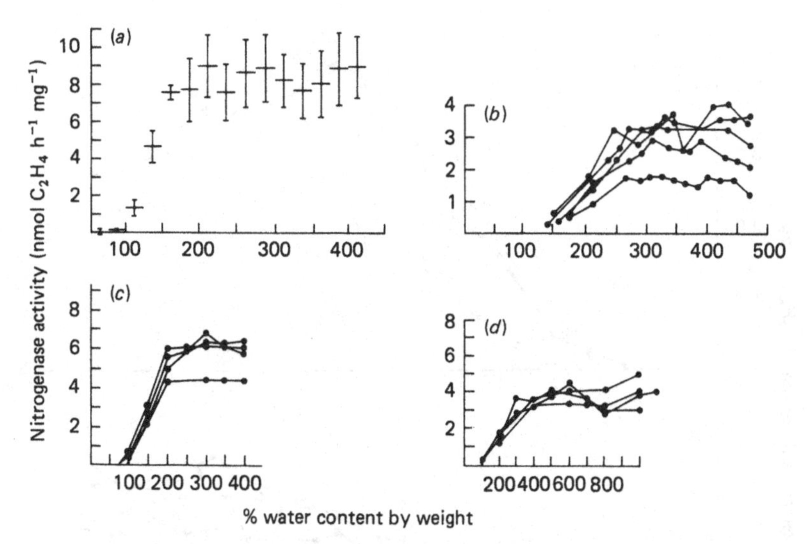

Fig. 73. The constant rate of nitrogenase activity at high to medium levels of thallus hydration in (*a*) *Stereocaulon paschale*, (*b*) *Peltigera praetextata*, (*c*) *P. rufescens*, and (*d*) *Collema furfuraceum*. Each line represents an individual replicate. The bars in (*a*) show the S.E.M.

Once a lichen has fully hydrated and the activity of the nitrogenase enzyme has stabilised, the response to changing levels of thallus hydration is remarkably constant in many species. The level of nitrogenase activity throughout a full drying cycle in *Stereocaulon paschale*, *Peltigera praetextata*, *P. rufescens* and *Collema furfuraceum* is given in Fig. 73. In each case there is negligible ethylene production below 100% thallus water content by weight but virtually constant rates from full thallus hydration down to 150%, 200%, 200% and 300% respectively. However, this relationship is not true for all species. Kershaw (1974) compares the responses of two other species of *Peltigera* with the linear response of *P. praetextata* and shows some differences between them (Fig. 74). These differences could reflect a degree of adaptation to a specific environment, with *P. canina*, for example, showing pronounced adaptation to a 'moist' environment. But neither *P. rufescens* nor *Collema*, both from markedly xeric environments, reflect any level of adaptation at all and accordingly an ecological interpretation as originally suggested by Kershaw (1974) now seems unlikely. However, this marked non-linear response in at least some species of *Peltigera* is important in another context and will be discussed further below.

The effects of long-term dehydration on nitrogenase activity have been examined by Huss-Danell (1977) and Kershaw and Dzikowski (1977) for

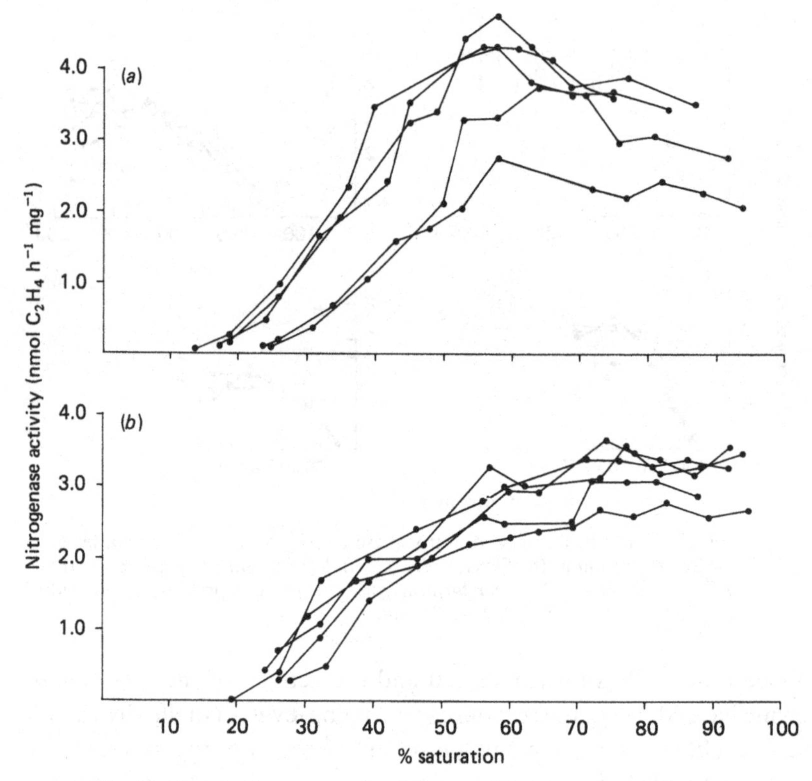

Fig. 74. The non-linear response of nitrogenase activity to thallus hydration in (*a*) *Peltigera evansiana* and (*b*) *P. canina*. Each line represents an individual replicate. (From Kershaw 1974.)

Stereocaulon paschale and *Peltigera polydactyla* respectively. In *P. polydactyla* the recovery rate of nitrogenase activity is rapid after a short drought period, but takes progressively longer as the drought period is extended. For example, after 3 days of storage air-dry, and only 4 h of soaking in the light, full nitrogenase activity was recovered compared with the equivalent rates in freshly collected material (cf. Fig. 75 and Fig. 76). After more extreme experimental levels of dehydration are achieved using calcium chloride, recovery of nitrogenase activity is still incomplete even after 4 h of soaking in the light (Fig. 76). After 66 days the lag phase is considerably more extreme, with 12 h of soaking in the light being required to induce full activity in the air-dry replicates. This treatment, however, fails to induce full recovery in the replicates dried using calcium chloride (Fig. 77).

It is important to recognise that the recovery of activity as a function of the length of a pre-soaking period also includes the effects of exposure to

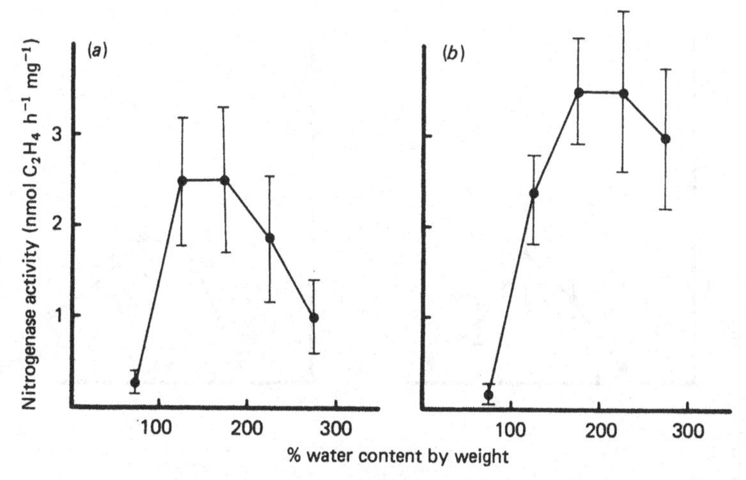

Fig. 75. Nitrogenase activity in the control replicates of *Peltigera polydactyla* at different levels of thallus moisture and 350 μmol m^{-2} s^{-1} illumination at 20 °C. (*a*) after soaking for 1 h; (*b*) after soaking for 4 h. Vertical bars show the S.E.M. of each 50% moisture class. (From Kershaw and Dzikowski 1977.)

light in the 12 h soaking period. MacFarlane *et al.* (1976) have emphasised the importance of light in the control of nitrogenase activity (see Section 4.2 below) and the full recovery of *Peltigera* after 66 days of air-dry storage could be partially due to the 12 h of illumination rather than simply being a rehydration effect. In addition to the direct effect of a period of drought on nitrogenase activity, it is also evident that the pattern of response to thallus moisture also changes. The normal pattern of response of nitrogenase activity in relation to different levels of thallus saturation shows a fairly constant initial rate until thallus dehydration is well advanced (cf. Figs. 75*b* and Fig. 77*b*). After a period of dehydration followed by a recovery period at insufficient hydration, the first two or three incubation periods show reduced levels of activity resulting in a very different response curve (cf. Figs. 75*a* and Fig. 77*c*, for example). This presumably reflects the continuing recovery of nitrogenase activity during the actual experimental period. It seems evident that *P. polydactyla* can quickly recover levels of nitrogenase activity after short periods of field drought. However, after lengthy dry storage under laboratory conditions it is essential that all experimental material should receive an identical pretreatment (MacFarlane and Kershaw 1977; Crittenden and Kershaw 1979). Equally after excessively dry summer periods, minimum recovery periods at full thallus hydration should be at least 12 h and sometimes may need to be longer.

In contrast to these findings, Huss-Danell (1977) reports no deleterious

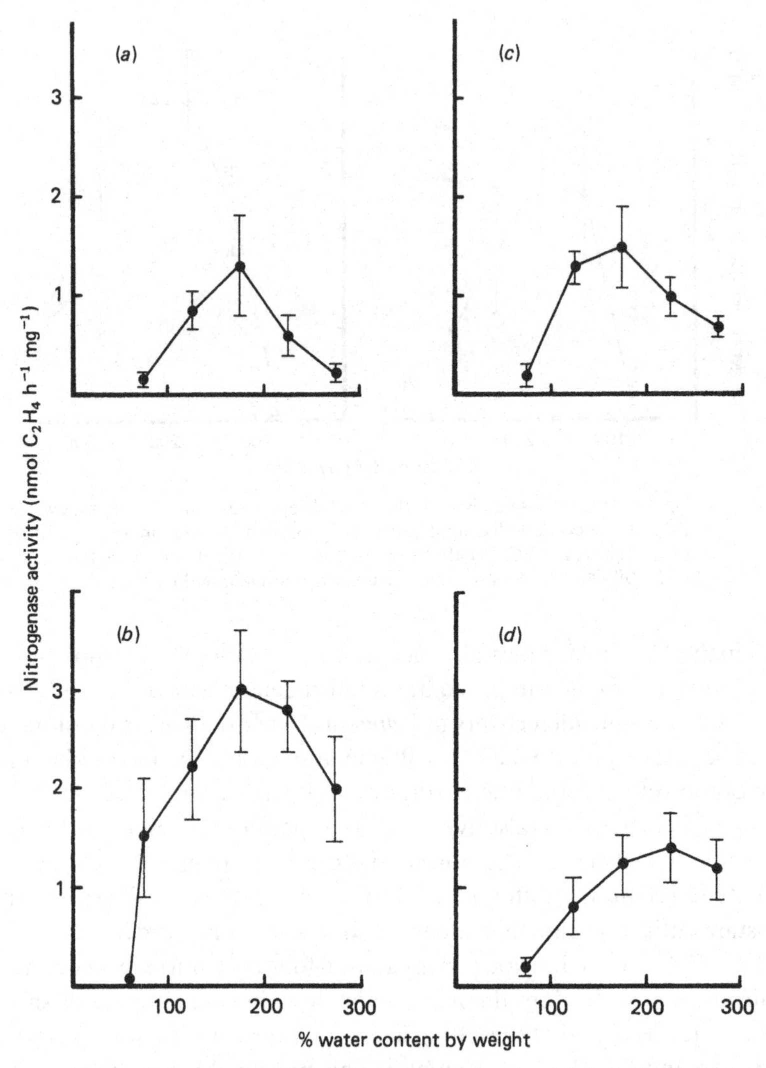

Fig. 76. Nitrogenase activity in *Peltigera polydactyla* at different levels of thallus moisture and 350 μmol m^{-2} s^{-1} illumination at 20 °C. (*a*) and (*b*) after 3 days of air drying and soaking for 1 and 4 h respectively; (*c*) and (*d*) after 3 days of drying over calcium chloride and soaking for and 1 and 4 h respectively. Vertical bars show the S.E.M. of each 50% moisture class. (From Kershaw and Dzikowski 1977.)

effects on nitrogenase activity from dry storage of *Stereocaulon paschale* for periods up to 75 weeks where thallus water content was from 3 to 10% of the oven-dry weight. Despite the very low rates of nitrogenase activity, particularly after 36 h of soaking, and the rather variable rates reported in the controls which make exact interpretation of the data difficult, it is

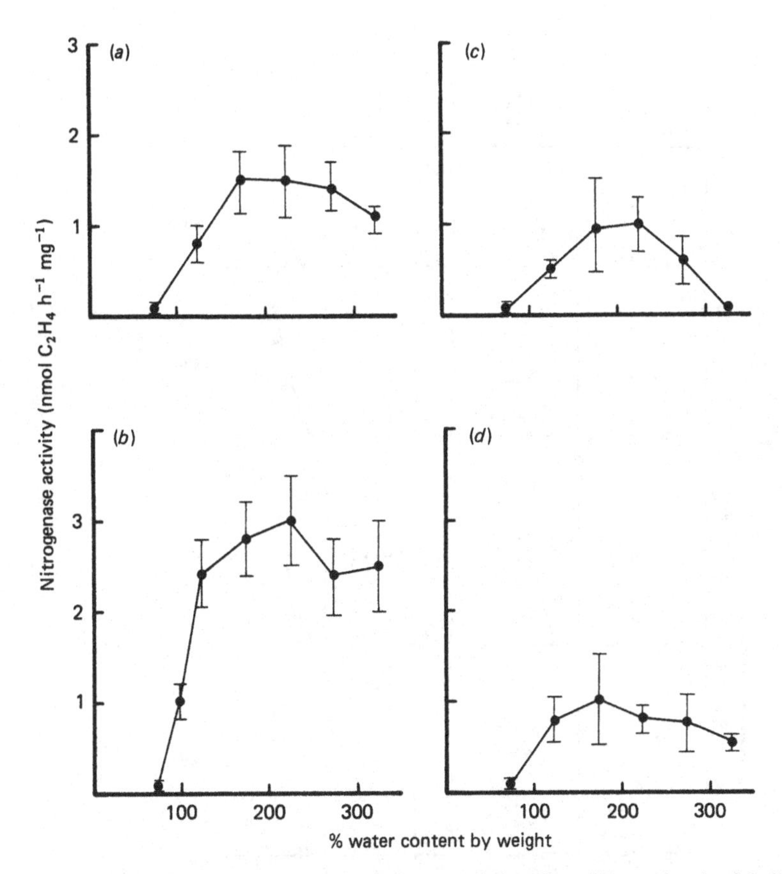

Fig. 77. Nitrogenase activity in *Peltigera polydactyla* at different levels of thallus moisture and 350 μmol m^{-2} s^{-1} illumination at 20 °C. (*a*) and (*b*) After 66 days of air drying and soaking for 4 and 12 h respectively; (*c*) and (*d*) after 66 days of drying over calcium chloride and soaking for 4 and 12 h respectively. Vertical bars show the S.E.M. of each 50% moisture class. (From Kershaw and Dzikowski 1977.)

certainly clear that after very long-term dry storage a considerable degree of activity is restored fairly rapidly after rehydration. This indeed may be a characteristic of this species, as Huss-Danell points out, and in fact Lange (1953) groups it with the most drought-resistant of the lichens he examined using the recovery of respiration rate as a criterion. This raises the possibility that each species may have a specific recovery time before full nitrogenase activity can be restored.

4.2 The interaction between nitrogenase activity and temperature

At high experimental temperatures, the constant rate of nitrogenase activity normally found during a drying cycle (see above) is

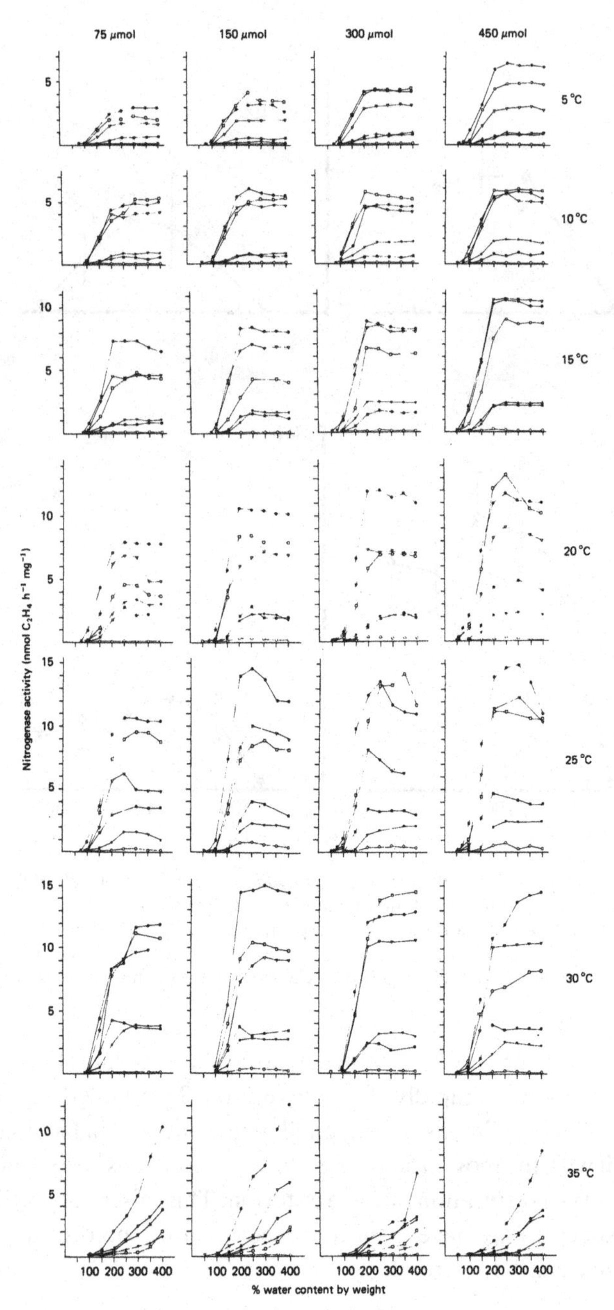

Fig. 78. Nitrogenase activity in *Peltigera praetextata* at all levels of thallus moisture, under 75, 150, 300 and 450 μmol m^{-2} s^{-1} illumination, and at 5, 10, 15, 20, 25, 30 and 35 °C. The data matrix was established for July 1975 (○), October 1975 (▲), December 1975 (●), March 1976 (□), May 1976 (△) and July 1976 (■). (From MacFarlane and Kershaw 1977.)

modified to a considerable extent. MacFarlane and Kershaw (1977) examined the response matrix of nitrogenase activity to temperature, light and season in *P. praetextata* and *P. rufescens* and showed that at 35 °C the rates were not initially constant at all, but declined rapidly (Fig. 78 and Fig. 79). More recently Kershaw and MacFarlane (1982) show an identical pattern of decline in *Collema furfuraceum* but at 30 °C (Fig. 80), and it appears that the nitrogenase enzyme in both these lichens is sensitive to high temperatures. It is ecologically significant that *Collema* is rather more sensitive than *Peltigera* (see below). MacFarlane and Kershaw (1977) confirmed the relationship with temperature and time of exposure by simply maintaining *P. praetextata* at 35 °C and at full thallus saturation for several hours, thus inducing a rapid decline in nitrogenase activity identical to the response in the drying curve (Fig. 81). The rapid decline of activity at high temperature is thus a function of the direct temperature inactivation of the nitrogenase enzyme over time, which can be restored subsequently by returning the experimental replicates to a more modest temperature regime (MacFarlane and Kershaw 1977).

The optimal temperature for maximum nitrogenase activity has been observed in a wide range of organisms to occur between 20 and 30°C. Waughman (1977) has found a temperature optimum of 20-30°C for several leguminous and non-leguminous plants, while Pattnaik (1966) and Fogg and Than-Tun (1960) reported a temperature optimum in the range 30-35 °C for free-living blue-green algae. Similar optimum temperatures for nitrogenase activity have been reported in a range of lichen species: for example, 20-25 °C in *Lichina confinis* and 35°C in *Peltigera rufescens* (Hitch and Stewart, 1973); 15 °C in *Solorina crocea* and *Nephroma arcticum* (Kallio, Suhonen and Kallio 1972); 16-21 °C in *Peltigera canina* (Maikawa and Kershaw, 1975); 30 °C in *Lobaria pulmonaria*, *Sticta weigelii* and *Leptogium cyanescens* (Kelly and Becker, 1975); 25-35°C in *Stereocaulon paschale* (Kallio 1973); and 20-30 °C for *Peltigera aphthosa* (Kallio and Kallio 1975, 1978; Kallio, Kallio and Rasku 1976; Englund, 1978). Unfortunately most of this work can be criticised for a variety of reasons and as a result the conclusions should be interpreted with caution.

The criticisms stem largely from the poikilohydric nature of lichens, coupled with the bivariate design of the experimental approach employed throughout all of the work. Traditionally, experimental replicates are held at thallus saturation whilst temperature is adjusted sequentially. If the interaction between thallus moisture and nitrogenase activity is constant over a wide range of thallus moisture at all experimental temperatures, then indeed this simple bivariate design will yield valid data. Unfortunate-

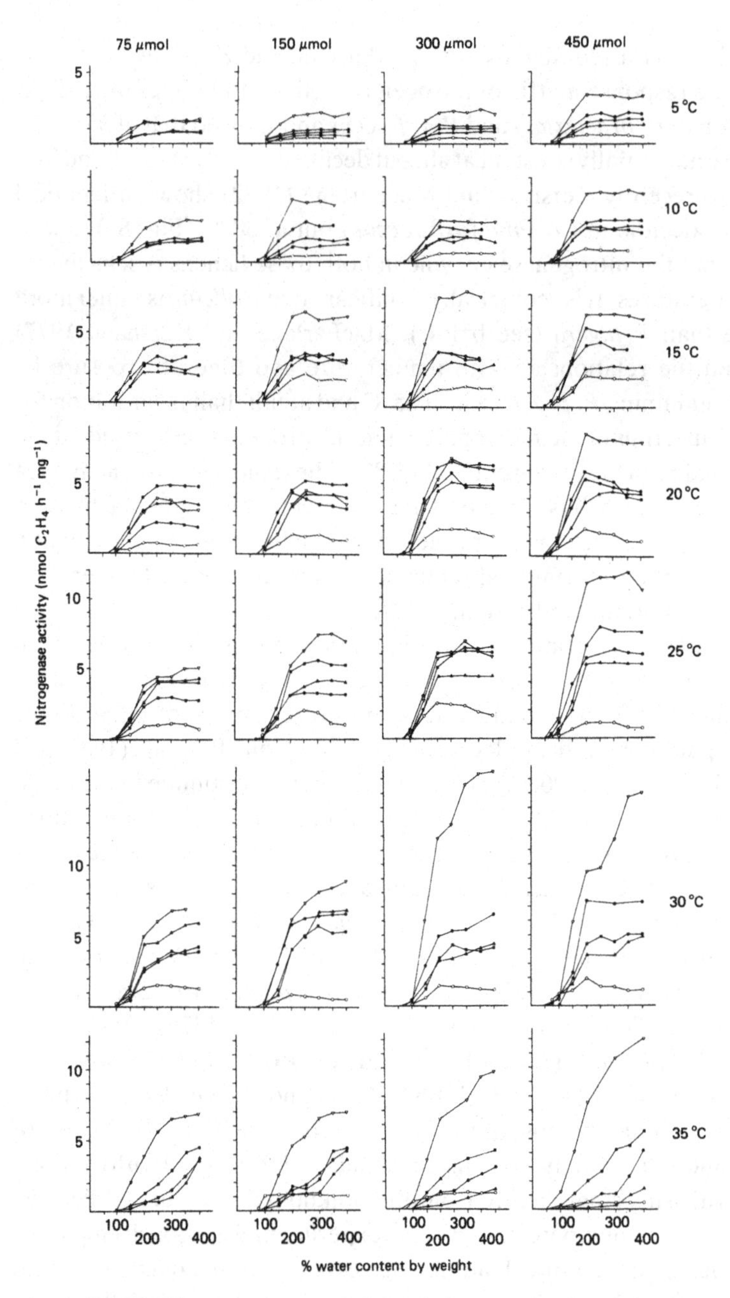

Fig. 79. Nitrogenase activity in *Peltigera rufescens* at all levels of thallus moisture, under 75, 150, 300 and 450 µmol m^{-2} s^{-1} illumination, and at 5, 10, 15, 20, 25, 30 and 35 °C. The data matrix was established for August 1975 (○), October 1975 (▲), December 1975 (●), May 1976 (△) and August 1976 (■). (From MacFarlane and Kershaw 1977.)

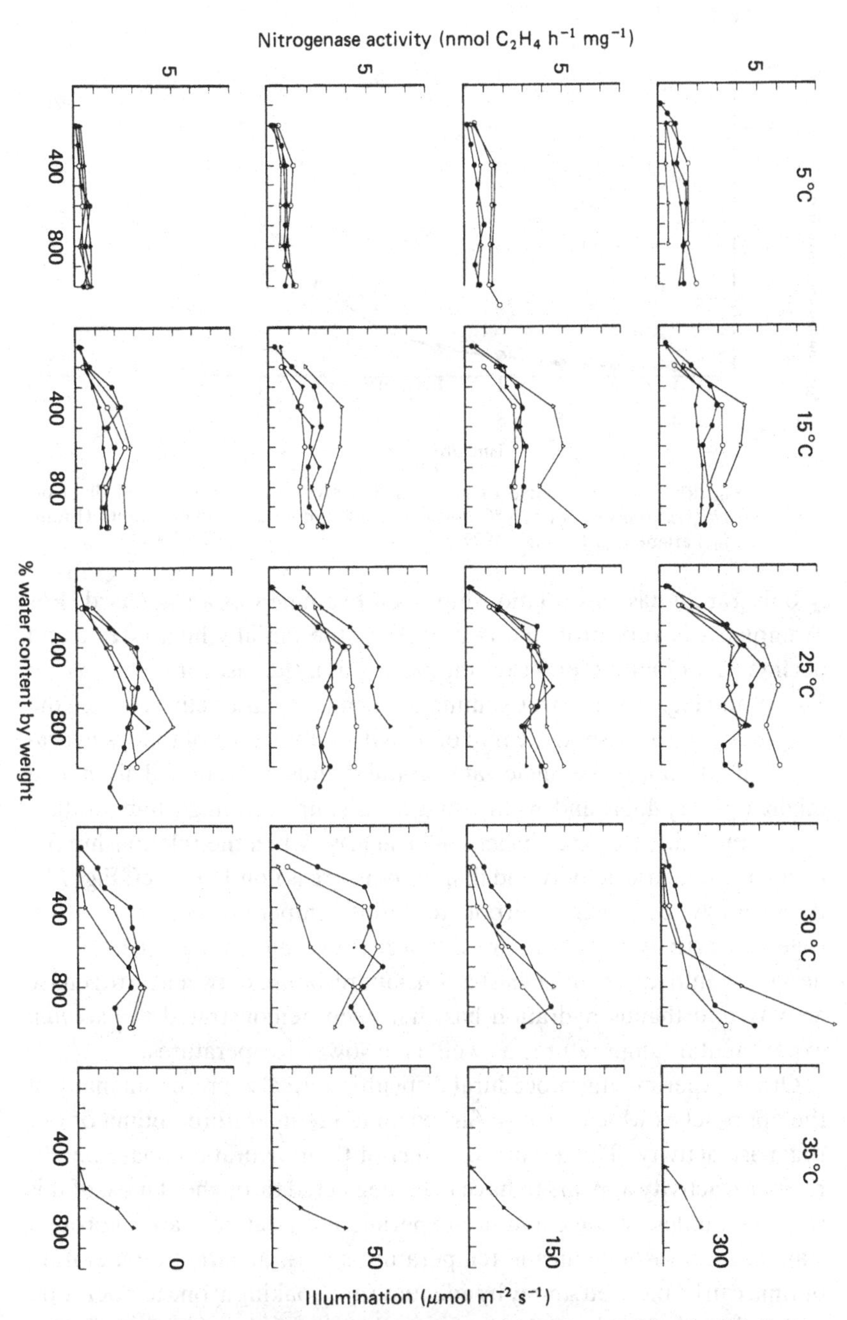

Fig. 80. The nitrogenase activity in *Collema furfuraceum* at all levels of thallus moisture, under 0, 50, 150 and 300 µmol m^{-2} s^{-1} illumination, and at 5, 15, 25, 30 and 35 °C. The response matrix was established for April (▲), July (△), September (○) and January (●). (From Kershaw and MacFarlane 1982.)

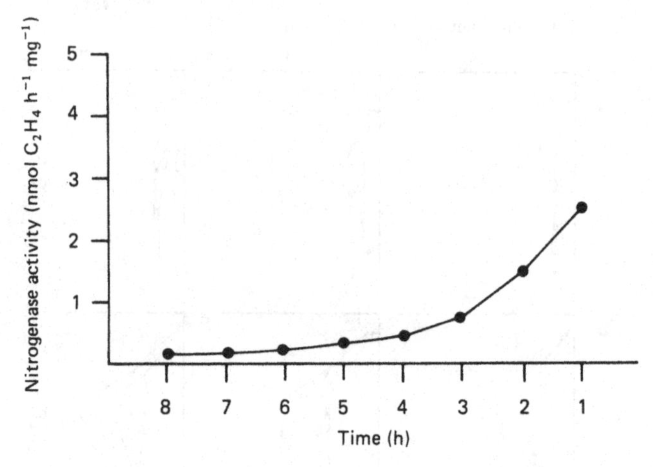

Fig. 81. The time course of nitrogenase activity in water-saturated thalli of *Peltigera praetextata* at 35 °C and under 300 μmol m^{-2} s^{-1} illumination. (From MacFarlane and Kershaw 1977.)

ly only rarely has this relationship been examined as a safeguard. For example, it is very probable that at 35°C the initially high rates could decline very rapidly over the drying period, as is the case for *Peltigera* (see Fig. 78 and Fig. 79). The true summation of nitrogenase activity is then the *integration* of the response curve of activity with degree of thallus hydration and not simply the single value at full thallus saturation. This indeed might be very high and would suggest a spuriously high temperature optimum. Similar discrepancies will also arise when the relationship between nitrogenase activity and thallus moisture is non-linear (cf. Fig. 74). Accordingly, data which purport to define temperature optima for nitrogenase activity in lichens are not acceptable unless the experimental design is multivariate or at least a constant response between nitrogenase activity and thallus hydration has first been demonstrated at *maximal* experimental temperatures as well as at lower temperatures.

Other experimental procedural difficulties are also present in many of the approaches which have so far examined temperature optima of nitrogenase activity. The important effect of the rehydration phase on nitrogenase activity appears to have been neglected throughout most of this work and unless storage and pre-experimental treatment are rigorously standardised throughout the temperature series, apparent temperature optima can be induced simply by adequate pre-soaking at one temperature with only a brief soaking of experimental replicates at other temperature combinations. In contrast, the data of Maikawa and Kershaw (1975), although they were safeguarded against the rehydration effect, can be

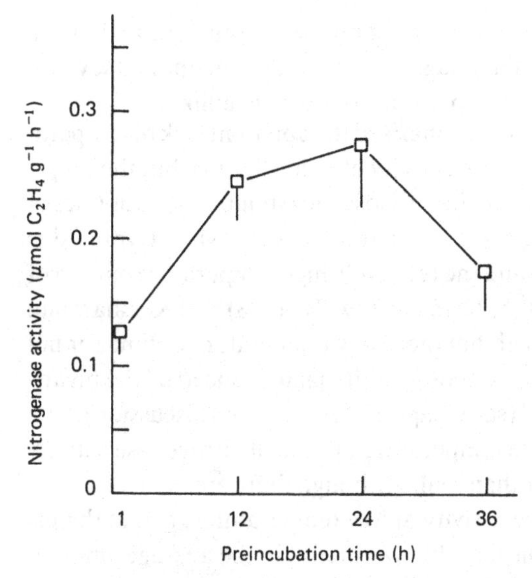

Fig. 82. The decline in nitrogenase activity in *Stereocaulon paschale* after long-term preincubation periods. The vertical bars show the S.E.M. (From Huss-Danell 1977.)

questioned because of the *continuous* experimental moisture regime which could induce a massive efflux of glucose from the phycobiont. Thus Tysiaczny and Kershaw (1979) and MacFarlane and Kershaw (1982) have shown that in some species of *Peltigera* the movement of glucose from the phycobiont is excessive at full thallus saturation. Under these conditions the energy supply required for nitrogen fixation could be seriously depleted, resulting in reduced levels of nitrogenase activity particularly at higher temperatures. Indeed Huss-Danell (1977) very effectively demonstrates the decline in nitrogenase activity in *Stereocaulon paschale* after 36 h of soaking (Fig. 82).

As a result of these criticisms we are left with a rather limited number of data sets derived from an adequate experimental design. They show a temperature optimum of 25 °C in *Stereocaulon paschale* (Fig. 92; Crittenden and Kershaw 1979) and *Collema furfuraceum* (Fig. 80; Kershaw and MacFarlane 1982), 25-30 °C in *Peltigera praetextata* (Fig. 78; MacFarlane and Kershaw 1977) and 30 °C in *P. rufescens* (Fig. 79; MacFarlane and Kershaw 1977). The geographical range of these species is from low arctic to temperate latitudes with corresponding temperature optima of 25 °C for the low arctic zone and 25-30 °C for the temperate species. The reliability of the low temperature optima reported for *Peltigera canina* (16-21 °C: Maikawa and Kershaw 1975), *Solorina crocea* and *Nephroma arcticum*

(15°C: Kallio *et al.* 1972) accordingly should be questioned, and in the absence of more critical data it is suggested here that an optimal level of nitrogenase activity below a temperature of 20°C is unlikely.

Kallio and Kallio (1978) have emphasised the apparent lack of temperature adaptation in the nitrogenase activity of arctic lichens, but the major contrast they made was between the supposed low temperature optima of net photosynthesis in the arctic species examined (which was interpreted as a high degree of adaptation) and the relatively high temperature optima of nitrogenase activity. However, Kershaw (1977*a*, 1978) and Kershaw and Smith (1978) have emphasised that these low temperature optima for net photosynthesis are often an erroneous result, again induced by the bivariate experimental design used (see Chapter 7 for a detailed discussion). The gross lack of adaptation in the temperature optima of nitrogenase activity is accordingly illusory rather than real. Although there are indeed universally low rates of nitrogenase activity at low temperatures, actual thallus temperatures are usually considerably warmer than the average ambient temperatures might suggest (Chapter 1). Adjustment of the nitrogenase activity temperature optima from a thallus temperature of 30°C down to 20°C would thus constitute a substantial advantage.

4.3 The interaction between nitrogenase activity and light

Although light is a general requirement for nitrogenase activity and there is a marked increase in activity with increasing light levels up to *c.* 75 μmol m^{-2} s^{-1} (Cox and Fay 1969), in *Peltigera praetextata* and *P. rufescens* there is apparently no further marked rate increase at light levels above 150 μmol m^{-2} s^{-1} (Fig. 83, and see also Fig. 78 and Fig. 80). Other earlier studies of the relationship between light and nitrogenase activity in both free-living and lichenized blue-green algae (Stewart 1965; Fogg and Stewart 1968; Henriksson *et al.* 1972; Kallio 1973; Hitch and Stewart 1973) also show little interaction between high light and nitrogenase activity. Stewart (1974) summarised the information on the effects of light intensity on terrestrial blue-green algae and suggested that there is a general, but not always a very direct, correlation between light intensity and nitrogen fixation in natural ecosystems. Thus the ability to fix nitrogen at low light intensities may reflect an adaptation to shade or darkness. As long as there are adequate supplies of ATP (which are apparently easily supplied by photophosphorylation under quite low levels of illumination) light is not a limiting factor. However, in the absence of photophosphorylation providing a source of ATP, nitrogenase activity is markedly reduced. There is

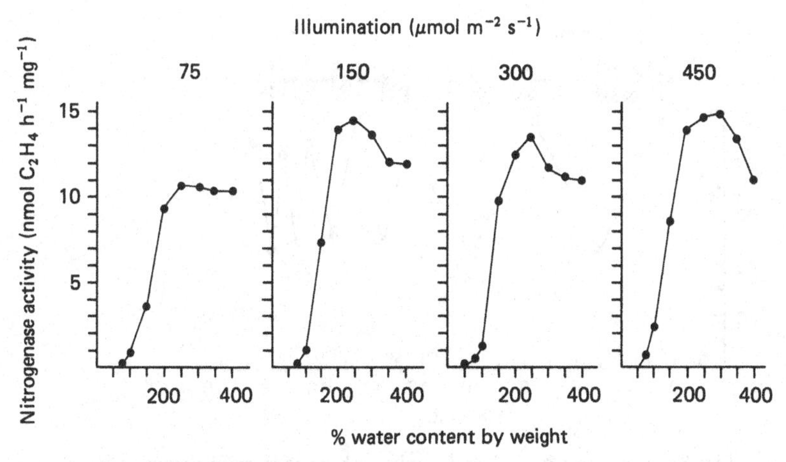

Fig. 83. The response of nitrogenase activity in *Peltigera praetextata* to the experimental level of illumination.

some discrepancy between the early studies of Kallio *et al.* (1972) and those of Hitch and Stewart (1973). Hitch and Stewart for example report that *Lichina confinis* and *Peltigera rufescens* continue to reduce acetylene in the dark, although at progressively decreasing rates, for up to 26 h. In marked contrast Kallio *et al.* (1972) found no nitrogenase activity in the dark in *Nephroma arcticum* and *Solorina crocea.*

MacFarlane *et al.* (1976) have re-examined this apparent contradiction by following the time course of nitrogenase activity in relation to the length of the dark period. When *Peltigera polydactyla* is transferred to the dark for as short a time as 1 min, there is a dramatic decrease in nitrogenase activity (Fig. 84). The initial value is typically 50% of the control value. Longer periods of darkness (Fig. 85) have similar effects. Equally remarkable is the immediate and complete restoration of full activity on returning the material to the light. With increasing dark periods of 2 h or more (Fig. 86) there is an increasing lag period before complete recovery of nitrogenase activity is observed on re-exposure to light and as the carbohydrate pool is restored. After 17 h in the dark there is very little immediate recovery of activity when light is restored, as presumably the carbohydrate pool is heavily depleted. Huss-Danell (1979) reports similar light/dark interactions with nitrogenase activity in *Stereocaulon paschale.*

From these results is is clear how the conflict between the findings of Hitch and Stewart (1973) and Kallio *et al.* (1972) arose. Different pretreatments were used in the two studies: Kallio used a long pretreatment (36-48 h) in the dark at the temperature conditions to be used, whereas Hitch and Stewart reported *in situ* measurements in the field. The probable degree of

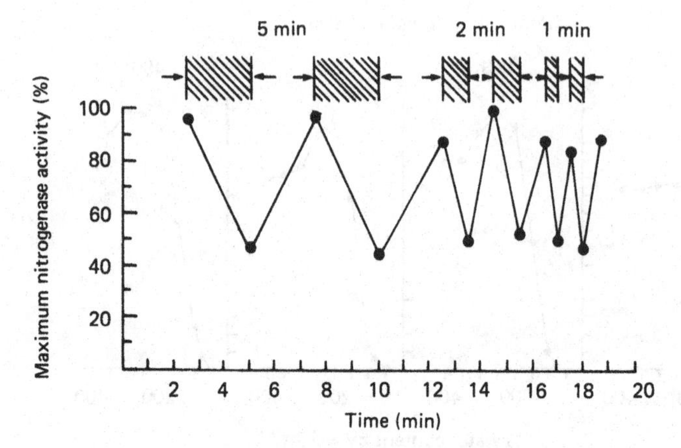

Fig. 84. The effect of 5, 2 and 1 min dark periods on nitrogenase activity in *Peltigera polydactyla*. (From MacFarlane *et al.* 1976.)

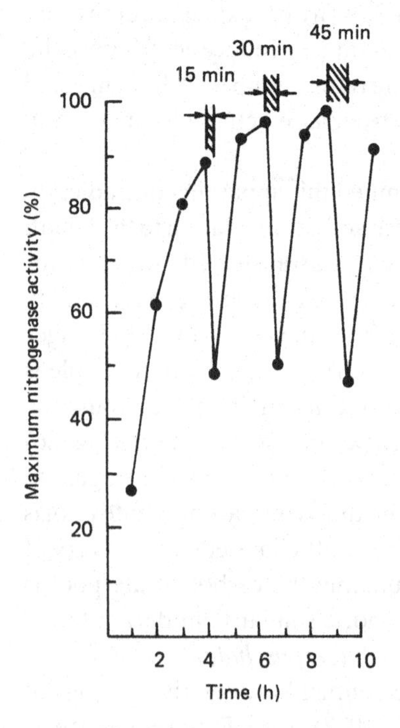

Fig. 85. The effect of 15, 30 and 45 min dark periods on nitrogenase activity in *Peltigera polydactyla*. (From MacFarlane *et al.* 1976.)

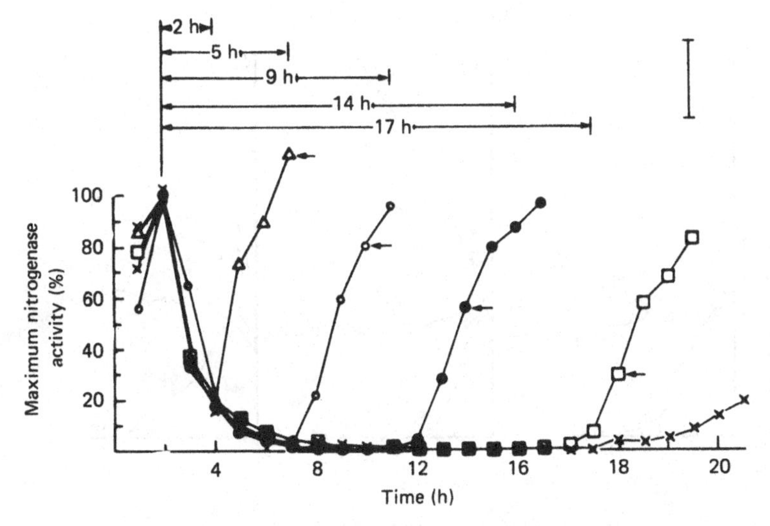

Fig. 86. The effect of long-term dark periods on nitrogenase activity in *Peltigera polydactyla*. The arrows mark the level of recovery of nitrogenase activity in each replicate after it has been returned to the light for 3 h. △, 2 h replicate; ○, 5 h replicate; ●, 9 h replicate; □, 14 h replicate; ×, 17 h replicate. The vertical bar shows the standard error. (From MacFarlane *et al.* 1976.)

depletion of thallus carbohydrates in Kallio's experiment after 36-48 h in the dark would be considerable and it is not surprising that no nitrogenase activity was detected.

This experiment also illustrates the importance of a dark/light regime in controlling the nitrogenase level in situ. For example, a typical night length at temperate latitudes might be 9 h. MacFarlane *et al.* (1976) show that under these conditions 6 h light would be required to return the lichen to its full level of nitrogenase activity. If the dark interval were extended to 15 h, a light period of 10 h would be required for maximum nitrogenase activity. Furthermore if reliable absolute rates of nitrogenase activity are to be determined experimentally for lichens, not only must the hydration response be examined and controlled but also the light pretreatment must be standardised. Experiments conducted after 9 h of pretreatment will give significantly different results from those after 15 h.

The most straightforward explanation of the rapid decline in nitrogenase activity in the dark is a switch in the source of ATP from photophosphorylation to oxidative phosphorylation. Weare and Benemann (1973) consider oxidative phosphorylation to be a source of ATP for nitrogenase activity in the light as well as in the dark. However, Stewart (1974) suggests that oxidative phosphorylation provides a basal level of

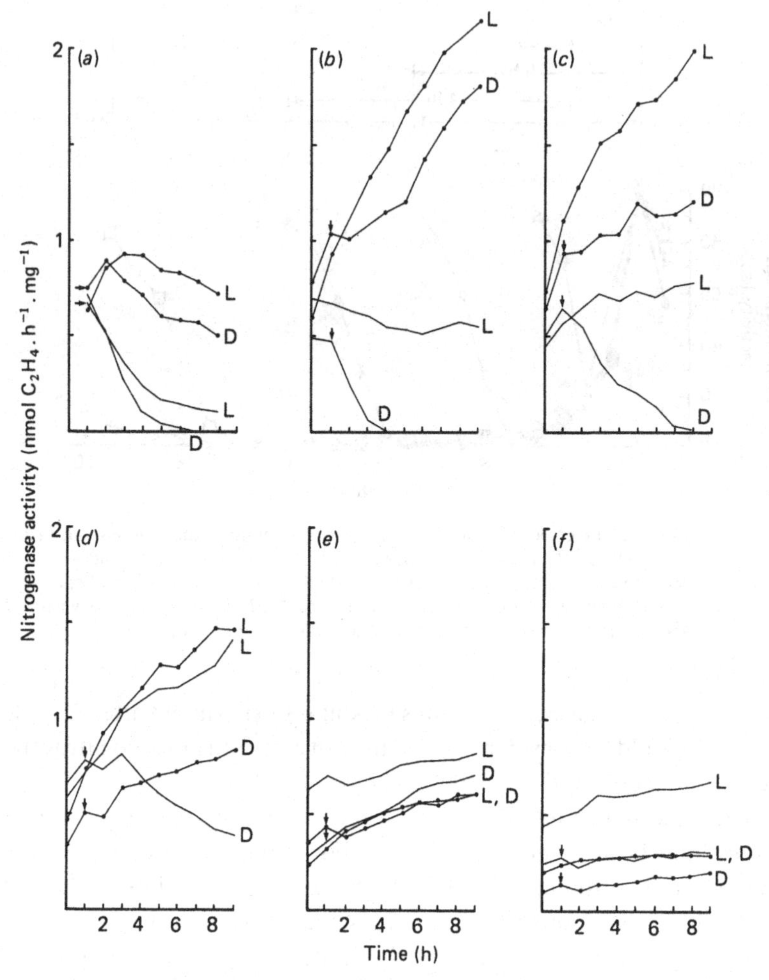

Fig. 87. The effect of light (L) and glucose upon nitrogenase activity in *Peltigera polydactyla* over a range of temperatures (●–●, 6% glucose replicates; —, water controls). The start of each 8 h dark period (D) in the control and treatment replicates is marked with an arrow. (*a*) 35 °C; (*b*) 30 °C; (*c*) 25 °C; (*d*) 20 °C; (*e*) 15 °C; (*f*) 5 °C. (From Kershaw *et al.* 1977.)

activity comparable with that found initially in the dark and that the additional ATP which enables the fixation by nitrogenase to proceed at a much greater rate in the light is provided by photophosphorylation. The results for *Peltigera polydactyla* are quite compatible with this latter interpretation. They show that 50% of the nitrogenase activity is lost immediately after transfer of the thallus to the dark and suggest that the remaining activity is supported by carbohydrate based oxidative phosphorylation. The decline in nitrogenase activity after several hours in the dark may represent a consequence of the depletion of the carbohydrate pool by a variety of cellular activities such as respiration, as well as the metabolic requirements for support of nitrogen fixation. If this indeed is the case, then the decline in nitrogenase activity could be substantially eliminated by the provision of an endogenous supply of carbohydrates or by lowering the experimental temperature, thus slowing the drain on the carbohydrate pool, particularly by metabolic demands other than nitrogen fixation. These potential relationships were confirmed by Kershaw, MacFarlane and Tysiaczny (1977) again using *P. polydactyla* as an experimental organism, and combining light and dark treatments with both temperature and an exogenous supply of glucose (Fig. 87).

The data illustrate two basic patterns: a response of nitrogenase activity to temperature, and the maintenance of nitrogenase activity in the dark at temperatures of 15 °C or less. The overall pattern of response to temperature shows an optimum for nitrogenase activity at 25-30 °C, which is in close agreement with the data of MacFarlane and Kershaw (1977) for *P. praetextata*. The data also indicate the sensitivity of the nitrogenase enzyme to high temperature in that the activities of the water controls in the light and dark decline very rapidly at 35 °C. Similarly, although the 6% glucose treatment initially stimulates the rate of acetylene reduction, both in the light and in the dark, there is eventually a marked decrease in nitrogenase activity. At 30 °C (Fig. 87*b*) no temperature inhibition of the enzyme is evident and there is a significant stimulation of the rates of nitrogenase activity by glucose for both light and dark treatments. This pronounced enhancement disappears as the experimental temperature is decreased (Fig. 87*d*). This suggests strongly that glucose uptake and utilisation are temperature dependent. As the experimental temperature is reduced still further, the rapid decrease in nitrogenase activity in the dark gradually disappears. Thus at 20 °C (Fig. 87*d*) the level of activity in the dark control has only declined to approximately 50% of the initial value. At 15 °C (Fig. 87*e*) the water control in both light and dark shows a progressive increase in nitrogenase activity during the 8 h dark period. Although the absolute rates are less at 5 °C (Fig. 87*f*), the same pattern of response is maintained.

The maintenance of nitrogenase activity in the dark controls at the lower temperatures again is consistent with the concept that activity is governed by energy availability from oxidative phosphorylation, and that there is a limited carbohydrate pool which becomes rapidly depleted at higher temperatures. It seems probable that other cellular processes which compete for energy could be differentially inhibited at lower temperatures, thereby allowing the rate of nitrogenase activity to continue in the dark. It is evident that in temperate latitudes, when periods of darkness are less than 12 h, nitrogen fixation could continue through the night, unimpaired by the absence of light, given that an adequate carbon reserve is available. At night, moist thalli would not dry further, temperatures typically would be 10-15 °C and a considerable proportion of the annual nitrogen budget could be accumulated under these conditions (see also Section 4.6 and Fig. 94 below).

Further evidence is given by Millbank (1977), who provides an extremely elegant data set which strongly supports the interpretation made here, of a switch to oxidative phosphorylation as the energy source for nitrogenase activity in the dark. He estimated the internal partial pressure of oxygen in saturated lichen thalli by implanting unprotected microelectrodes for brief experimental periods. Experiments with the main thallus of *Peltigera aphthosa* with electrodes implanted in the *Coccomyxa* layer show partial pressures of oxygen (P_{O_2}) at or slightly below atmospheric levels under various conditions of illumination, falling quite gradually in the dark as oxygen is consumed by respiration (Fig. 88). The pattern in the cephalodia, though, is totally different, showing an extremely rapid and striking decline of internal P_{O_2} (Fig. 89). Millbank simply interprets the difference as a function of the very active general metabolic state of *Nostoc* in the cephalodia of *P. aphthosa* (Hitch and Millbank 1975*a*,*b*). A more specific interpretation, however, is that an immediate switch to oxidative phosphorylation as the energy source of nitrogenase activity results in the very dramatic lowering of partial pressures of oxygen in the cephalodia. This decline is much greater in amplitude than the equivalent response in the algal layer of *P. canina* and correlates with the very high frequency of heterocysts found in the cephalodia of *P. aphthosa* as compared with the thallus of *P. canina* (Tables 4 and 5 in Hitch and Millbank 1975*b*).

4.4 The seasonal dependency of the basic level of nitrogenase activity

Seasonal variation of nitrogen fixation in lichens has been reported by Hitch and Stewart (1973), Kallio and Kallio (1975) and by Horne and Goldman (1972). All these studies measured nitrogenase activity *in situ* and the rate changes observed were directly related to light, temperature and moisture. Since these environmental parameters *have a marked seasonal distribution*, there was a corresponding seasonal distribution of nitrogenase activity. Huss-Danell (1978) and Crittenden and Kershaw (1979) examined replicate thalli under rigorously controlled experimental conditions and showed a cessation of activity in the winter as a result of

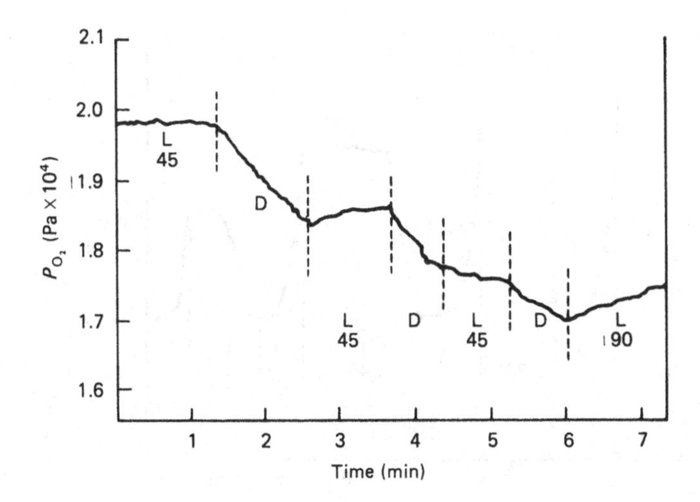

Fig. 88. The response of an oxygen microelectrode implanted in the *Coccomyxa* layer of *Peltigera aphthosa*. D, darkness; L, thallus illuminated at energies shown (W m^{-2}). (From Millbank 1977.)

snow cover (see below). It seems probable that many lichens will experience a comparable decline in nitrogen fixation capabilities during such periods. This will not be true for epiphytic species however, as Kershaw and MacFarlane (1982) have shown for *Collema furfuraceum* (Fig. 80).

The data for *C. furfuraceum* equally prevent any generalisation about the decline of nitrogenase activity during mid-summer. MacFarlane and Kershaw (1977) have examined the seasonal levels of nitrogenase activity in *Peltigera praetextata* and *P. rufescens* to temperature, light and moisture (Fig. 78 and Fig. 79). The data again emphasise the constancy of response at higher levels of thallus hydration, except at high temperatures where there is the characteristic response due to heat inactivation of the nitrogenase enzyme. The very limited reponse to increasing light intensity and the standard temperature optimum of 25-30 °C are also evident in the data. Of outstanding interest, however, is a marked decline in nitrogenase activity during mid-summer in both species. This is particularly conspicuous in *P. praetextata* but only obvious in 1975 in *P. rufescens* after an abnormally hot and dry summer. By August 1975 the level of nitrogenase activity in *P. praetextata* was barely detectable and in *P. rufescens* was severely reduced.

In temperate regions after exceptionally high summer temperatures which presumably exceed the normal range of adaptation of a species, a significant reduction in the level of nitrogenase activity may be present. The data for *Collema* presented above, however, point to the lack of

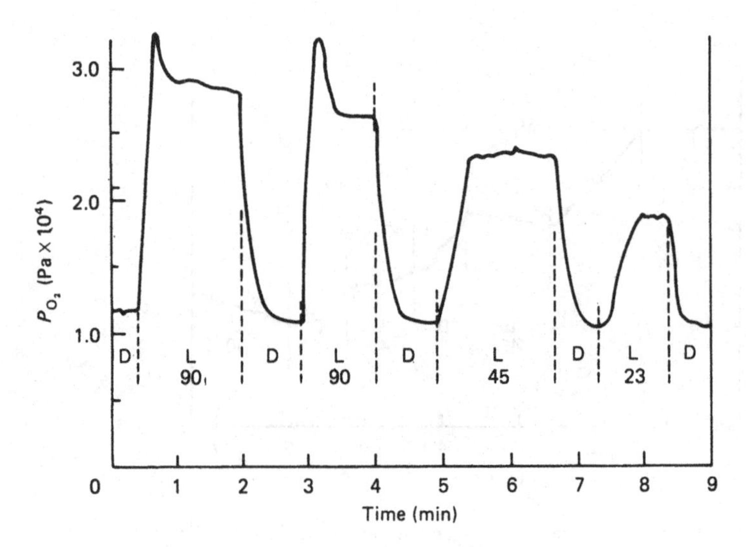

Fig. 89. The response of an oxygen microelectrode implanted in the cephalodium of *Peltigera aphthosa*. D, darkness; L, thallus illuminated at light energies shown (W m^{-2}). (From Millbank 1977.)

generality of this relationship and it is quite probable that in a less extreme summer period in Ontario rates of nitrogenase activity would not be affected. Equally, populations of *P. rufescens* or other species from a more southerly location could exhibit a full level of adaptation to high summer temperatures and may not be affected at all. The data do emphasise, though, the potential importance of thallus temperature in the field as a stress factor. A full discussion of temperature stress on metabolic activity will be postponed until later (Chapter 7). Nevertheless, it becomes quite clear that there is considerable potential for large-scale seasonal variation in the nitrogenase activity of lichens in addition to any diurnal fluctuations. The magnitude of both these fluctuations is in turn modified to an extent by the thallus environment of the day, the week or indeed the season, and the importance of a full multivariate response matrix incorporating light, temperature, hydration and season of the year is re-emphasised. The interaction with snow cover is particularly pronounced.

4.5 The interaction between nitrogenase activity and snow cover

There has been considerable recent interest in the role that cyanophilic lichens may play in the overall nitrogen balance of ecosystems, especially in the extensive and lichen-rich systems of northern Canada, Russia and Scandinavia. As a result, the effect of prolonged snow cover on nitrogenase activity has been closely examined by a number of workers

(Alexander and Kallio 1976; Kallio *et al.* 1976; Crittenden and Kershaw 1979). The data for northern studies consistently show a complete absence in winter of nitrogenase activity, and in each case temperatures around 0 °C have been implicated as the key environmental factor which affects the cold-labile nitrogenase enzyme. This interpretation has been supported by the *in vitro* studies of Dua and Burris (1963) and Haystead, Robinson and Stewart (1970), who document the cold-labile nature of the nitrogenase enzyme in *Clostridium* and *Anabaena* respectively. However, this interpretation presents an apparent paradox since there is also an extensive literature which clearly demonstrates that under experimental conditions nitrogenase activity continues at temperatures around 0 °C although at a reduced level (e.g. Fogg and Stewart 1968; Horne 1972; Kallio *et al.* 1972; Hitch and Stewart 1973; Maikawa and Kershaw 1975; Huss-Danell 1978).

Stewart (1977) has suggested that since a number of studies have shown an immediate increase in nitrogenase activity upon transference of experimental organisms from lower to higher temperatures, nitrogenase resynthesis is an unlikely explanation for the rapid increase in activity. Thus the nitrogenase enzyme *in vivo*, unlike that of a cell-free extract, is not in fact cold-labile, and Kershaw and MacFarlane (1982) have indeed shown that nitrogenase activity continues unimpaired throughout the winter in *Collema furfuraceum* when thallus temperatures can be as low as −40 °C (see below). An explanation for the lack of nitrogenase activity under snow cover must accordingly be sought in the depletion of carbohydrate pools.

MacFarlane and Kershaw (1980*b*) examined the influence of snow cover on the rates of nitrogenase activity in *Peltigera praetextata* and *P. rufescens*. In accordance with the existing literature, only very low rates of nitrogenase activity could be detected after several months of snow cover. After storage in the dark at 2 °C, very low initial rates of acetylene reduction were in evidence, at all temperatures for both experimental light levels (300 and 0 μmol m^{-2} s^{-1}). However, after 2 weeks of storage at a 12 h photoperiod of 300/0 μmol m^{-2} s^{-1} and 15/10 °C, the rate of nitrogenase activity increased almost twofold for all experimental temperatures and both experimental light levels (Fig. 90). Furthermore, material stored in a hydrated state for 8 days under a combination treatment of either a 2/0 or 15/10 °C thermoperiod and either in the dark or a 350/0 μmol m^{-2} s^{-1} photoperiod showed an extremely clear-cut recovery pattern (Fig. 91). The rate of acetylene reduction increased in replicates stored at a 12 h thermoperiod of 15/10 °C day/night temperature and a 12 h photoperiod of 350/0 μmol m^{-} s^{-1} when compared with the response on day 1, and reached

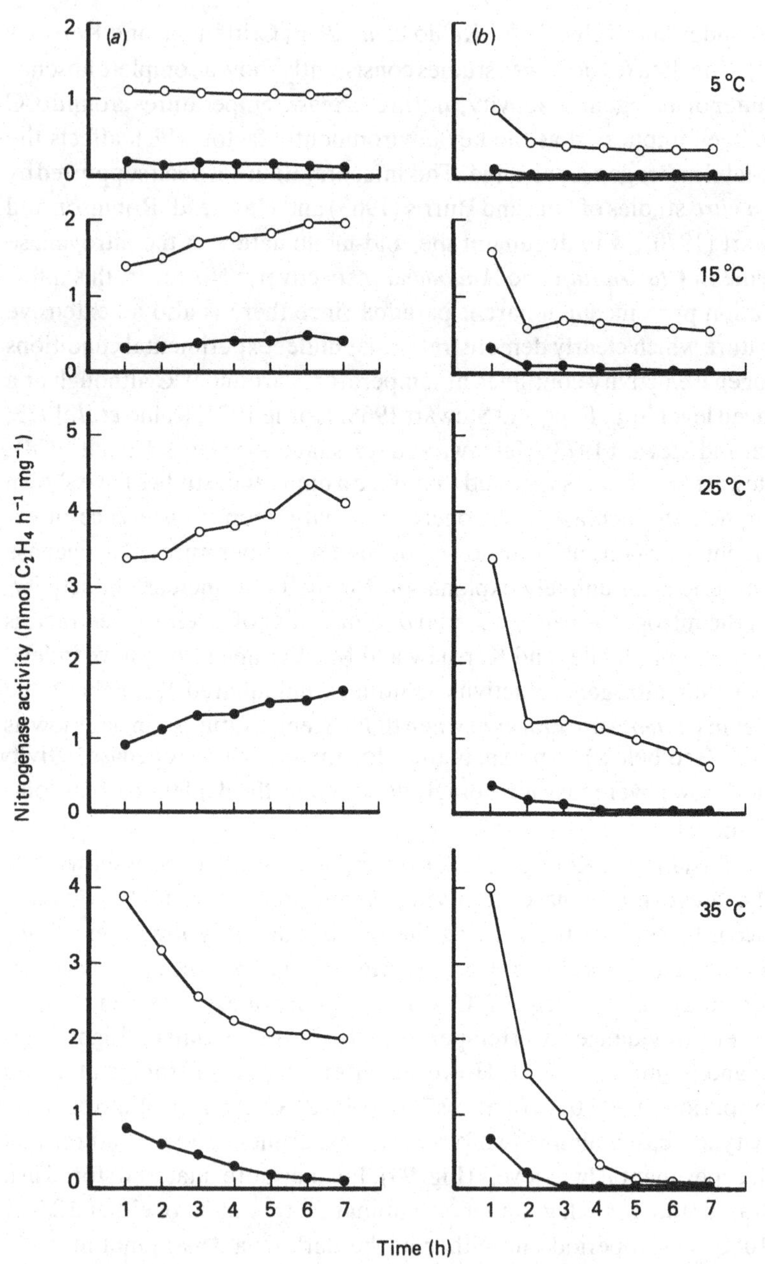

Fig. 90. Nitrogenase activity in fully hydrated replicates of *Peltigera rufescens* at 5, 15, 25 and 35 °C, under (*a*) 300 and (*b*) 0 μmol m^{-2} s^{-1} illumination. ●, activity immediately following collection from under snow and with a standard 12 h light pretreatment; ○, activity after an experimental 2 week storage period at 15/10 °C day/night temperatures and a 300 μmol m^{-2} s^{-1} photoperiod of 12 h. Each point is the mean of four replicates, with a standard error of less than 0.5 nmol C_2H_4 h^{-1} mg^{-1}. (From MacFarlane and Kershaw 1980*b*.)

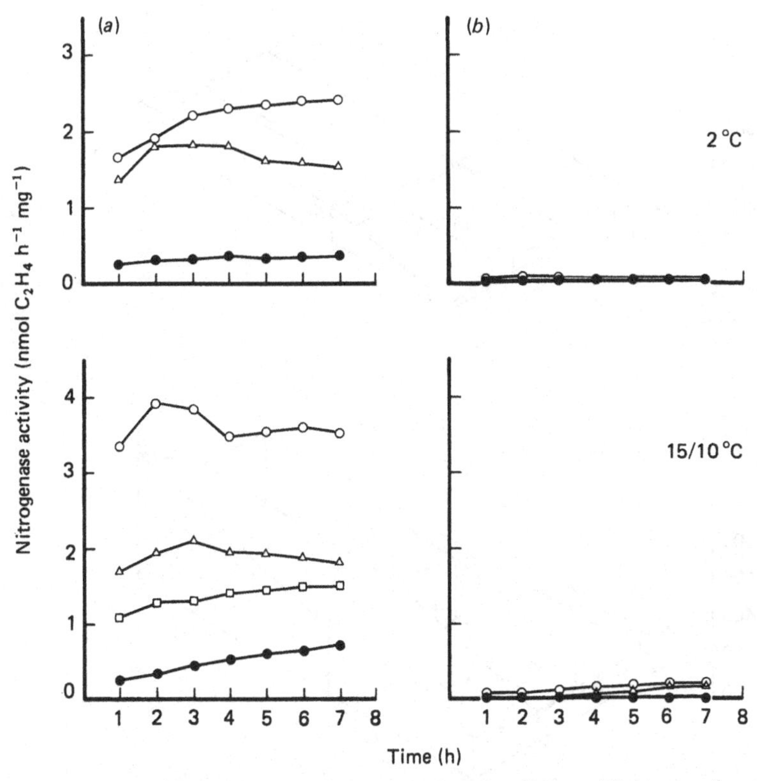

Fig. 91. Nitrogenase activity in fully hydrated replicates of *Peltigera rufescens* at 25°C and 300 μmol m^{-2} s^{-1} illumination. The rates during recovery were examined in experimental material stored under (*a*) a 12 h photoperiod of 350 μmol m^{-2} s^{-1} or (*b*) continuous darkness, at either 2/0 °C or 15/10 °C. Each point is a mean of four replicates, with a standard error of less than 0.5 nmol C_2H_4 h^{-1} mg^{-1}. ○, day 8; △, day 4; □, day 2; ●, day 1. (From MacFarlane and Kershaw 1980*b*.)

a maximum of 3.9 nmol C_2H_4 h^{-1} mg^{-1} after 8 days. The rate of recovery at 2 °C and a 12 h photoperiod of 350/0 μmol m^{-2} s^{-1} is again much slower than that observed at 15/10 °C under identical illumination. Irrespective of storage temperature, lichen thalli kept in continuous darkness exhibit very low levels of nitrogenase activity and again these low levels can be directly attributed to the experimental pretreatment. Whilst temperature is intimately involved in the speed of recovery, the prime requirement appears to be light.

These interactions are also evident in the data of Huss-Danell (1977) and Crittenden and Kershaw (1979) (Fig. 92). *Stereocaulon paschale* collected from under the snow in March shows extremely low to zero rates of nitrogenase activity, but significantly the rates at 35 °C are maximal, con-

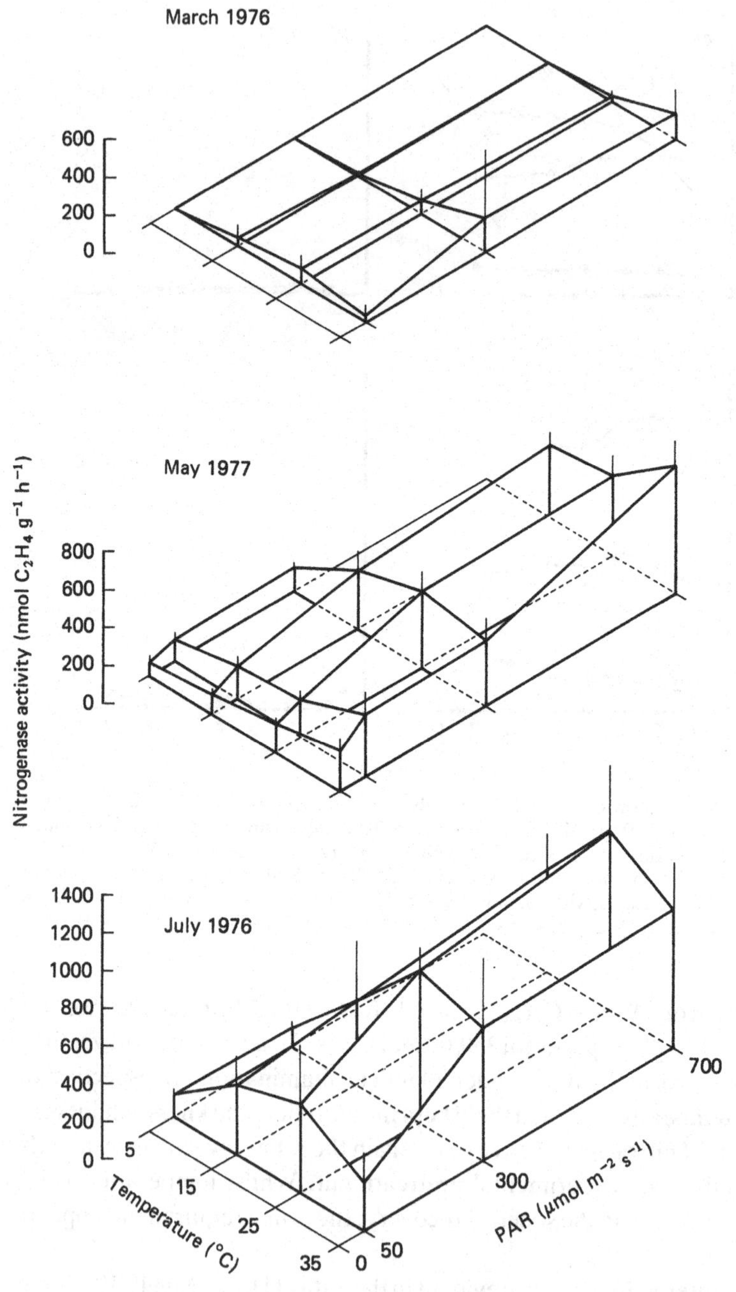

Fig. 92. Effects of factorial combinations of photosynthetically active radiation (PAR) and thallus temperature on nitrogenase activity in collections of *Stereocaulon paschale* made at different times in the year. Mean values are plotted together with 95% confidence limits (*n* ranges from 7 to 36). (From Crittenden and Kershaw 1979.)

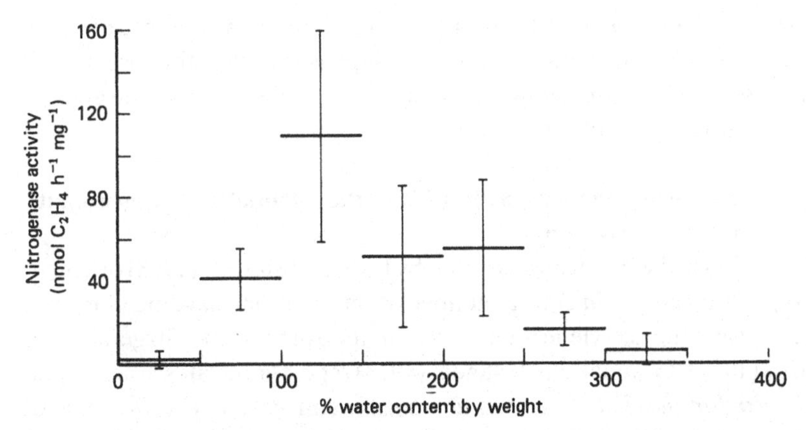

Fig. 93. The partial recovery of nitrogenase activity in *Stereocaulon paschale* during the experimental incubation period at 25 °C and under 300 μmol m^{-2} s^{-1} illumination. The vertical bars show the S.E.M. (From Crittenden and Kershaw 1979.)

firming the control of the velocity of recovery by temperature. This starts to occur in the experimental replicates actually during the incubation period at 25 °C (Fig. 93). In May, shortly following snow-melt and thus after a brief period of full thallus hydration from the snow-melt, whilst under full radiation, rates of nitrogenase activity increase substantially with, again, a marked interaction with high light at 35 °C. By July, full levels of activity are evident (Fig. 92). Thus in low-arctic conditions with limited rainfall and low ambient temperatures the velocity of recovery of nitrogenase activity can be quite slow. In contrast however, when Huss-Danell incubated at 15 °C under 170 μmol m^{-2} s^{-1} illumination, *S. paschale* that had shown very low or zero levels of nitrogenase activity under the snow, he found that rates increased substantially after 24 h of treatment.

The continued nitrogenase activity in *Collema furfuraceum* throughout the full winter period in northern Ontario further emphasises the absence of low-temperature inhibition of nitrogenase in the intact thallus. *C. furfuraceum* is a species abundant on the north side of balsam poplars (*Populus balsamifera*) in northern Ontario. It always occurs more than 1 m above ground level and therefore is always above snow level in mid-winter. The nitrogenase response matrix to moisture, temperature and light, throughout the year, shows a remarkable degree of constancy (Fig. 80). The data confirm the ubiquitous nature of temperature optima around 25 °C even for a low arctic species, the very limited response to increased levels of illumination, and the constant response of nitrogenase activity at higher levels of thallus hydration, all of which have already been discussed above. The lack of any seasonal variation in winter in the absence of snow cover is of particular interest, further emphasising that the loss of nitrogenase activity under snow is indeed due to the depletion of the energy pool supplying the nitrogenase reaction.

4.6 Nitrogenase activity in the field and the potential for nitrogen input into the ecosystem

Since the initial observations by Fogg and Stewart (1968) of lichen nitrogenase activity *in situ*, a number of observations have been made, largely using the acetylene reduction technique of Stewart, Fitzgerald and Burris (1967). Granhall and Selander (1973) report good rates of activity in *Stereocaulon paschale*, *Nephroma arcticum* and *Peltigera scabrosa*; and Crittenden (1975) similarly reports good activity in *P. aphthosa* and *P. canina* (see also Horne 1972; Schell and Alexander 1970). More recently several attempts have been made to examine rates of nitrogenase activity in the field over a period of time, particularly in terms of the potential contribution of nitrogen to arctic ecosystems (Kallio and Kallio 1975; Kallio *et al.* 1976; Huss-Danell 1977; Crittenden and Kershaw 1979). The majority of this work suffers to a considerable extent from lack of the realisation that field temperatures measured with thermometers are wildly inaccurate and are unlikely to bear much relationship to thallus temperature. Similarly, thallus temperature inside an incubation bottle under even modest levels of radiation is elevated to a considerable extent, modifying both rates of thallus nitrogenase activity and moisture, especially during long incubation periods (> 1 h). The microclimate data are often incomplete and in the absence of a full response matrix generated under carefully controlled combinations of environmental conditions, interpretation of changing rates of nitrogenase activity in the field, over 24 h or longer periods, is speculative at best.

Crittenden and Kershaw (1979) exhaustively examined the response matrix of nitrogenase activity in *Stereocaulon paschale* to moisture, temperature, light and time of year, (Fig. 92 above), in addition to the rates of activity in the field. Concurrently field measurements of thallus moisture, temperature and illumination were also taken. The field data agree well with the rates of activity that would be expected on the basis of the response matrix. For example on 10-11 August following a period of heavy rain, with full cloud cover, both thallus moisture and thallus temperature, measured by embedded microthermocouples, were almost constant (Fig. 94). As a result there was a marked diurnal fluctuation of nitrogenase activity solely induced by changes in the level of photosynthetically active

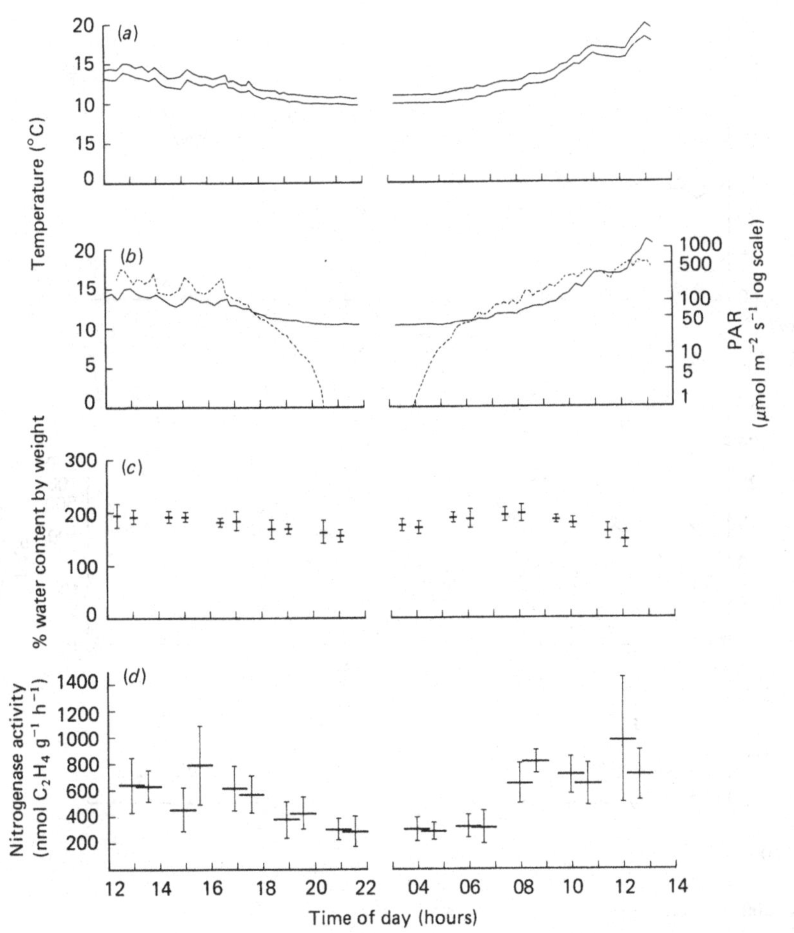

Fig. 94. Microclimatic variables and nitrogenase activity in a mat of *Stereocaulon paschale* on 10-11 August 1976. (*a*) Maximum and minimum thallus temperatures within the lichen mat (range of five signals). (*b*) Thallus temperature (—) and photosynthetically active radiation (PAR: note log scale) (----) within the incubation bottles. (*c*) Mean thallus water content within the lichen mat and of thalli before incubation for acetylene reduction assays (percentage of oven-dry weight). (*d*) Mean nitrogenase activity. Mean values (n = 10) are plotted together with 95% confidence limits. Heavy rain commenced at *c*. 15.30 hours on 9 August and measurements were begun at 12.00 hours the next day. (From Crittenden and Kershaw 1979.)

radiation (PAR). This particularly striking data set is consistent with the decline of nitrogenase activity in the dark (MacFarlane *et al.* 1976), which apparently can be maintained unimpaired throughout the night at 15 °C (Kershaw *et al.* 1977). It was not until 08.00 hours with PAR values exceeding 100 μmol m^{-2} s^{-1} that nitrogenase activity was sustained at full

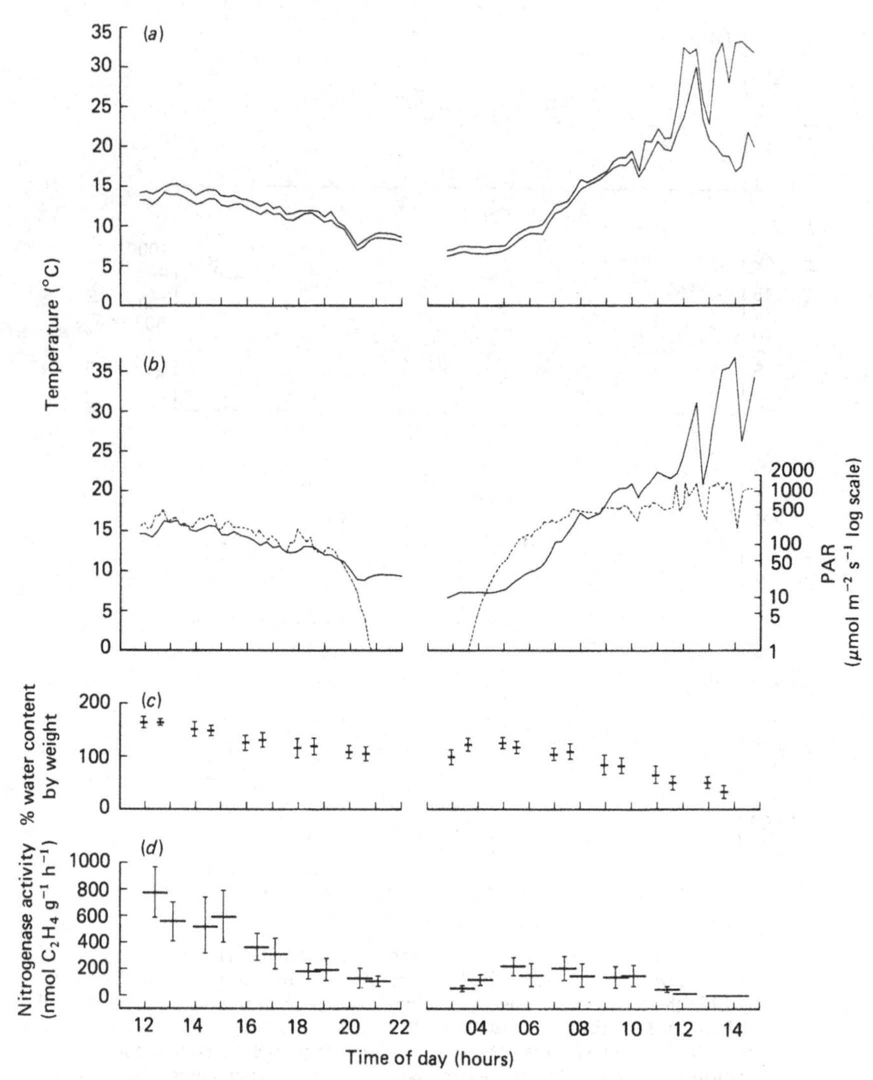

Fig. 95. Microclimatic variables and nitrogenase activity in a mat of *Stereocaulon paschale* on 6-7 August 1976. Rainfall commenced at *c*. 02.30 hours on 6 August and measurements were begun at 11.30 hours. For further explanation see legend to Fig. 94. (From Crittenden and Kershaw 1979.)

rates. The experimental run was discontinued after temperature control inside the incubation bottles was lost under increasing levels of radiation. The control of nitrogenase activity by thallus moisture content is well illustrated by the data of 6-7 August during increasing radiation conditions (Fig. 95). By 18.00 hours low levels of radiation and declining tempera-

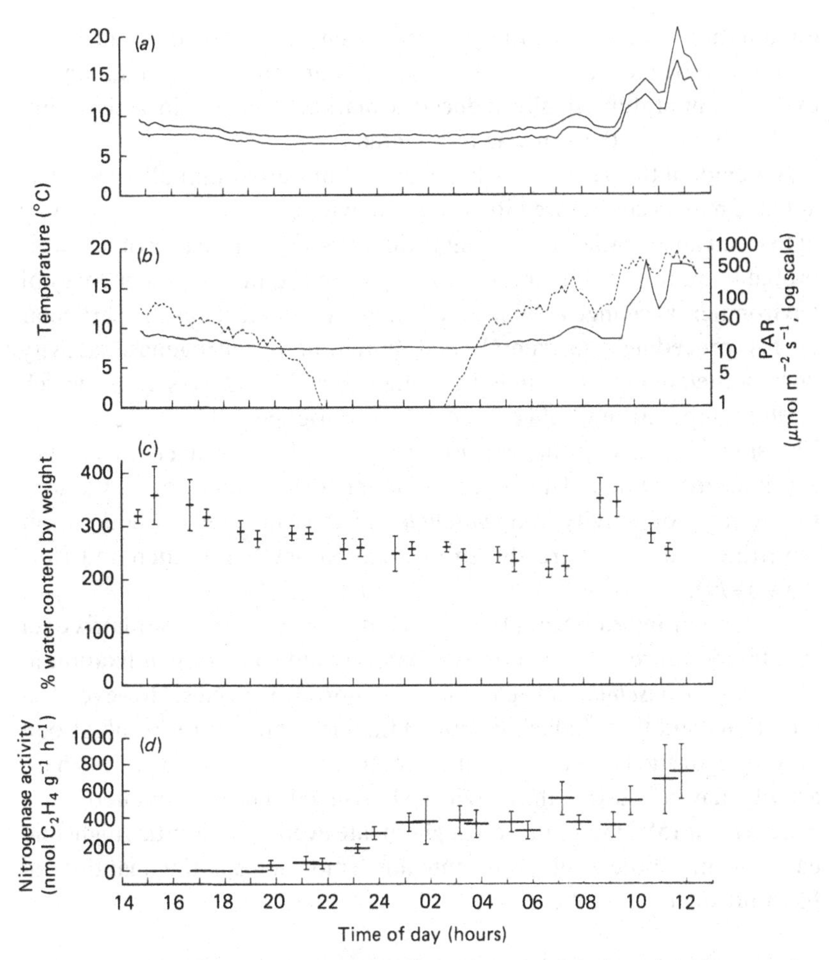

Fig. 96. Microclimatic variables and nitrogenase activity in a mat of *Stereocaulon paschale* on 24-25 June 1976. Heavy rain commenced at *c*. 09.15 hours on 24 June and measurements were begun at 14.30 hours. For further explanation see legend to Fig. 94. (From Crittenden and Kershaw 1979.)

tures produced low levels of nitrogenase activity which were maintained until 04.00 hours. As light and temperature levels increased the following morning there was a rise in the level of activity generated until *c* 08.00 hours after which a steady decline of thallus moisture reduced activity to zero as ambient temperatures inside the incubation bottles rose steeply.

The data for 24-25 June (Fig. 96) are particularly important in that heavy rain after 12 days of drought occurred at 09.15 hours but no ethylene production was detectable for almost 10 h despite isothermal temperatures of approximately 8 °C, at full thallus hydration, but under modest levels of illumination. The actual initiation of nitrogenase activity occurred during darkness and reached a maximum by 02.00 hours. Increasing radiation and thallus temperature finally induced a marked increase in activity immediately before the run was terminated.

It is evident that once the lichen mat is saturated by rainfall a considerable lag may occur before nitrogenase activity is recovered and certainly light is not an essential component in this instance. The extent of this lag is probably quite variable and is dependent on the previous sequences of environmental change to which the lichen is exposed. Thus the particular events preceding the rainfall and initiation of nitrogenase activity documented in Fig. 96 included a drought period of 12 days during which thallus temperature would exceed 45 °C for brief periods on several days. The slow recovery of nitrogenase activity probably reflects this rather extreme environmental history, since under normal laboratory conditions steady rates of activity in *S. paschale* are evident after 2-3 hours from rewetting under non-stressful storage conditions (Crittenden and Kershaw 1979).

Crittenden and Kershaw (1979) conclude from the response matrix data that temperature is the major rate determinant for nitrogen fixation in *Stereocaulon paschale* and indeed for most northern species. However, full radiation conditions inducing optimal thallus temperatures result also in rapid dehydration of the thallus and although maximal rates of nitrogenase activity may be present, the activity is short-lived. The more limited levels of activity at 15 °C integrated throughout late evening, the entire night and early morning potentially contribute much more substantial quantities of fixed nitrogen to the lichen.

4.7 The fate of fixed nitrogen in the lichen thallus

Millbank and Kershaw (1969) have shown that the cephalodia of *Peltigera aphthosa*, which contain *Nostoc* as a phycobiont, actively fix nitrogen and are particularly suitable for experimental manipulation as they allow the subsequent fate of the fixed nitrogen to be examined. Using $^{15}N_2$ as a tracer and by dissecting off the cephalodia from the thallus, the translocation of the fixed $^{15}N_2$ in the cephalodia to the remainder of the lichen thallus can be readily determined. The data showed that the level of ^{15}N in the cephalodia increased linearly over the first few days of exposure to a ^{15}N enriched atmosphere, thereafter becoming constant (Fig. 97). In contrast the level of nitrogen in the remainder of the thallus, with *Coccomyxa* as the phycobiont, continued to increase with time in a linear fashion. The cephalodia can be considered as a nitrogen-fixing system with

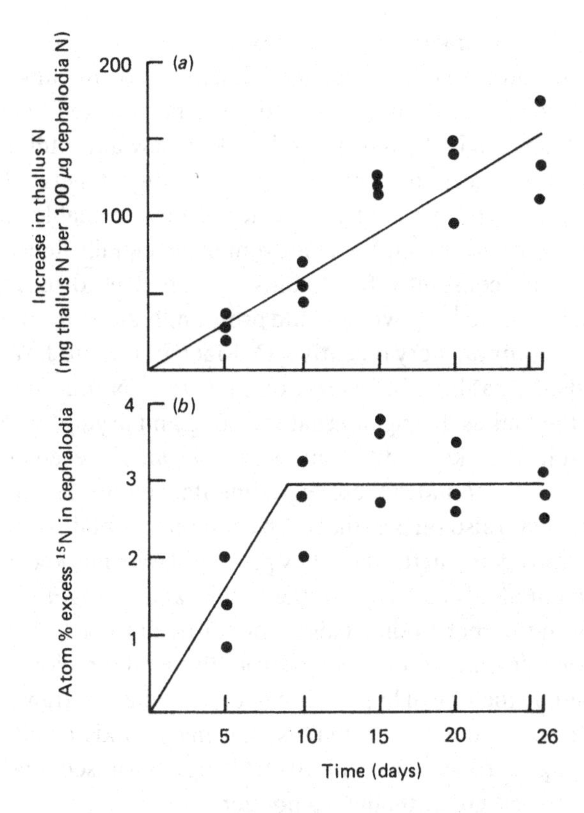

Fig. 97. (*a*) Uptake of elementary nitrogen by the thallus of *Peltigera aphthosa* at 12 °C. (*b*) $^{15}N_2$ enrichment of *Peltigera aphthosa* cephalodia at 12 °C when exposed to nitrogen gas enriched to 30% $^{15}N_2$. (From Millbank and Kershaw 1969.)

the rate of fixation equal to the rate of secretion. The lapse time before equilibrium is related to the fixation rate, whereas the level of $^{15}N_2$ at equilibrium is related to the size of the metabolic pool. Thus at 25 °C the amount of $^{15}N_2$ enrichment in the cephalodia levelled off after only 4 days, in contrast to 10 days at 12 °C ; the apparent difference between pool sizes reflects the use of 95% $^{15}N_2$ and 30% $^{15}N_2$ respectively in the two experiments (Millbank and Kershaw 1969). Thus although 19-28% of the nitrogen fixed by a culture of *Nostoc* isolated from the lichen *Collema tenax* is released into the medium (Henriksson 1951), a very large proportion of the nitrogen fixed by *Nostoc* in the intact thallus is released, presumably under the influence of the fungal component.

The subsequent division of $^{15}N_2$ in the thallus between the mycobiont and the *Coccomyxa* phycobiont shows that only a small proportion of $^{15}N_2$ actually appears in the green alga. Since the thallus on average contains *c*.

7% algal nitrogen, the quantity of $^{15}N_2$ expected to accumulate on a proportional basis can be readily calculated (Kershaw and Millbank 1970). In fact *Coccomyxa* receives about one-twentieth of the expected amount of $^{15}N_2$. This had been originally interpreted by Kershaw and Millbank as contradictory to any concept of symbiosis where sharing of metabolites is implicit. However, in the light of the evidence now available for the environmental control over metabolite movement an equally likely interpretation lies in the constantly full thallus hydration used during the experimental exposure to $^{15}N_2$, which could profoundly affect the metabolic balance of *Coccomyxa*. Very recently J.D. MacFarlane and J.W. Millbank (unpublished data) have in fact re-examined the $^{15}N_2$ throughput to *Coccomyxa* in the thallus during alternate wetting and drying cycles and have shown that it is markedly different, with *Coccomyxa* receiving substantial quantities of ^{15}N under these experimental conditions. Stewart, Rowell and Rai (1980) also suggest that each component obtains a proportion of the fixed nitrogen, but the data they present show a marked decline of $^{15}N_2$-enrichment after day 5 in both the thallus and *Coccomyxa* replicates, as well as in the cephalodia. This is very difficult to interpret but it seems quite probable that after its release from the algal symbiont $^{15}N_2$ is rapidly absorbed by the fungal hyphae in the cephalodia and transported throughout the remainder of the thallus. During periods of alternate wetting and drying, a proportion of this fixed nitrogen is subsequently also made available to the green thallus component.

There is some apparent controversy as to the form of released label from the phycobiont. Millbank (1974*b*, 1976) examined the nature of the substances released from the cephalodia of *Peltigera aphthosa*, *Placopsis gelida* and *Lobaria amplissima*, and from the dissected cortex of *Peltigera polydactyla*. The dissected cephalodia or cortical tissue were placed on moist filter paper and the products released allowed to accumulate over a 2 week period. The results showed that peptides with molecular weights of the order of 1100 accumulated in the filter paper and had closely similar composition to those released by the lichen algae when cultured separately. In contrast Stewart *et al.* (1980) consider that the fixed nitrogen in the cephalodia of *Peltigera aphthosa* is transported to the surrounding fungal hyphae as ammonium ions. Thus the level of glutamine synthetase activity in lichens with blue-green symbionts is extremely low (Stewart and Rowell 1977; Sampaio *et al.* 1979; Stewart *et al.* 1980). This infers that because of the resultant slow amino acid synthesis within the phycobiont, the accumulating ammonium ions are released and assimilated by the fungal component. The suggestion is that the ammonium is largely converted to alanine

(Stewart *et al.* 1980). They also found that the reduced level of glutamine synthetase activity is rapidly lost when the algal symbiont from *Blasia pusilla* in particular is cultured in isolation, and it seems evident that the reduction of glutamine synthetase activity is specifically induced by the state of symbiosis. The apparent discrepancy between these data and those of Millbank is, however, largely a result of his emphasis on the release of organic nitrogen: the release of inorganic nitrogen in the form of ammonium ions was not included in Millbank's assay. It should also be borne in mind that some simple organic nitrogen could be released from cellular tissue simply as a result of the surgical technique used by Millbank. It seems most likely that the majority of the nitrogen fixed by the algal component is released as ammonium, as Stewart and his co-workers have shown, with a residual amount released as simple peptides. The subsequent mobility of fixed nitrogen may well be also heavily influenced by wetting and drying cycles.

It is usually assumed that nitrogen fixed by lichens constitutes an important component of the input nitrogen-flux, particularly in northern ecosystems. Unfortunately there is little reliable information to substantiate this belief. Skujins and Klubeck (1978) have shown that despite a good level of $^{15}N_2$-fixation in blue-green algal and lichen crusts on arid soils in Utah, only a minor fraction of the nitrogen fixed enters the soil for further use by plants. The majority is rapidly lost through the activities of denitrifying organisms. This may not be the case in northern systems, where low soil temperature may reduce the activity of denitrifying bacteria, but evidence for this is lacking. Redistribution of fixed nitrogen by decomposition processes is also usually assumed to be an active process, potentially transferring nitrogen fixed by the lichen to the remainder of the system. However, it is equally possible that at least some of the nitrogen released by the decomposing parts of a lichen thallus is immediately reabsorbed by the younger, adjacent thallus.

Release of nitrogen occurs then, largely during the initial few minutes of each hydration period or ultimately by a relatively catastrophic event such as fire or herbivory (Millbank 1982). As a result of these uncertainties, current estimates of the nitrogen input of ecosystems are probably wildly unrealistic. Certainly many estimates have been derived from field studies using ethylene production as a measure of nitrogenase activity, and although classically a 3 : 1 ratio of ethylene production to actual nitrogen fixation is acceptable under univariate laboratory conditions, it is most unlikely to be applicable over a full range of light intensities and temperatures or even from season to season (Millbank 1981, 1982; Millbank and

Olsen 1981). Even more complexity is apparent when the marked lag phase following snow-fall or a long dry period (see above) is considered. Consequently, predictive modelling of nitrogen fixation without information from a full matrix of seasonal response to temperature, light and moisture as well as the potential reduction of activity during thallus rehydration periods is of little value.

5

Photosynthesis in lichens: its measurement and interaction with thallus hydration

Because of the continuous exchange of carbon dioxide by respiratory and photosynthetic processes, the direct measurement of gross photosynthesis can only be achieved by the use of $^{14}CO_2$. The difficulties and expense of operating with $^{14}CO_2$, however, make it unattractive in most experimental situations in ecology and as a result the vast majority of the data which are available on gas exchange in lichens are measurements of dark respiration and net photosynthesis using a variety of other methods. This discussion will be largely restricted, then, to gas exchange measurement utilising infrared gas analysis, one of the most commonly used techniques.

Net photosynthetic rates are controlled by a number of external environmental factors as well as by several internal factors. The important external factors which are influenced directly by the environment are the degree of thallus hydration, the intensity and quality of the ambient illumination, the operating thallus temperature, the thallus ionic environment, and day-length as a function of the time of year. Some of these parameters are partially correlated and the measurement of the photosynthetic response to even a single factor is accordingly complex. In addition the rate response is usually non-linear, further complicating experimental design. Unfortunately a considerable proportion of the published work on net photosynthetic rates has steadfastly ignored this marked degree of non-linearity of response and has established data using a univariate experimental framework. Thus, for example, experimental replicates are held at a single level of hydration, usually full saturation, whilst the experimental temperature is changed sequentially. However, the metabolic response to the level of thallus hydration is often non-linear and to extrapolate the response to temperature at full thallus saturation indiscriminately can lead to considerable error (see below). In an analogous way, net photosynthetic response has largely been examined only for a single time period and the measured response is assumed to hold true for the remainder of the year. We are now aware that in a number of species this is not true and in fact the majority of the species which so far have been examined critically show considerable seasonal changes in both the capacity and the pattern of their gas exchange responses. Other serious deficiencies in the storage and pretreatment of experimental replicates may also be involved in much of the early work and it becomes necessary at this point to evaluate critically the methods that have been used for gas exchange measurement. At the same time it is worthwhile to discuss storage methods and the pretreatment of experimental replicates, before a synthesis of the findings can properly be made.

5.1 The measurement of gas exchange

The exact measurement of any physical parameter is fraught with difficulties and the whole science of mensuration even in physics has proceeded in a step-wise fashion from some 'pretty good measures' of the last century to the refinement of a further few decimal places in the 1980s. The measurement of net photosynthesis in a meaningful way has also become more precise, even though there are physical difficulties which are superimposed on a normal level of biological variability and which in turn can be further modified by seasonal variation.

5.1.1 *Flow systems*

The standard approach has been to use an infrared gas analyser (IRGA), usually with an open flow system, which measures changing levels of carbon dixoide in a temperature-controlled and illuminated cuvette. There has been a tremendous amount of effort expended to obtain absolute measures of net photosynthetic and respiratory activity, by ensuring, for example, very adequate mixing of air in the cuvette to eliminate any carbon dioxide limitation of photosynthesis by the establishment of a carbon dioxide gradient. Equally, adequate flow rates over a leaf surface have been ensured in order to reduce the potential boundary-layer diffusive resistance to carbon dioxide, which could again limit the development of the full rate of photosynthesis. These concepts are particularly relevant to measurement of net photosynthesis in higher plants, especially where high rates of gas exchange are involved. For a full discussion the reader is referred to Jarvis (1971).

These exhaustive efforts were perhaps significant as part of the International Biological Programme project examining comparative productivity in different ecosystems. Even then, however, they may have reflected excessive technical zeal, since subsequent data handling during the development of productivity models often coupled these absolute measures to much less accurate values obtained for additional parameters. Recent technological advances have considerably improved the accuracy and reliability of temperature measurement and flow rate control, for example, but not the flexibility of the system, and even the most complex flow systems are still not capable of handling more than a few replicates at a time. Realistic estimates of standard errors to be attached to the absolute measures of net photosynthesis are thus largely absent. Although the absolute values of net photosynthesis that are available are significant in terms of productivity, considerable ecological information is also contained in *the seasonal pattern of response*, irrespective of the absolute rates. Such investigations demand a multivariate experimental design, particularly if seasonal changes are to be examined, and such data are still fragmentary. The advantage of a flow system is that it allows the examination of gas exchange – even in plants which have a high photosynthetic capacity – with a good level of accuracy, but it has to be evaluated within the framework of the small number of replicates that can be examined at any one time. An adequate estimate of population variability, or frequent and repeated sampling throughout the year, is difficult if not sometimes impossible. The measurements using a flow system, as a result, tend to be restricted to a limited univariate form rather than a full multivariate design.

5.1.2 *Closed loop and dish systems*

The alternative IRGA system which has also been used for some time, is a closed loop system where the changing level of ambient carbon dioxide is monitored over short time periods before photosynthetic rates become limited by the lowered ambient carbon dioxide concentration. This basic closed loop system can be easily modified to accommodate up to 25 replicates within a single exposure period by replacing the closed IRGA-cuvette loop with a number of sealed glass cuvettes which have a side arm covered by a rubber septum. Samples of the air before and at the end of an incubation period under a known temperature and level of illumination are then injected into a nitrogen/carbon dioxide standard 'carrier' gas which circulates through the reference and sample tube of an IRGA (Larson and Kershaw 1975*b*; Atkins and Pate 1977; Clegg, Sullivan and Eastin 1978). The differential between the reference and sample tube is recorded as a sharp peak and the difference between the peak height at the beginning and end of an incubation period can then simply be quantified by calibration against nitrogen/carbon dioxide gas standards. Between each exposure period the glass cuvettes are fully vented and the lichen replicate weighed, thus readily allowing a multivariate experiment (net photosynthesis, all moisture levels, one light level and one temperature), replicated up to 25 times, to be run as a standard procedure.

There are two potential problems involved in the use of closed incubation dishes: the absence of any gas mixing, and the declining level of ambient carbon dioxide concentration. Jarvis (1971) has stressed the importance of well-ventilated cuvettes to minimise boundary-layer carbon dioxide resistance (r_a). There is a finite resistance to the diffusion of a gas in completely still air or laminar flow conditions and accordingly the boundary layer offers a measurable impedance to the flux of carbon dioxide required to maintain photosynthesis at a maximum rate for the given conditions. Typical values for r_a in the field under windy conditions range from 0.1 to 0.3 s cm^{-1}. In well-ventilated cuvettes r_a values of 0.5 - 2.0 s cm^{-1} are desirable when plants with high photosynthetic capacities are involved, to avoid significant reduction of net photosynthetic values due to carbon dioxide limitation. Green and Snelgar (1981*b*) have examined the problem in detail and have concluded that r_a represents only a small proportion of the unusually high total diffusive resistance (Σr) reported for lichens. Thus Snelgar, Green and Beltz (1981*b*) report a value for Σr of 50-300 s cm^{-1} and Collins and Farrar (1978), 136 s cm^{-1}. Green and Snelgar (1981*b*) suggest a value of r_a for still air conditions of 7.0 s cm^{-1}, corresponding to a rather unrealistic boundary-layer depth of 1.03 cm. However, their values of r_a are all derived using a value for d (diameter of lichen) of 5 cm and utilise the formula given by Monteith (1964).

$$r_a = 1.3\sqrt{\left(\frac{d}{u}\right)}, \tag{7}$$

where u is the wind speed at a specific height.

Usually experimental replicates are considerably smaller than 5 cm and the estimates of r_a given by Green and Snelgar should be considered as extreme values. Accordingly r_a, the boundary-layer resistance to carbon dioxide diffusion, becomes a small proportion of the internal diffusive resistance term and lack of ventilation in closed dishes where the lichen rarely is capable of net photosynthetic rates greater than 6.0 mg CO_2 h^{-1} g^{-1}, is not a serious concern.

The restriction of photosynthesis by low ambient levels of carbon dioxide, however, presents a much more serious problem. Larson and Kershaw (1975*b*) showed that in *Alectoria ochroleuca*, *Cetraria nivalis* and *Parmelia caperata*, gas exchange rates are independent of carbon dioxide concentration between *c*. 200 and 350 ppm. If the ambient level of carbon dioxide did not drop below 200 ppm then no significant limitation of net photosynthesis occurred. However, it was emphasised that the pattern of response shown by these three species was not necessarily a generalised pattern for all lichens and each new experimental species should be separately examined. Green and Snelgar (1981*b*), in contrast, have suggested that the response of lichens to carbon dioxide concentration is similar to that of higher plants and may result in apparently depressed rates of net photosynthesis if incubation periods are not kept very short. Lange (1980) and Lange and Tenhunen (1981) express a similar concern but also fail to distinguish clearly between the effects of ambient carbon dioxide concentration at full thallus saturation and the effects at optimal water content. A full discussion of this will be postponed until the effects of the degree of thallus saturation are examined below, but in essence, for some lichens, the internal resistance to carbon dioxide diffusion is so high at full levels of hydration that the rates of net photosynthesis will indeed be depressed. Experimentation in a closed dish using long incubation periods with decreasing levels of ambient carbon dioxide can lead to some errors of measurement (see below).

It is re-emphasised that before the use of a closed dish injection approach for the measurement of gas exchange in any particular experimental species, it is essential to establish the level at which carbon dioxide limitation occurs. Some species may require quite short incubation periods to keep ambient carbon dioxide levels close to or above 300 ppm.

5.2 Variation between replicate thalli, and data presentation

As a direct result of the inadequate levels of replication inherent in the use of an IRGA flow system, little effort has been made to provide a comprehensive estimate of between-replicate variation. Largely, such potential variation is often deliberately obscured by utilising net photosynthetic rates expressed on a relative scale as a percentage of the maximum. Not only does this approach negate the effort put into the accurate measurement of net photosynthesis, it also ignores extremely valuable information which may relate, for example, directly to seasonal capacity changes or ecotypic differences. This is seen very clearly from the data presented by Kershaw (1977*b*, *c*) on the clear summer/winter change in photosynthetic capacity which occurs in *Peltigera praetextata*. This is a marked attribute of this species (see p. 189 for a detailed discussion) but its presence would be completely obscured by adjusting the absolute values to relative values.

The use of net photosynthetic values expressed as a percentage of the maximum appears to have developed as an alternative to exact control of

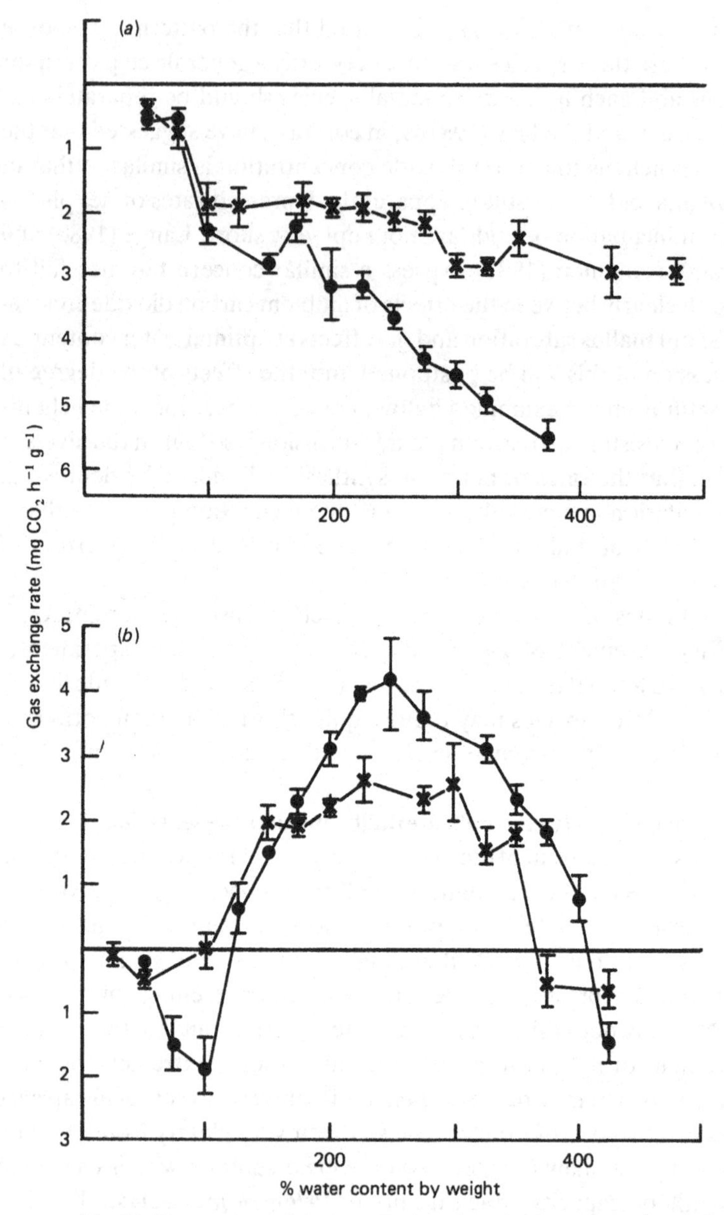

Fig. 98. The response of photosynthetic and respiration rates to changes in thallus moisture (% of oven-dry weight), in young (•) and old (×) thallus replicates at 25°C and (*a*) 0 or (*b*) 300 μmol m^{-2} s^{-1} illumination. The vertical bars show the S.E.M. for each moisture class. (From Kershaw 1977*b*.)

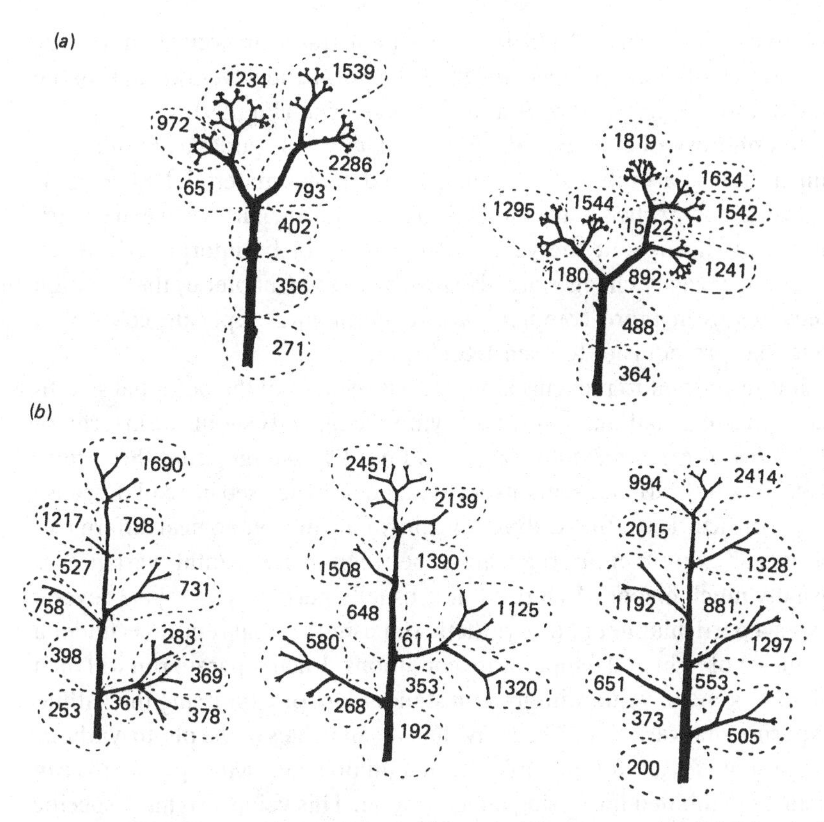

Fig. 99. Photosynthetic activity (as ^{14}C cpm) of the upper portion (2.0-2.5 cm) of (*a*) two specimens of *Cladonia stellaris* and (*b*) three specimens of *C. rangiferina*. (From Nash *et al.* 1980.)

between-replicate variation, or the even more difficult control of between-collection variation. Kärenlampi, Tammisola and Hurme (1975), Kallio and Kärenlampi (1975), Kershaw (1977*b*) and Nash, Moser and Link (1980) have all documented the considerable range of photosynthetic, respiratory or net photosynthetic rates which occur in young or old parts of many lichen thalli. In *Peltigera polydactyla* for example (Kershaw 1977*b*), mean maximum rates of net photosynthesis in the young thallus lobes are 4.2 mg CO_2 g^{-1} h^{-1} contrasting with 2.6 mg CO_2 g^{-1} h^{-1} for older parts of the thallus. Similarly respiration rates of the younger thalli have a mean maximum of 5.6 mg CO_2 g^{-1} h^{-1} compared with 3 mg CO_2 g^{-1} h^{-1} in older material (Fig. 98). These differences are large and can induce considerable scatter in the data unless very careful control is maintained over the selection of replicates for each experiment. Such scatter, however, can be easily hidden by expressing the net photosynthetic or respiratory rates on a relative scale. Nash *et al.* (1980) have demonstrated an even greater range of photosynthetic capacity in the upper, central and lower portions of the podetia of *Cladonia stellaris* and *C. rangiferina* (Fig. 99).

Net photosynthetic rates of whole thalli or whole podetia are thus very much a function of the ratio of young to old thallus material. Unless this is rigorously controlled between replicates and particularly between experiments or throughout the year, the data cannot be interpreted. When seasonal capacity changes are also involved (see Chapter 6) the situation becomes even more complex, with experimental replicate control an essential prerequisite to each IRGA run.

It is important to recognise that the assessment of the potential significance of a seasonal change in photosynthetic capacity should be in terms of the *between experimental variance* and not by assuming that the *within*-experimental variance remains constant and can be used instead. Because of the difficulties outlined above of controlling morphological differences between experimental replicates, between-experimental variance is usually much higher. As a result it is usually necessary to repeat several times a particular net photosynthetic run using several replicates under a given set of light and temperature conditions. Each separate run will then provide a mean value with its *own* standard error expressing the within-experimental variation. The individual mean values of the photosynthetic maximum (P_{max}), for example, from a number of such experiments can then be combined into a single mean value. This value also has a specific standard error which, however, expresses variation *between individual experiments* characteristic of a particular time of year, experimental temperature and level of illumination. If this *experimental standard error* is subsequently exceeded at a different time of the year then the *seasonal* change is indeed significant statistically.

The use of relative thallus moisture scales equally can hide differences that may express very significant ecotypic adaptive features. Fig. 100 shows quite readily the marked differences in the water-holding capacity of two ecotypes of *Peltigera aphthosa*, but only when this is expressed in absolute terms. In other instances, however, it is acceptable to express thallus moisture on a relative scale to allow a comparison of net photosynthetic responses in different species to thallus hydration. Coxson and Kershaw (1983*b*) encountered severe problems in attempting to express the thallus water content of *Rhizocarpon superficiale* on an absolute basis, throughout a complete drying cycle. Different experimental replicates behaved quite differently in their drying response, as a result of the very specific architecture of each rock and lichen replicate (Fig. 101). As a result

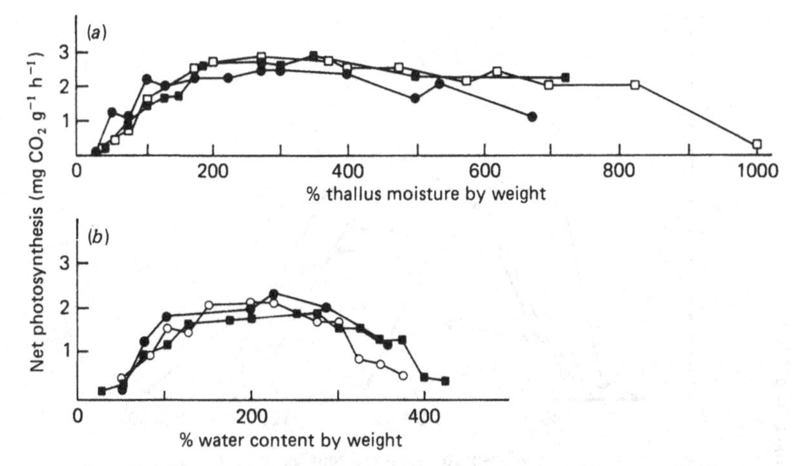

Fig. 100. The contrasting water-holding capacities of (*a*) sun and (*b*) shade ecotypes of *Peltigera aphthosa*, which hold up to 1000% and 400% thallus moisture by weight, respectively, in absolute terms. Each line represents a replicate experiment. (From MacFarlane and Kershaw 1980*a*.)

the calculation of mean net photosynthetic rates within each specific moisture class often gave meaningless results. The use of a relative scale in this instance at least allowed a good estimate of variation to be attached to the net photosynthetic values.

Larson (1980) has advocated some further form of statistical testing of a gas exchange response matrix and develops a simple model to assess the relative advantages of the particular response patterns of species of *Umbilicaria* to their environmental parameters of temperature, moisture and light. Such an ambitious approach is perhaps premature since even a simple model should utilise the specific parameters defining the operating environment of the lichen rather than standard meteorological data which are not necessarily relevant. Furthermore, to test the significance of seasonal changes demands at least one degree of freedom which in turn requires a full response matrix for a 2 year period. This is still a daunting prospect even with a computer-assisted, discrete sampling, closed loop IRGA system.

5.3 Storage and pretreatment of experimental material

Because of the rather inhospitable habitat of some lichens, where rigorous temperatures or low availability of nutrients and moisture eliminate higher plant competitors, it has been almost casually assumed that *all* experimental material can be kept in equally inhospitable storage conditions, particulary when dry, without any detriment to the lichen. Storage

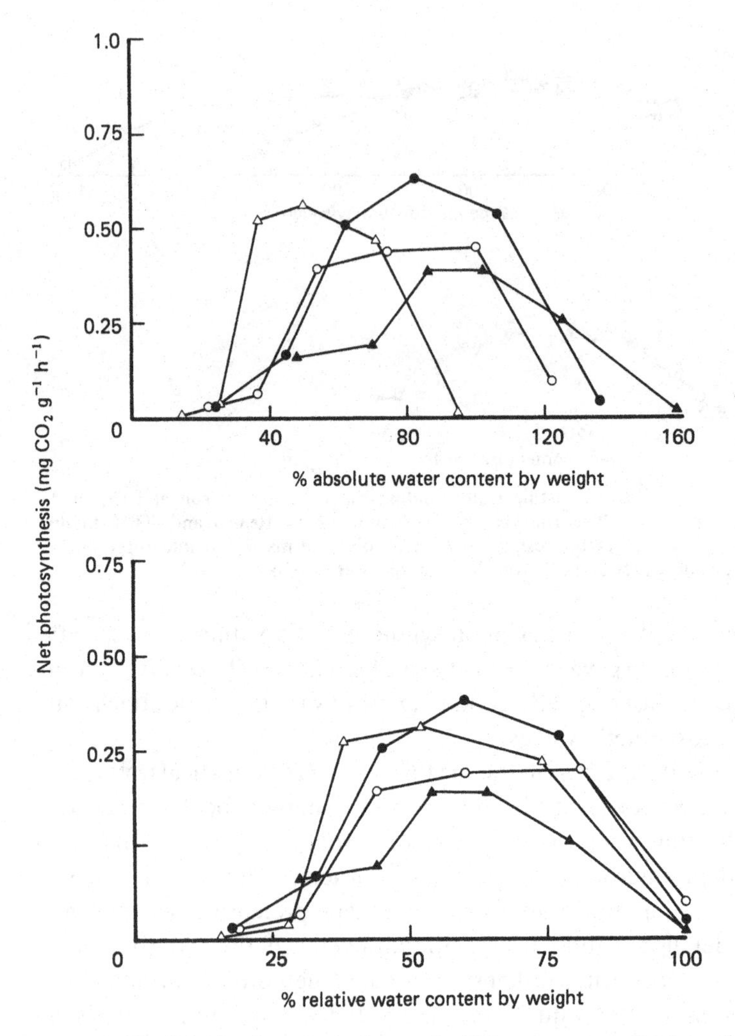

Fig. 101. Net photosynthetic rates for *Rhizocarpon superficiale* in relation to the degree of thallus hydration expressed as absolute and relative water contents. Each line represents a replicate experiment. (From Coxson and Kershaw 1983*b*.)

conditions reported in the literature range from deep-freezing to dry storage on a bench top, often without any examination of possibly deleterious effects, or even whether a metabolic change has been induced by the storage conditions so that experimental data subsequently obtained no longer reflect the metabolic pattern which originally existed in the field. It has been shown that a change from field temperature conditions to a different regime in a growth chamber can, for example, induce net photosynthetic acclimation to the new storage temperature (Kershaw 1977*b*,*c*;

Larson and Kershaw 1975*b*; etc.). Furthermore a response to a new temperature regime can take place rapidly even in air-dry thalli which contain only 10% moisture by weight. In a comparable way some lichen species can also adjust their photosynthetic light saturation point when stored air-dry under low levels of illumination, thus maintaining a full level of photosynthetic capacity under the new lower light environment (Kershaw and MacFarlane 1980). High light and continuous light can be detrimental to certain lichen species (Kershaw and MacFarlane 1980) as can low- or high-temperature storage (MacFarlane and Kershaw 1978, 1980*a*; Larson 1980).

It is now evident that each experimental species must be stored under light, temperature and daylength conditions as close as possible to the corresponding levels in the field. In the absence of this level of storage control, experimental results can be of limited worth and may have to be interpreted with caution. Equally, it is always advisable to allow a 4 or 5 day period of storage under controlled conditions before any replicate material is used experimentally, as collection in the field and transportation to the laboratory often results in the lichen material being enclosed in the dark in some suitable container, effectively modifying the current field daylength. The potential effects of this phenomenon can be simply avoided by a short pre-experimental dry-storage period to re-establish any pattern of diurnal environmental control which may have been important in the field.

In an analogous way, the same pretreatment for all replicates immediately before an experiment can also be essential. Some lichens completely rehydrate and achieve full rates of gas exchange over a short time period whilst others take longer. Nitrogenase activity particularly has both a distinct rehydration phase and a phase where marked depletion of carbohydrate reserves occurs during the night (see above). The possibility of an efflux of carbon dioxide on rehydration or a higher rate of respiration for a period immediately after rehydration can also be important. A consistent pre-soak period under defined illumination levels avoids the possible interaction of such uncontrolled metabolic states with the experimental variables at the start of an experiment.

5.4 The interaction of net photosynthetic rates and the level of thallus hydration

As early as 1927, Stocker demonstrated that the photosynthetic capacity of *Umbilicaria pustulata* and *Lobaria pulmonaria* at first increased with decreasing thallus water content before finally declining at lower levels of hydration. Ried (1960*a*,*b*) emphasised that the results obtained by Stocker (1927) reflected the high resistance to gas exchange by the water-saturated thalli and extended similar observations to a number of additional species which were characteristic of a series of habitats ranging from semi-aquatic to exposed xeric situations. He concluded that there was a definite relationship between the compactness of thallus structure, the available capillary spaces for gas diffusion, and whether photosynthesis was thus severely limited at full thallus hydration. It should be noted that in common with other workers, Ried carefully removed the surface film of water and defined 100% thallus saturation as a function solely of water held internally in the thallus. This is in contrast to the more strictly ecological approaches where it is recognised that a surface water film is a normal occurrence under field conditions during and after rain. Furthermore thallus saturation is defined by Ried on a 0-100% relative scale which, although allowing ready comparison within his own data set, sometimes has to be re-interpreted for comparison with data where moisture content has been expressed in absolute terms. Ried's data show that *Umbilicaria cylindrica*, for example, has an optimum rate of net photosynthesis at 100% relative saturation with only a slight reduction of rate at the higher levels of thallus 'super-saturation'. Ried interprets this pattern as a significant ecological correlation, which at first sight appears to be quite a reasonable hypothesis. However, there now seems little doubt that decreased rates of net photosynthesis at maximum levels of thallus hydration are largely due to the high impedance to carbon dioxide diffusion together with the consistently high respiration rates associated with full thallus water content. The clarity of the correlation with habitat implied by Ried is not necessarily always apparent.

Thus, Snelgar *et al.* (1980), for example, demonstrate an almost linear decline of respiration rate with decreasing moisture content in *Pseudocyphellaria colensoi* and *P. dissimilis* (see also Chapter 8). As a result, net photosynthetic rates can be completely respiration-dominated at full thallus hydration, particularly at higher temperatures. Conversely many other species have a more or less constant respiration rate during a drying cycle and in these instances depression of net photosynthetic rates at full hydration will be controlled almost entirely by diffusive resistances to carbon dioxide uptake alone.

The major resistances between the atmosphere and the actual incorporation of carbon dioxide into the photosynthetic pathway of higher plants have been discussed exhaustively by Chartier and Catský (1970), Jarvis (1971), Jones and Slatyer (1972), Körner, Schul and Bauer (1979), Landsberg and Ludlow (1970), and Prioul, Reyss and Chartier (1975), whilst Green and Snelgar (1981*a*,*b*) and Snelgar *et al.* (1981*a*,*b*) give a particularly useful account for lichens. The major resistances are summarised in Fig. 102 for both upper and lower surface diffusion. The boundary-layer resistances, r_a, are due to a still-air zone adjacent to the lichen surface, through which movement of carbon dioxide (or water molecules) is solely by diffusion; turbulent transfer processes are absent. This is a variable resistance, the size of which is controlled by wind speed. The potential limitations of a closed loop system, as a function of the lack of ventilation and the resultant high value of this boundary-layer resistance, have been discussed above. It was concluded that the resistance term r_a is usually quite small in relation to the internal resistance of a lichen thallus, which controls to a considerable extent the rate of diffusion of carbon dioxide to the chloroplast.

These internal resistances reflect the anatomy as well as the detailed membrane and cellular construction of a lichen. They can be generally categorised as three components: The upper and lower cortex offer an internal resistance to carbon dioxide diffusion which also functions as a variable resistance under the control of the thallus hydration level but probably also in terms of its detailed anatomy and particularly as a result of ecotypic adaptations (see Chapter 9). The medulla offers a further diffusive and variable resistance to the pathway from below the thallus, the size again being specifically controlled by anatomy and the level of thallus hydration. Finally there is a carboxylation resistance term, r_c, which reflects the direct demand for water molecules and carbon dioxide by the actual uptake mechanisms. (Fig. 102) Thus there is a considerable degree of interaction between the degree of thallus hydration, the level of internal resistance to carbon dioxide diffusion, and the resultant pattern of net photosynthetic response to thallus hydration.

Snelgar *et al.* (1981*a*) have now confirmed by measurement that depression of photosynthesis at full thallus saturation is indeed largely due to carbon dioxide diffusive resistance. This resistance and the pattern of gas exchange with changing thallus water content for *Pseudocyphellaria amphisticta* and *P. homoeophylla* are given in Fig. 103 and Fig. 104. *P. amphisticta* holds over 500% water content by weight with only a small increase in the carbon dioxide diffusive resistance at full thallus saturation. There is a correspondingly small decrease in net photosynthesis. At low levels of thallus hydration there is a sharp rise in the diffusive resistance of carbon dioxide which corresponds with a sharp decrease in the net photosynthetic rate. In contrast *P. homoeophylla* shows a very large increase in carbon dioxide diffusive resistance at higher levels of thallus hydration,

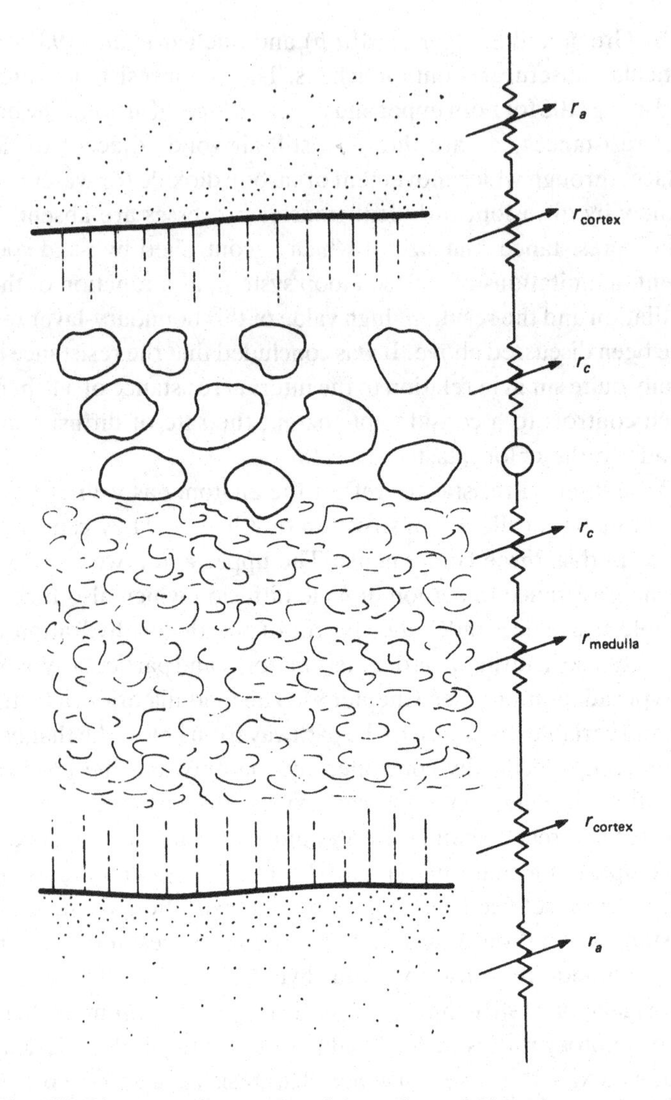

Fig. 102. Diagrammatic summary of the major variable resistances to carbon dioxide diffusion in a generalised lichen thallus: r_a, boundary-layer resistance; r_{cortex}, cortical resistance; $r_{medulla}$, medullary resistance; r_c, carboxylation resistance.

with the resistance reaching values comparable to those at low water contents (Fig. 104). Snelgar *et al.* (1981*a*) conclude that there are three identifiable phases in the six species which they examined: very high resistance at low water contents, which they suggest is most readily ex-

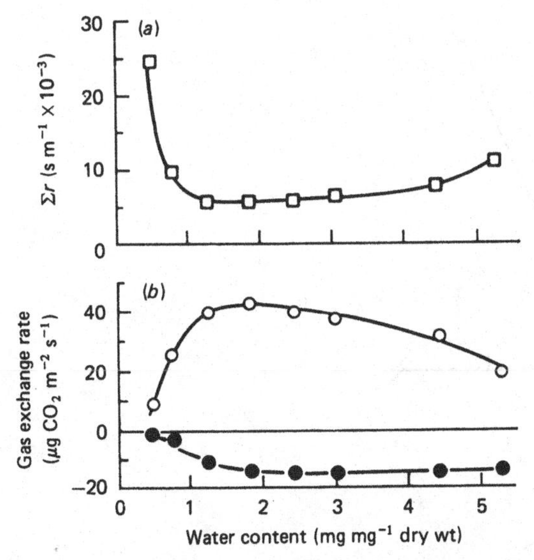

Fig. 103. (*a*) Diffusive resistance to carbon dioxide (Σr) and thallus water content in *Pseudocyphellaria amphisticta*. (*b*) Net photosynthetic (○) and respiratory (●) rates in response to thallus hydration. (From Snelgar *et al.* 1981*a*.)

plained by increased carboxylation resistance; low resistance over a range of intermediate water contents during which there are maximal rates of gas exchange; and an increase in resistance at high water contents, which can be quite small or large depending on the species. Although respiration rate is little affected by carbon dioxide diffusive resistance at high levels of thallus hydration they suggest that there is little doubt that, when a lichen is close to carbon dioxide saturation, then increased respiration must depress net photosynthetic rates. However, the increase in carbon dioxide diffusive resistance will also contribute to a decrease in net photosynthetic rate at full thallus hydration. Snelgar *et al.* (1981*a*) explain the differences between species in terms of the presence or absence of tomentum or pseudocyphellae. They also point to a close correlation of each lichen with a specific habitat, and a considerable degree of ecophysiological adaptation is implied.

Lange and Tenhunen (1981) further substantiate these findings with an elegant experiment using *Ramalina maciformis*. Replicate thalli were exposed to varying ambient carbon dioxide concentrations and the time course of net photosynthesis during thallus drying was then monitored (Fig. 105). At natural ambient carbon dioxide concentrations (320-360 ppm) there was a marked depression of net photosynthetic rates at full thallus saturation, to approximately 75% of the maximum rates generated

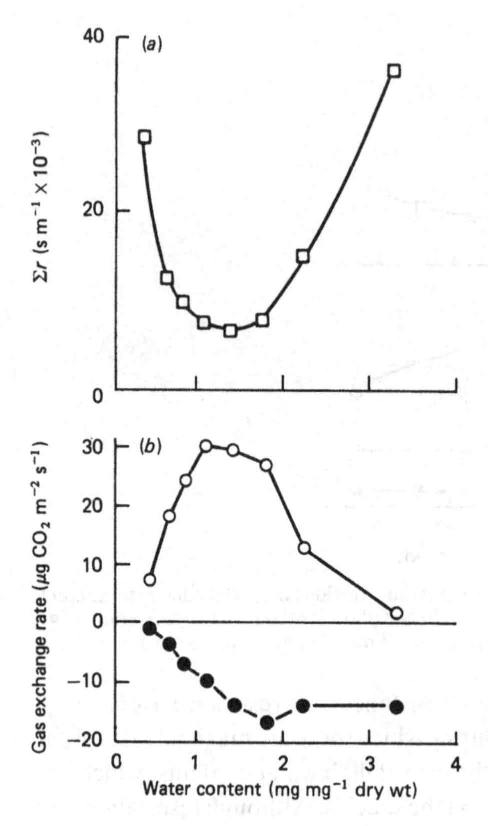

Fig. 104. (*a*) Diffusive resistance to carbon dioxide (Σr) and thallus water content in *Pseudocyphellaria homoeophylla*. (*b*) Net photosynthetic (○) and respiratory (●) rates in response to thallus hydration. (From Snelgar *et al.* 1981*a*.)

at the optimum thallus water content. At lower ambient levels of carbon dioxide this depression at full thallus hydration increases still further. At higher levels the depression of net photosynthetic rates at full thallus saturation becomes less marked and, with 1600 ppm carbon dioxide it is, in most replicates, entirely eliminated. Thus the effect of the increased internal diffusive resistance can be eliminated by supplying an atmosphere with excess carbon dioxide; the slow diffusion rate at thallus saturation is counteracted by a large increase in the carbon dioxide gradient. These results are particularly significant in relation to the discussion above of the problems of closed dish incubations and the development of low carbon dioxide levels during long incubation periods.

Larson and Kershaw (1975*b*) examined the point at which net photosynthesis became limited by low carbon dioxide concentrations in closed dish incubations and found for *Alectoria ochroleuca*, *Cetraria nivalis* and

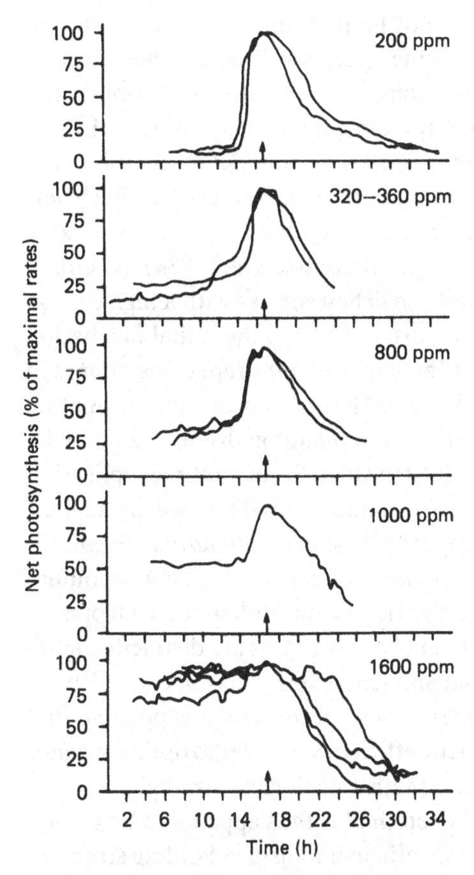

Fig. 105. Time course of relative rates of net photosynthesis as the thallus of *Ramalina maciformis* dries under different ambient concentrations of carbon dioxide. Under 1600 ppm carbon dioxide the depression of net photosynthesis at full thallus hydration is virtually eliminated. Each line represents a replicate experiments. Arrows indicate time of maximal carbon dioxide uptake. (From Lange and Tenhunen 1981.)

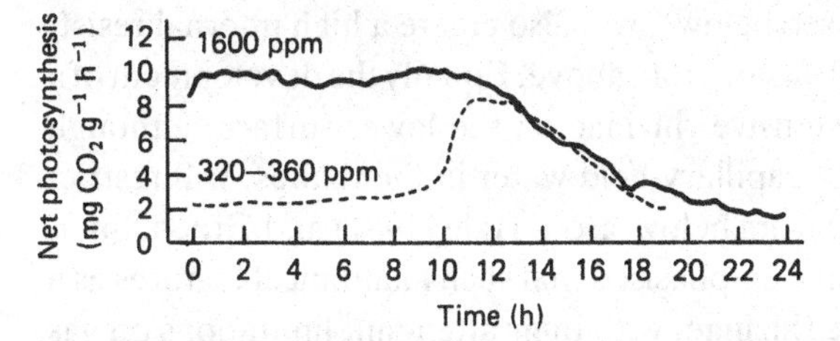

Fig. 106. Time course of absolute rates of net photosynthesis during drying of thalli of *Ramalina maciformis*. Under 1600 ppm ambient carbon dioxide, net photosynthetic rates are significantly higher from full thallus hydration down to the optimum level. (From Lange and Tenhunen, 1981.)

Parmelia caperata that rates were not limited until *c.* 200 ppm carbon dioxide. It is now apparent that some species will be affected by sub-ambient carbon dioxide concentrations at full thallus hydration if the thallus internal resistance is markedly altered by the level of thallus hydration. Under these conditions using a closed loop IRGA method, net photosynthetic rates will be underestimated. However, at optimum levels of thallus hydration no reduction in net photosynthetic rates will probably be evident in most species until *c.* 200 ppm (Coxson *et al.* 1983*a*). As Green and Snelgar (1981*a*) point out, in those lichen species with a high carbon dioxide diffusive resistance it is important to keep the initial incubation periods as short as possible until the experimental replicates approach their optimum levels of thallus hydration. The problem of underestimating net photosynthetic rates particularly at full thallus hydration can then be avoided to a considerable extent whilst the benefits of multiple replication are exploited. The data of Lange and Tenhunen (1981), however, suggest that where internal resistances are very high, as in *Ramalina maciformis* for example, full photosynthetic potential is not achieved even at optimum water content (Fig. 106). Under an enriched atmosphere of 1600 ppm of carbon dioxide, optimum rates of net photosynthesis are distinctly higher than those achieved under normal ambient levels. Coxson *et al.* (1983*a*) also report similar patterns in other lichen species and it is possible that some species sacrifice photosynthetic efficiency in order to optimise water economy, as a more successful alternative ecological strategy.

The evolution of an effective lichen thallus thus appears to be a compromise between the provision of an efficient moisture-holding structure for the maintenance of the normally aquatic algal component, and one which does not at the same time limit free gas exchange. The presence of a massive paraplectenchymatous upper cortex which will protect the algal cells from too rapid or excessive desiccation, and incidently from the destructive effects of too intense solar radiation (Rundel 1978*c*; Kershaw and MacFarlane, 1980; see below), will also create a high internal resistance to carbon dioxide diffusion from above. Equally the development of a thick medulla and/or extensive rhizinae on the lower surface, although increasing the amount of capillary-held water in the thallus, will restrict carbon dioxide diffusion from below. Both Hale (1981) and Green, Snelgar and Brown (1981) have emphasised that such anatomical features as a thick medulla and dense rhizinae, with their attendant limitations on gas diffusion, can be further modified by cyphellae, pseudocyphellae or the development of a pored epicortex. As Hale (1981) points out, until very recently the function of vegetative structures in lichens has been inter-

Table 7. *Carbon dioxide exchange in* Sticta latifrons *complete thalli and separately through upper and lower surfaces, during incubations at 16-20°C and 150 μmol m^{-2} s^{-1} (saturating intensity)*

Thallus surface assayed	Net photosynthetic rate $\mu g\ CO_2\ m^{-2} s^{-1}$	Mean	%	Respiration $\mu g\ CO_2\ m^{-2} s^{-1}$	Mean	%
Intermediate water content (0.8-1.0)						
Top	−1.9,−2.8,4.9	0.1	0.2	−6.6,−28.5,−5.7	−13.6	19
Bottom	62.2,59.9,56.5	59.5	99.8	−72.3,−42.9	−57.6	81
Complete thallus	40−60			−30		
High water content (1.7-2.0)						
Top	0,0,0,−4.3	−1.1	0	−13.6,−8.5,−18.4,−19.8	−15.1	16
Bottom	17,11.3,48,22.6	13.9	100	−72.3,−84.8,−63.6,−107.4	−82.0	84
Complete thallus	15-40			−40 to 50		

preted largely by analogy or purely by speculation. It is now evident that the morphology of lichens plays a direct role in their water economy (Larson and Kershaw 1976; Larson 1979, 1980) and there seems little doubt that anatomy can have a major effect on the carbon dioxide diffusion rate within the thallus. Green *et al.* (1981) examined carbon dioxide exchange in *Sticta latifrons* using a split chamber which allowed separate analysis of the gas exchange at the upper and the lower surface of the thallus. The results show that gas exchange occurs very largely through the lower surface, providing strong circumstantial evidence for the central role of cyphellae in both the uptake of carbon dioxide during photosynthesis and the uptake of oxygen during respiration (Table 7). Hale (1981) similarly concludes, from his extensive scanning electron microscope studies, that the large foliose and fruticose lichens in particular, have evolved a number of specialised structures that enhance the rate of gas exchange without simultaneously removing the advantages of a massive cortex and medulla on the water economy of the lichen. These structures range from the rather elaborate form of cyphellae via the much simpler pseudocyphellae to the simple pores and cortical cracks which are usually well developed in the foliose genera that have lobes broader than 5 mm (Hale, 1981). Thus it becomes apparent that in a lichen thallus the diffusion of carbon dioxide is inextricably linked to the thallus anatomy and water economy. It is not unexpected then, that there need not be a simple relationship between the ecology of a lichen and the level of thallus hydration at which optimum rates of net photosynthesis are developed.

It has been widely assumed that there is a close correlation between the water relations, net photosynthetic optima and ecology of a lichen. Those species which show little if any increase in internal carbon dioxide diffusive resistance at full thallus hydration grow in habitats of a mesic nature, whilst those lichens from a more xeric habitat have net photosynthetic optima which occur at lower thallus moisture levels. This concept is implicit in the work of Ried (1960*a*, *b*). Kershaw and Rouse (1971*a*) and Kershaw (1972) similarly suggest such a close relationship but their results are questionable because of the extremely low flow rates used (see Larson and Kershaw 1975*b* for discussion). However, Kershaw and Smith (1978) and Snelgar, Brown and Green (1980) again emphasise a correlation between the ecology of a lichen and the pattern of net photosynthetic response to the degree of thallus hydration. In some instances such a correlation may indeed be true but in other cases identical response patterns may be developed in quite contrasting environments.

For example, *Bryoria americana* subsp. *canadensis*, an epiphytic species, *Collema furfuraceum* a corticolous species, and *Peltigera aphthosa*, a terricolous species, are all characteristic of the northern boreal forests of Canada. All three species have an identical net photosynthetic response pattern to level of thallus hydration (Fig. 107). Their microenvironmental situations, however, contrast totally. *Bryoria* is characteristic of the upper branches of black spruce and is entirely dependent on precipitation and atmospheric humidity for its transient supplies of moisture. Evaporation will be consistently rapid and the thallus will be at very low levels of hydration for considerable time periods. *Collema* is a characteristic lichen on the north-facing trunks of balsam poplar, particularly in drainage tracks on the bottom 4 m of the trunk (Kershaw and MacFarlane 1982). Its habitat is thus reasonably xeric but considerably less extreme than the upper canopy of black spruce. For example, evaporation rates on the northerly aspect will be limited by reduced levels of solar radiation and will be 'buffered' to an extent by the bark of the tree. *Peltigera aphthosa* is equally widely distributed in the same regions of Canada but is always restricted to deep moss ground cover directly under the dense canopy provided by the lower branches of black spruce (Kershaw and MacFarlane 1980). The thallus accordingly is rarely if ever exposed to direct solar radiation and in

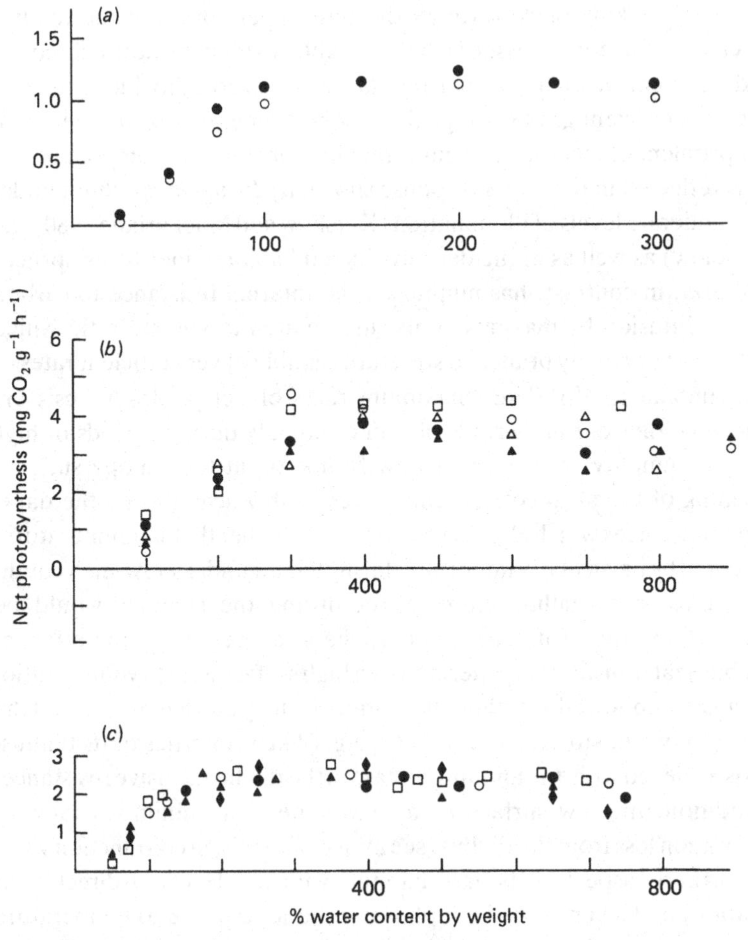

Fig. 107. The net photosynthetic response to the degree of thallus hydration in (*a*) *Bryoria americana*, (*b*) *Collema furfuraceum* and (*c*) *Peltigera aphthosa*. Different symbols represent replicate experiments.

fact is quite sensitive to even medium levels of illumination (see p.248). Only during extremely dry periods does the thallus dry out completely and then only over a period of many days.

The constant net photosynthetic rate which is maintained from full thallus saturation down to quite low levels of hydration and which shows an identical pattern in all three species, in fact reflects the three quite different ecological strategies adopted by each species. *Peltigera aphthosa* has developed a thick medulla covered below with an extensive tomentum allowing very efficient water storage and with the majority of the gas exchange presumably taking place through the thin upper cortex of the thallus. Effective water storage tissue is thus available without the normally correlated problem of high internal resistance to carbon dioxide diffusion. However, efficient gas exchange through the upper cortex has the attendant problem of insufficient light screening for the algal component and this is reflected in the stress response shown by *Peltigera aphthosa* under even moderate levels of illumination (Kershaw and MacFarlane 1980; and also below) as well as in the densely shaded habitat under black spruce.

Bryoria, in contrast, has minimised the internal resistance to carbon dioxide diffusion by maximising its surface area to volume ratio which results in a very finely branched structure capable of very efficient rates of water uptake (p.45). Thus maximum rates of net photosynthesis are achieved either during periods of rain or equally during periods of high ambient humidity, with mist or dew formation at the canopy surface. Screening of the algal component is presumably achieved by the dark, pigmented cortex which also serves to induce a higher thallus temperature. This could be particularly important during winter under snow-melt conditions. Excessive thallus temperatures during the summer would be avoided by virtue of its upper canopy habit as well as by the efficient sensible heat transfer also generated by a high surface area to volume ratio. *Collema* has followed a further alternative strategy and has maximised the amount of water stored by virtue of the gel-like properties of its thallus. This is achieved with a minimum increase in the thallus diffusive resistance. In addition, the low surface area to volume ratio minimises rates of evaporation loss from the thallus (see above) and this, in conjunction with, the northerly aspect of its bark habitat (with low levels of direct solar radiation) and a very broad net photosynthetic response to temperature allows very effective use of periods of rain (Kershaw and MacFarlane 1982).

Thus very different but equally effective alternative ecological strategies have been adopted by lichens. They can provide a direct correlation between rates of net photosynthesis generated at full thallus hydration and the relative mesic nature of the habitat, and this is seen repeatedly in many species. The correlation can, however, be inverse, with maximal photosynthetic rates at full hydration occurring in some lichens from xeric habitats. Equally *Peltigera canina* and *P. polydactyla*, both from a fairly mesic habitat, have very low rates of net photosynthesis at thallus saturation and have both adopted a quite different strategy in order to optimise carbon gain (see below).

5.5 Extreme levels of thallus hydration as an environmental stress

The poikilohydric nature of lichens is correlated, in most instances, with an ability to withstand periods of extremely low levels of thallus hydration that is coupled with rapid and effective reactivation of metabolic activity following rewetting. Such a correlation is not surprising but in some cases the period of desiccation may be both prolonged and extreme. Lange (1969) and Lange *et al.* (1970*b*), for example, have documented the remarkable level of adaptation achieved by *Ramalina maciformis*, a characteristic lichen of the Negev Desert. Even after 51 weeks at a thallus water content of 1%, respiratory activity is initiated extremely rapidly by exposure to water vapour in a fully saturated atmosphere (Fig. 108). As the thallus continues to imbibe water vapour, photosynthetic rates rise above compensation after 24 h. Spraying the thallus with a mist of water droplets yields a further increase in gas exchange. The stress effect of 38 weeks of dehydration is less marked and it would appear that *Ramalina* is little affected by periods of desiccation to which it is normally exposed in its desert environment. Ried (1960*b*) found a similarly rapid and full recovery of net photosynthetic activity in *Rhizocarpon geographicum* after 27-30 days at 60% relative humidity (Fig. 109). In contrast, however, *Lecidea soredizodes* required a full hour of hydration to recover its normal level of net photosynthesis, and *Aspicilia lacustris* and *Dermatocarpon aquaticum* only achieved partial recovery even after 6 days of soaking. *Verrucaria elaeomelaena* showed no restoration of photosynthetic capacity at all and Ried (1960*b*) interprets its sensitivity to dehydration as the dominant factor controlling its ecology. Thus even exposure to 24 h drying at 40% relative air humidity resulted in a considerable reduction of net photosynthetic rate which had not recovered 7 days after resoaking (Fig. 110). It is evident, then, that a few aquatic or semi-aquatic lichens are very sensitive to low levels of thallus hydration, although the majority of lichens can withstand dehydration on a regular basis and some species from desert environments can indeed tolerate very long periods of extreme desiccation.

The inverse correlation is also important: aquatic lichens can survive prolonged periods of submersion very successfully, whereas the majority of lichens are rapidly killed in such an environment. Ried (1960*b*) again shows that *Verrucaria* is understandably adapted to an aquatic environment whereas *Rhizocarpon* and particularly *Parmelia saxatilis* rapidly lose their photosynthetic capacity within a few days of immersion (Fig. 111). These findings have led to the conclusion that a drying period is an essential component in the maintenance of the integrity of a lichen thallus.

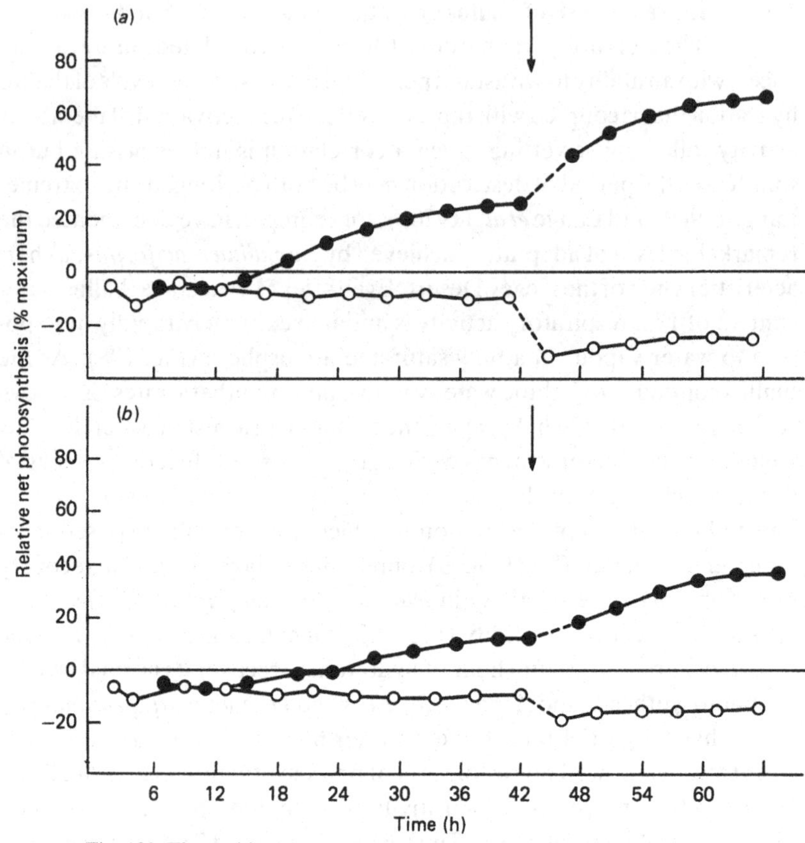

Fig. 108. The rapid recovery of metabolism in *Ramalina maciformis* on exposure to moist air after (*a*) 38 and (*b*) 51 weeks of desiccation. Filled and open circles are photosynthesis and respiration respectively. The arrow indicates the application of a mist of water droplets. (From Lange 1969.)

5.6 The requirement for alternate wetting and drying periods

The few attempts to maintain lichens under laboratory conditions have only been successful when thalli were alternately rehydrated and then allowed to dry (Pearson 1970; Dibben 1971; Harris and Kershaw 1971).

Farrar and Smith (1976) present a very useful summary of the events that immediately follow the rehydration of a lichen thallus. As well as there being an appreciable emission of carbon dioxide (which is discussed elsewhere: see Chapter 8), there is a very rapid recovery of physiological processes including photosynthesis and nutrient uptake. Active uptake of phosphorus, for example, proceeds within 10-20 s of rewetting, with labelled phosphorus detectable in the insoluble fraction after only 10 s. Farrar and Smith emphasise that the main effects of drying are potential

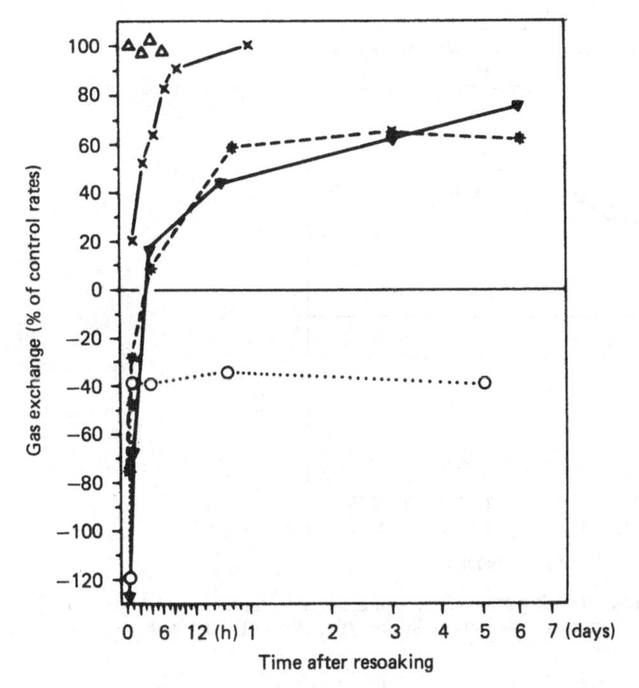

Fig. 109. The contrasting relative rates of recovery after 27-30 days of dehydration at 60% relative humidity, expressed as a percentage of normal levels of gas exchange in control replicates, in *Rhizocarpon geographicum* (△), *Lecidea soredizodes* (×–×), *Aspicilia lacustris* (*----*), *Dermatocarpon aquaticum* (▼–▼), *Verrucaria elaemomelaena* (○····○). (From Ried 1960*b*.)

damage to both macromolecules and membranes, the integrity of which is very dependent on water molecules. Removal of water molecules from membranes will usually result in increased membrane permeability (Simon, 1974). Farrar and Smith (1976) in fact demonstrate the leakage of ^{14}C- and ^{32}P-labelled compounds as well as potassium ions immediately following rewetting. They suggest a two-phase process. During the first phase, which lasts only a few minutes, there would be a rapid efflux of solutes whilst membrane integrity was being re-established. The second phase, they suggest, is also due to incomplete membrane integrity; following hydration the membrane may need some repair or at least energy input to maintain control over permeability. E.H. Nieboer (personal communication) similarly has observed an efflux of ions immediately following thallus rehydration. These ions are then subsequently taken up again.

Thus the rehydration phase demands, at least, some metabolic expenditure since most lichens experience wetting and drying cycles on a fairly continuous basis. Farrar (1976*b*) has emphasised the importance of a

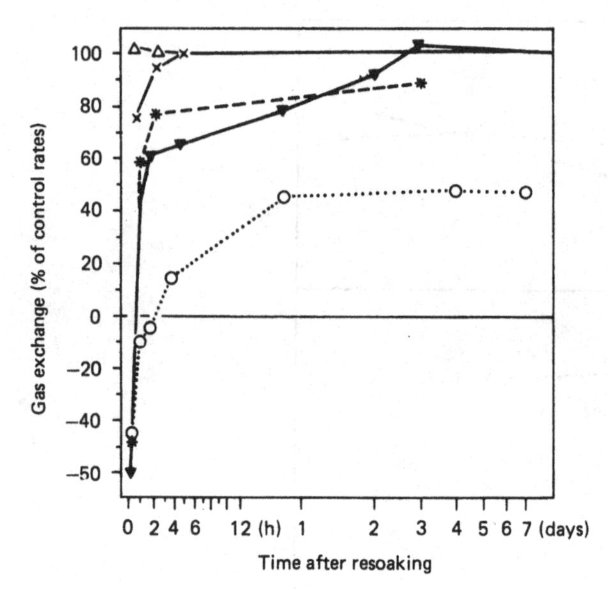

Fig. 110. The contrasting relative rates of recovery after 24 h of dehydration at 40% relative humidity. Symbols as in Fig. 109. (From Ried 1960*b*.)

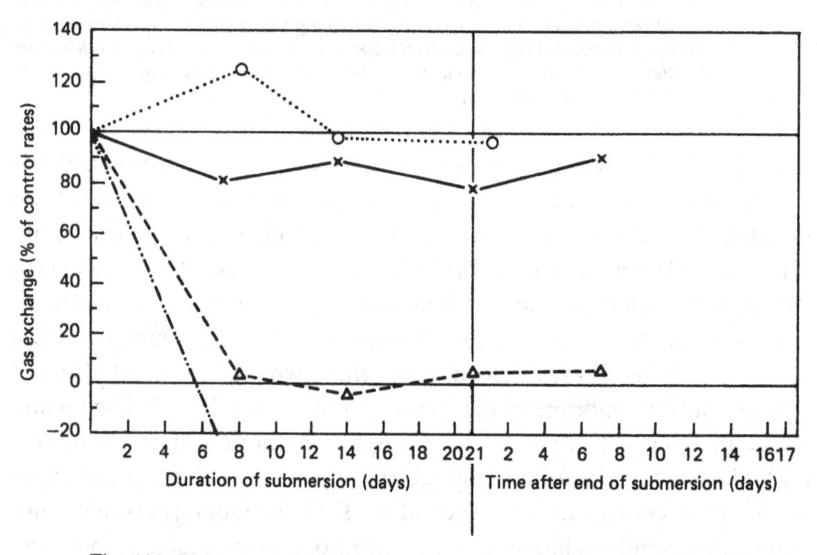

Fig. 111. The contrasting stress levels induced by immersion in water for 1-3 weeks, expressed as a percentage of the initial rates of gas exchange. △----△, *Rhizocarpon*; ×–×, *Lecidea*; ○····○, *Verrucaria*; –··–, *Parmelia saxatilis*. (From Ried, 1960*b*.)

substantial polyol pool and has hypothesised that it acts as an environmental buffer. In the absence of an immediate and substantial pool of respiratory energy the lichen would be forced to draw on insoluble materials. As photosynthetically fixed carbon accumulates directly into polyols and respiratory loss is directly from them, reaction to periods of stress can be largely restricted to this area of metabolism rather than affecting processes such as protein turnover and polysaccharide synthesis. This hypothesis has some attractive aspects, although some of the experimental methodology lacks proper controls. Nevertheless there is still an appreciable demand on the energy resources of the lichen as a whole and the hypothesis of an 'environmental buffer' does not explain the central importance of alternate wet and dry periods. This still requires an explanation. Farrar (1976*a*) suggested simply that continuous water saturation was detrimental, but his experimental methods included continuous daylight or darkness. Kershaw and MacFarlane (1980) have shown subsequently that the continuous light regime was more important physiologically than continuous moisture over such a limited experimental period.

Tysiaczny and Kershaw (1979) and MacFarlane and Kershaw (1982) have advanced an alternative hypothesis that although the relative amount of photosynthetically fixed $^{14}CO_2$ appearing in the mycobiont as mannitol remains constant over a range of experimental temperatures, there is a marked interaction with the level of thallus hydration. At full levels of thallus saturation in *Peltigera praetextata* a considerable proportion of the fixed carbon appears in the mycobiont, whereas the requirements of the algal component are met at much lower levels of thallus hydration when rates of photosynthesis are still high but only a small proportion of labelled photosynthate appears as mannitol in the fungal component. Alternate wetting and drying cycles thus ensure an adequate provision of fixed carbon for each biont and hence maintain the physical integrity of the lichen itself:

The carbon fixed by the algal component of a *Peltigera* appears in glucose which is transported to the mycobiont where it is converted into mannitol. The ratio of labelled mannitol to labelled photosynthetic products accordingly gives a good relative measure of the rate of transport of glucose and its final incorporation to mannitol within the fungal partner. In *Peltigera praetextata* approximately 40% of the total soluble $^{14}CO_2$ label constantly appears in mannitol irrespective of the experimental temperature (Fig. 112). At 5 °C the proportion is a little less, which could simply represent a slower rate of chemical conversion of glucose to mannitol. This reaction is presumably enzyme mediated, and could be limited at lower

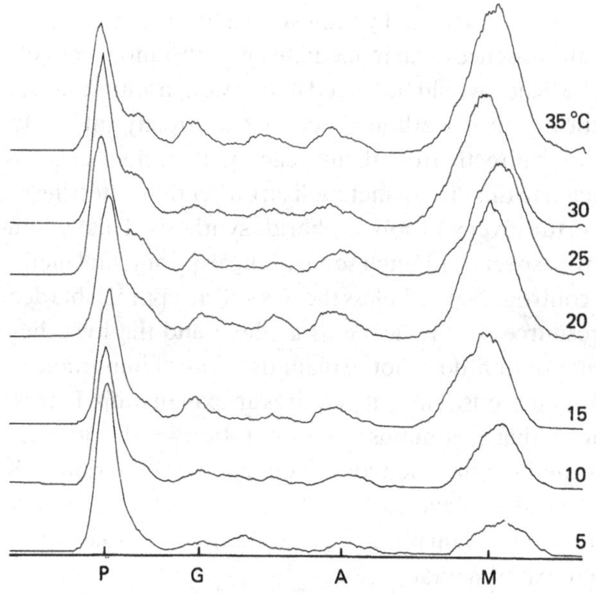

Fig. 112. Strip chromatograph scan of the ethanol soluble fraction of *Peltigera praetextata* replicates collected in July, and exposed to $^{14}CO_2$ at seven experimental temperatures. P, phosphates; G, glucose; A, alanine; M, mannitol. (From Tysiaczny and Kershaw 1979.)

temperatures. The transport of glucose to the fungal partner thus appears to be largely independent of temperature and accordingly is a passive rather than an active process.

There is, however, a very marked interaction with thallus moisture (Fig. 113). The incorporation of label into mannitol changes proportionally as the thallus dries out. Initially at full thallus hydration approximately 30% of the label is incorporated into mannitol. Subsequently at approximately 275% thallus water content by weight, and as the net photosynthetic rate increases to its maximum, there is a 50/50 division of total label between the two bionts. As the thallus dehydrates still further the proportion changes again with a marked decrease in the amount of label appearing in mannitol but with high rates of net photosynthesis still being maintained. At 100% of thallus water content by weight 86% of the products of photosynthesis are retained by the algal component, and finally at 50% thallus water content by weight, with some photosynthetic activity still present virtually no transport of glucose and conversion to mannitol are evident at all (Fig. 113). In *Peltigera polydactyla* there is a similar pattern of response to temperature with approximately 50% of the total label appearing in mannitol above 10°C (MacFarlane and Kershaw, 1982). At full thallus hydra-

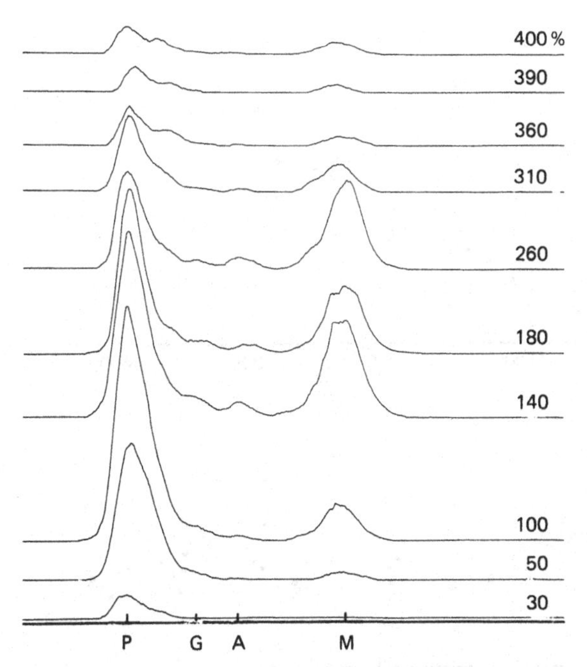

Fig. 113. Strip chromatograph scan of the ethanol soluble fraction of *Peltigera praetextata* replicates exposed to $^{14}CO_2$ at different levels of thallus hydration. Each scan comprises five replicates with an averaged percentage thallus moisture by weight. P, phosphates; G, glucose; A, alanine; M, mannitol. (From Tysiaczny and Kershaw 1979.)

tion again 50-60% of the label appears in mannitol, but below 300% thallus moisture by weight the proportion falls rapidly as the net photosynthetic rate rises to an optimum (Fig. 114*a*). Thus there is a changing balance in the translocation of glucose during the drying cycle, with the mycobiont obtaining the major proportion of its carbon requirements near full thallus hydration whilst the algal component retains and presumably utilises the glucose formed as an immediate photosynthetic product at medium and lower levels of thallus hydration. At least for these two species from a relatively mesic environment, the drying cycle appears to be an essential feature of their internal metabolic balance, but it remains to be seen whether a similar explanation is valid for lichens with green phycobionts from similar habitats. If this is so then it offers a plausible explanation for the requirement of alternate wetting and drying cycles.

However, *Peltigera rufescens* and particularly *Collema furfuraceum* have adopted, in part, an alternative strategy: The proportion of label retained in the insoluble photosynthetic products in *Collema*, presumably in the algal component, is approximately 75% of the total. The transfer of

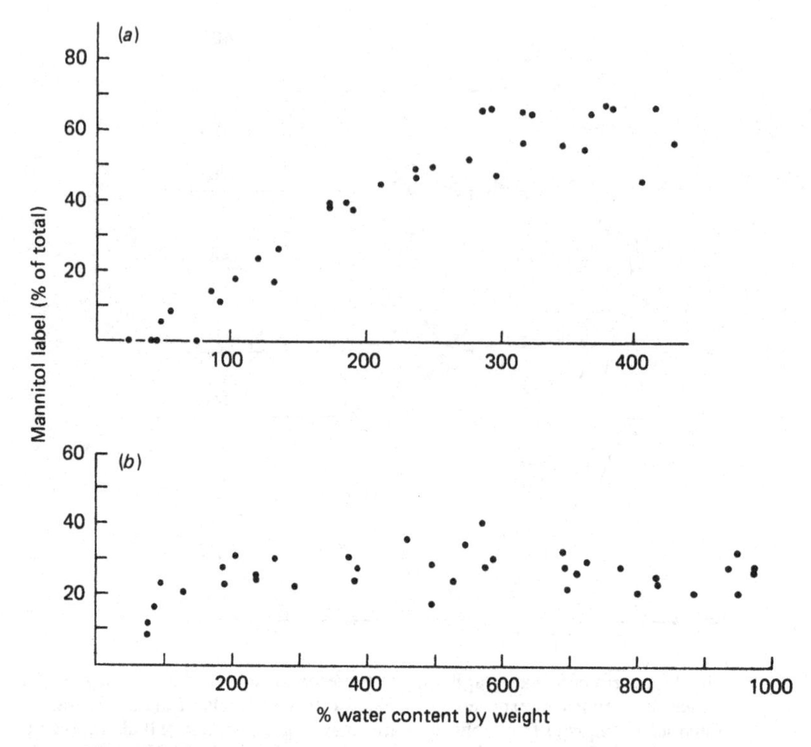

Fig. 114. The amount of label in the soluble fraction appearing in mannitol, as a percentage of the total soluble label, expressed as a function of thallus moisture in (*a*) *Peltigera polydactyla* and (*b*) *Collema furfuraceum*. Each point represents a single experimental replicate at a specific level of thallus hydration. (From MacFarlane and Kershaw 1982.)

the smaller proportion of soluble label into mannitol stays at a constant rate down to very low levels of thallus hydration (Fig. 114*b*), and not only represents an adaptation to a xeric environment but also again ensures an adequate provision of carbohydrate to the mycobiont throughout each drying cycle (MacFarlane and Kershaw, 1982). It thus appears that different proportions of photosynthate *are retained or exported* by blue-green phycobionts. There is not a continuous and massive efflux of photosynthate from all phycobionts, and equally, the *rate* of transport is closely controlled by the relative level of thallus hydration and this in turn reflects the xeric or mesic nature of the habitat. These concepts are integrated into a number of alternative strategies allowing each biont a share of the carbon resources and thus maintaining the integrity of the lichen but at the same time allowing a diversity of species throughout a wide range of environments. It seems probable that lichen species with *Trebouxia* or other algae as their phycobiont will have similar environmental control over the transport of their specific photosynthetic products, but experimental corroboration of this is required before a general explanation of the essential nature of drying cycles can be confirmed.

6

The interaction between net photosynthesis, light and temperature

There has been considerable progress over the last 5 years in understanding the interaction between light and the rate of photosynthesis in free-living algae. This interest was, to an extent, promoted by environmental concerns over the nitrogen and phosphorus loading of freshwater lakes or rivers, related to inadequate sewage disposal, and also by a widespread interest in ocean productivity. It has been shown over the last decade that several algal genera and species have developed a number of photosynthetic strategies in response to their varying light environments. Of fundamental importance are the seasonal or diurnal changes in photosynthetic capacity which develop in response to the corresponding fluctuations in illumination levels. Particularly excellent treatments are given by Harris (1978) and Prézelin (1981), for phytoplankton.

Unfortunately, until very recently these detailed syntheses of physiological concepts which have been applied to the ecology of free-living algae have had no counterpart in the literature on lichens. The potential for growth and cell division by the phycobionts of lichens appears to be considerably less than that of free-living algae; certainly their light environment is often much more constant. The possible importance of the level of illumination received by a lichen has been either largely ignored or relegated to a rather cursory examination. This has certainly been true for the work of Kershaw and his co-workers (Kershaw 1977*b*,*c*; Kershaw and Smith 1978; Kershaw and MacFarlane 1980, 1982; Kershaw and Watson 1983) but is also evident in the summaries provided by Smith (1962), Farrar (1973) and Hale (1967). Lange (1969) and Kallio and Heinonen (1971), and particularly Lechowicz and co-workers (Lechowicz and Adams 1973, 1974; Lechowicz, Jordan and Adams 1974), provide some information but their data appear to have had little impact other than a general realisation that light saturation at a specific level of illumination may vary between species from contrasting environments. However, the more recent reports of photosynthetic capacity changes in some lichens, induced by seasonal changes in their ambient levels of illumination (Kershaw and MacFarlane 1980; Kershaw and Watson 1983; Kershaw and Webber 1984) now point to

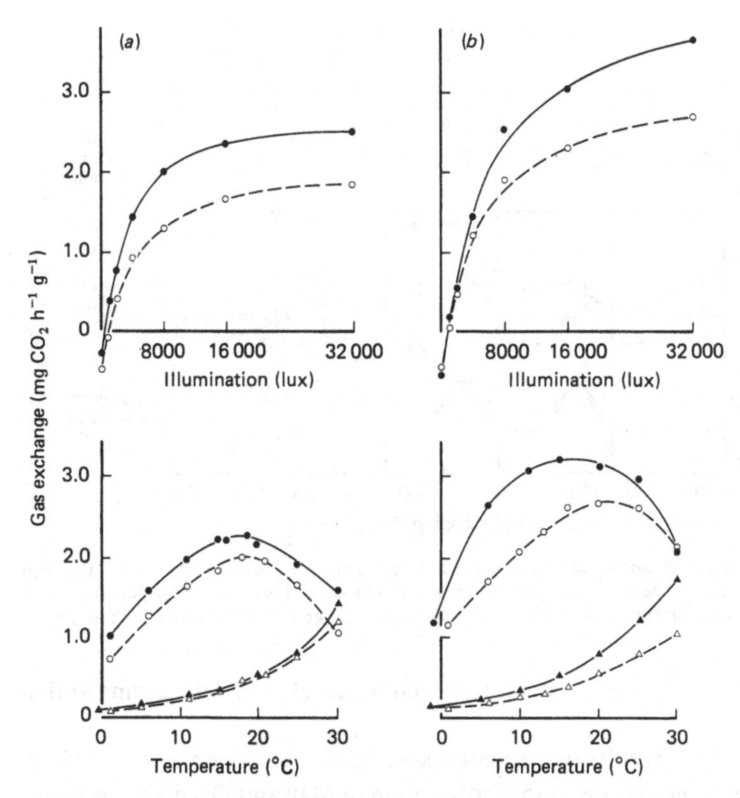

Fig. 115. Gas exchange in (*a*) *Evernia prunastri* and (*b*) *Ramalina farinacea*, expressed as seasonal photosynthetic-illumination curves (above) and seasonal temperature optima (below). Circles, photosynthesis; triangles, respiration; open symbols, winter rates; filled symbols, summer rates. (From Stålfelt 1939.)

an urgent need for a careful re-examination of the interaction between photosynthesis and illumination for a range of lichen species.

The intent here is first to examine the in-depth analysis that has been achieved for lichens using a response-matrix approach, to summarise the data available for free-living algae and finally to relate this summary, expressed as photosynthetic-illumination (PI) curves, to what little similar information is currently available for lichens.

6.1 Seasonal photosynthetic capacity changes in lichens: the response matrix approach

Although Stålfelt (1939) carefully documented capacity changes in both respiration and net photosynthesis, surprisingly his work had little impact on the scientific community as a whole, probably because the

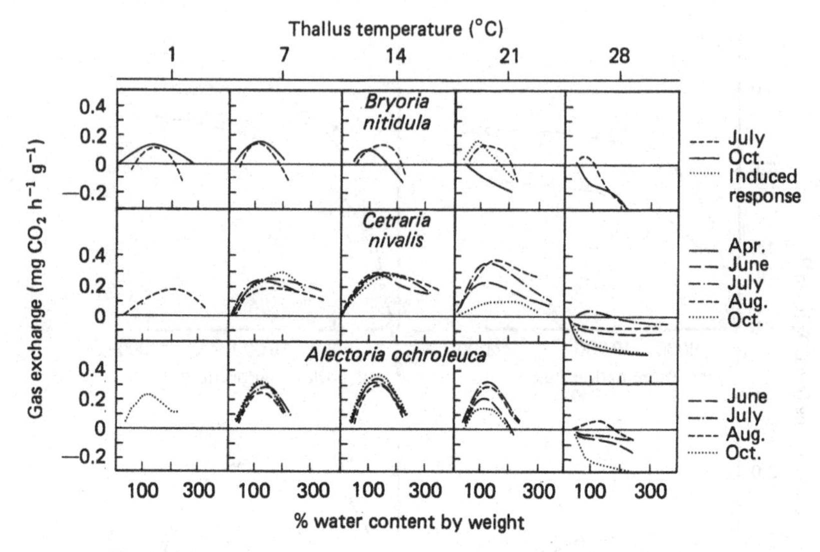

Fig. 116. Seasonal net photosynthetic patterns in (*a*) *Bryoria nitidula*, (*b*) *Cetraria nivalis*, and (*c*) *Alectoria ochroleuca*. The data show a considerable degree of temperature acclimation in the summer to 21 and 28 °C. (From Kershaw 1983.)

results were difficult to interpret given the level of understanding at that time.

Two of the results of particular significance are given in Fig. 115. In *Evernia prunastri* (Fig. 115*a*) respiration in May and December remains constant whilst there is a marked photosynthetic capacity increase in summer. Conversely in *Ramalina farinacea* (Fig. 115*b*) there is both a change in respiratory and photosynthetic rates and an apparent shift in the net photosynthetic temperature optimum. The statistical significance of these differences is difficult to assess, although the close agreement between the winter/summer respiratory rates of *Evernia* suggests that there was good experimental control over between-replicate variation. When examined together with the range of data that is now currently available, documenting capacity changes in lichens, there is no doubt that Stålfelt described actual events which have only just become more fully understood.

Rather more recently Larson and Kershaw (1975*c*,*d*,*e*) and Kershaw (1975*b*) have documented net photosynthetic capacity changes in three low-arctic lichens. Although the response matrix in each case is not complete, at least the net photosynthetic response to the degree of thallus hydration for a range of temperatures and throughout the seasons was examined. The data for *Bryoria nitidula*, *Alectoria ochroleuca* and *Cetraria* *nivalis* (Fig. 116) all show a closely similar seasonal pattern of response. This was interpreted as acclimation of the photosynthetic optimum during the summer period to the higher environmental temperatures prevailing in the field during June and July. Thus in *Cetraria* net photosynthetic rates at 28 °C in April and October do not reach compensation, are slightly higher in June and August, but do increase sufficiently to allow a significant carbon gain in July. At 21 °C there is a comparable sequential increase and decrease of net photosynthetic rates, again providing optimum rates in July. There is no significant change in rates at lower temperatures and the overall effect of the capacity changes in *Cetraria* is to maintain a constancy of photosynthetic gain up to 21 °C despite marked seasonal environmental temperature fluctuations. *Alectoria ochroleuca* shows a closely comparable seasonal response pattern, with full photosynthetic capacity maintained at 21 °C and positive carbon gains being achieved in early summer even at 28 °C. In *Bryoria nitidula* there is the same basic pattern, with constancy of photosynthetic capacity shown up to 20 °C but with a significant winter decline in rates at 15 °C. These slightly lower temperature optima correlate well with the exposed ridge-top habitat of this species (Kershaw 1975*b*).

There has been considerable comment on the potential for photosynthetic 'acclimation' in plants and exactly what does or does not constitute photosynthetic acclimation: Larson (1980) summarised the pattern of change in a number of lichen species and emphasised that the term 'acclimation' refers to an adaptation resulting in the constancy or homeostatic control of some vital process (Prosser 1955).

Since net photosynthetic rates are being measured, any capacity change should be quite independent of any concurrent respiratory changes. However, a number of other net photosynthetic capacity changes have been documented in lichens and the response patterns are now considerably more complex than the original work supposed (Larson and Kershaw 1975*c*,*d*,*e*; Kershaw 1975*b*). Some of these additional patterns do in fact involve respiratory changes and to make a critical examination of the apparent constancy of net photosynthetic capacity demands a factorial matrix that is developed several times throughout the spring and summer period. This allows at least a limited number of degrees of freedom, enabling the potential for acclimation of net photosynthetic optima to be ascribed to an environmental variable such as changing temperatures. Thus there is a marked capacity change in photosynthesis at the onset of winter in both *Peltigera praetextata* and *P. polydactyla*. The maximum winter rate was originally reported as being approximately half that of the

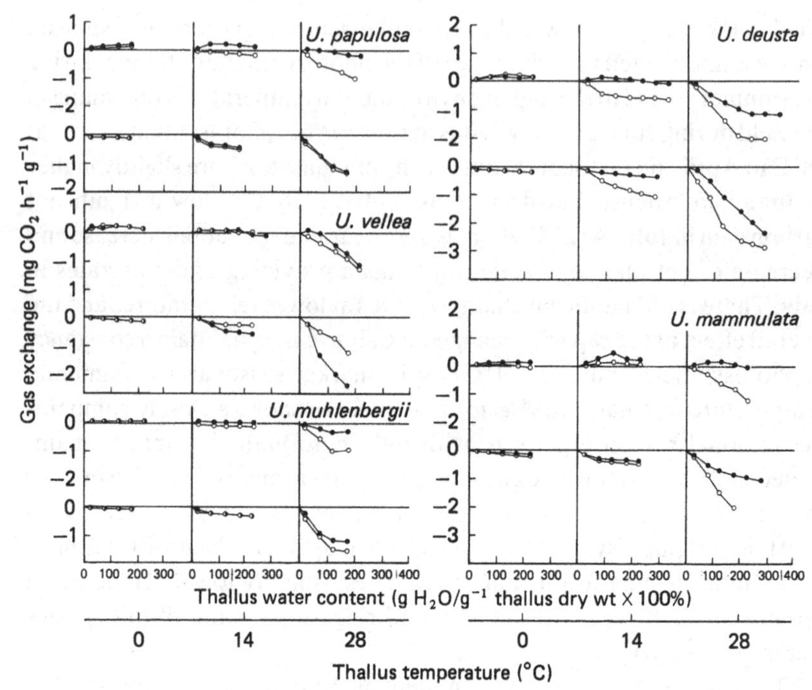

Fig. 117. The pattern of seasonal changes of photosynthetic and respiratory capacity in *Umbilicaria* species (●, summer; ○, winter). (From Larson 1980.)

spring, summer and autumn rates (Kershaw 1977*b*,*c*), and achieved without any significant change in dark respiration rate. Throughout the non-winter period the temperature optimum for net photosynthesis apparently closely corresponds to the prevailing ambient temperature, reflects homeostatic control of photosynthesis and was initially presented as a straightforward example of acclimation. More recently, however, this has been shown to be only partly true (see below).

In *Peltigera* then, there are quite different sets of photosynthetic events during the winter compared with the remainder of the year. Larson (1980) similarly presents a very complete summary of the net photosynthetic and respiratory changes in *Umbilicaria* spp. which occur with the onset of the winter period and which could be correlated with the ability of these species to withstand freezing temperatures. The pattern of events for *Umbilicaria* is summarised in Fig. 117. The winter state may be signalled by (1) an increase in winter respiration rate (as in *Umbilicaria deusta* and *U. mammulata*), with or without a corresponding photosynthetic capacity change; (2) a decrease in winter respiration rate as in (*U. vellea*, and also in *Cladonia rangiferina*) (Tegler and Kershaw 1980); or (3) a marked change in photosynthetic capacity at all temperatures without any concurrent change in respiration rate (as found in *Peltigera praetextata*, *P. polydactyla* (Kershaw 1977*b*) and *Umbilicaria papulosa* (Larson 1980)). Finally, during the winter period there may be no apparent change in carbon dioxide gas exchange characteristics, as is found in *Collema furfuraceum* (Kershaw and MacFarlane 1982).

Other patterns of photosynthetic capacity change have also been reported: Kershaw and MacFarlane (1980) examined the ability of *Peltigera praetextata* and *P. scabrosa* to acclimate to low or high light levels and maintain a given level of photosynthetic capacity irrespective of the tree canopy status. The photosynthetic acclimation to low light in air-dry experimental material of *P. scabrosa* is very clear-cut (Fig. 118). Maximum net photosynthetic rates of approximately 3.5 mg CO_2 h^{-1} g^{-1} are evident in material freshly collected in the field in mid-summer and are subsequently maintained in experimental material stored at 20 μmol m^{-1} s^{-1} for 7 days. However, by day 20 maximum net photosynthetic rates examined at 300 μmol m^{-2} s^{-1} increase significantly to 5.7 mg CO_2 h^{-1} g^{-1}. Further acclimation to low light did not occur after an additional week of low light treatment. With the return of replicates to a high-light regime of 350 μmol m^{-2} s^{-1} there is an extremely rapid change of net photosynthetic rate back to the initial response level. Concurrently, respiration rate also changes significantly. Initially, maximum rates of 2.2 mg CO_2 h^{-1} g^{-1} are achieved at 25 °C and thallus saturation (Fig. 118). These rates increase significantly to 3.5 mg CO_2 h^{-1} g^{-1} after air-dry storage under 20 μmol m^{-2} s^{-1} illumination and return rapidly to 2.2 mg CO_2 h^{-1} g^{-1} when experimental replicates are subsequently transferred to high-light conditions. The pattern of net photosynthetic change in *P. praetextata* is very similar. However, the concurrent change in respiration rate during acclimation to low light is very different from the direction of change seen in *P. scabrosa*. With adaptation to low light, respiration rates fall from maximum values of more than 3 mg CO_2 h^{-1} g^{-1} down to 1.8 mg CO_2 h^{-1} g^{-1}, whereas in *P. scabrosa* respiration rates increase significantly under low-light storage.

In *Stereocaulon paschale* Kershaw and Smith (1978) have shown yet another pattern, where there is no significant seasonal change in net photosynthetic temperature optima except at 35 °C where a considerable degree of capacity change in mid-summer is evident. This limited adjustment of photosynthesis to high summer temperatures maintains a constancy of photosynthetic carbon dioxide uptake at 35 °C, but the temperature acclimation is *restricted* to these summer conditions only. Kershaw and MacFarlane (1982) have suggested that the term 'restricted-acclimation'

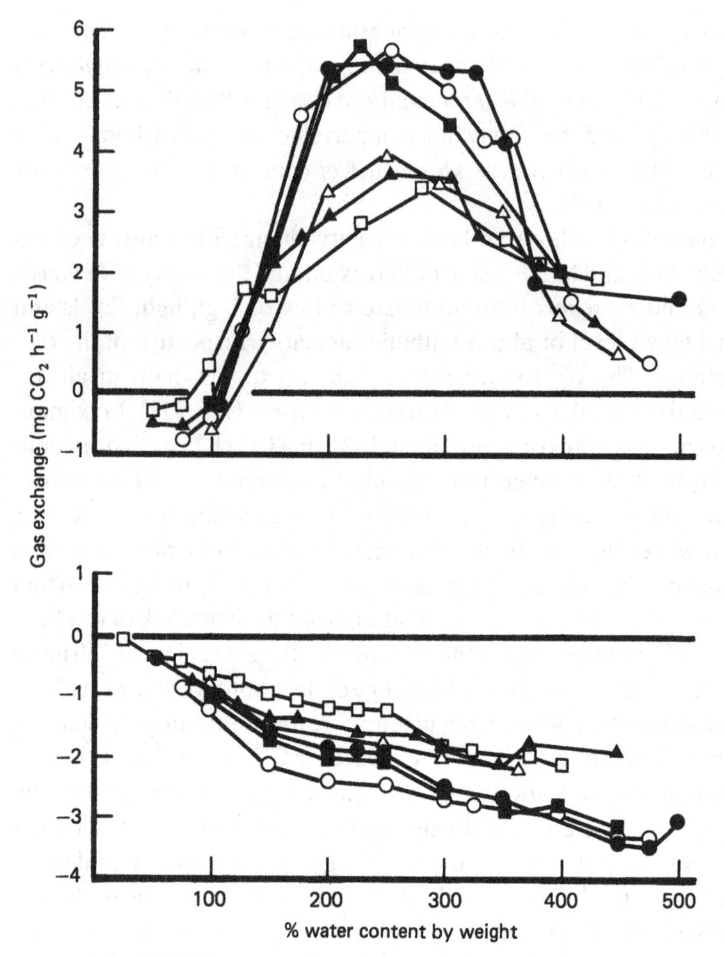

Fig. 118. Adjustments of net photosynthesis and respiration in *Peltigera scabrosa* during air-dry storage under contrasting light regimes. □, initial response pattern in freshly collected material; ▲, response after 7 days; ○, after 20 days; ●, after 28 days; ■, after 32 days storage under 20 μmol m^{-2} s^{-1} illumination. On day 28 replicates were moved to 350 μmol m^{-2} s^{-1} illumination and examined after 2 days (△). Standard error: 0.5 mg CO_2 h^{-1} g^{-1}. (From Kershaw and MacFarlane 1980.)

be used for such cases in which there is a marked capacity change across all experimental light levels only at the higher range of summer temperatures.

6.2 The interaction between photosynthesis and illumination in algae

Harris (1978) and Prézelin (1981) both give comprehensive accounts of the structure of algal chloroplasts, their range of photosynthetic pigments and characteristic absorption spectra, the membrane location

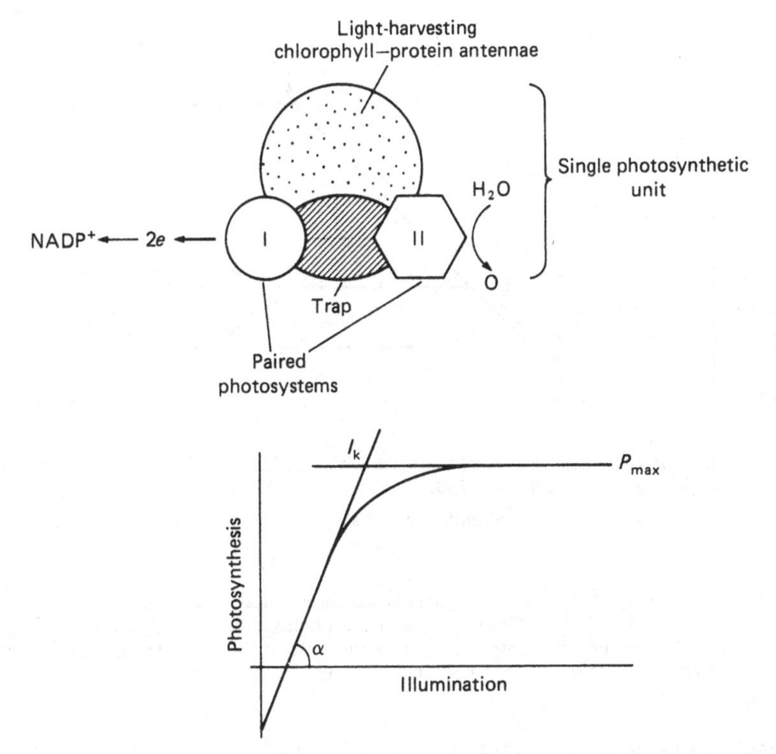

Fig. 119. A diagrammatic representation of the photosynthetic process simplified to its essential components. Below is shown the photosynthetic-illumination (PI) curve, expressing photosynthetic efficiency (α), light saturation point (I_k) and the maximum rate of photosynthesis (P_{max}.)

of photosystems I and II, etc., as well as seasonal capacity changes. Thus, the photosynthetic process can be viewed as a number of photosynthetic pigments which constitute a light harvesting mechanism (chlorophyll *a* is of particular importance together with accessory pigments which broaden the width of the absorbence range of light energy). The interaction with light can be viewed in terms of 'antennae' (chlorophyll protein complexes) that focus energy quanta at energy 'traps' which in turn redistribute the energy to paired reaction centres (photosystems 1 and 2: PS1 and PS2) for the production of electrons. A single photosynthetic unit (PSU) comprises the antennae together with their traps and reaction centres. An electron transport chain results in ATP formation, the reduction of NADP and finally the reduction of carbon dioxide molecules which then enter the Calvin cycle for the production of sugars. This simplified conceptual framework is based on the model originally proposed by Hill and Bendall

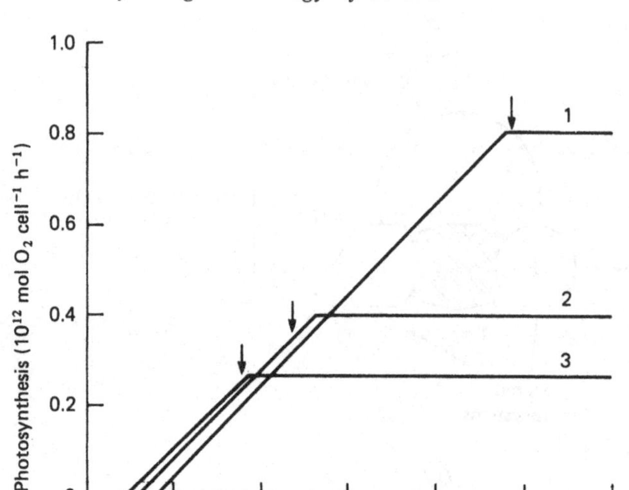

Fig. 120. PI curves from green algal populations in Lake Ontario. 1, 21 °C July; 2, 12 °C June; 3, 8 °C October. The mean photosynthetic efficiencies and P_{max} values are not correlated. P_{max} is a function of temperature; I_k (arrows) is therefore a function of temperature. (From Harris 1978.)

(1960) and Thornber *et al.* (1977) and is shown diagrammatically in Fig. 119 together with the equivalent photosynthetic illumination (PI) curve.

Herron and Mauzerall (1972) and Prézelin (1981) interpret the form of the PI curve simply in terms of light absorption and the size and number of the reaction centres. Since the number of reaction centres, the number of PSUs and the cellular chlorophyll content are all inter-related, the initial slope of the PI curve (α) is a function of the amount of chlorophyll and accessory pigments in the cells, together with their absorption characteristics. The value of the photosynthetic maximum, P_{max}, is therefore completely independent of illumination level and is simply a function of the number of reaction centres modified by the rate constant for the single rate-determining step which controls the overall throughput in the Calvin cycle. It follows then that P_{max} is particularly dependent on temperature, which will control the maximum reaction rate through the limiting step. The initial slope of the PI curve is accordingly a direct measure of photosynthetic efficiency. I_k, the saturating level of illumination, is usually, but not always, temperature dependent.

For example, Harris (1978) presents PI curves for green algal populations in Lake Ontario in which P_{max} is correlated with temperature whilst photosynthetic efficiency remains constant. I_k in this instance is a function of temperature (Fig. 120). Harris (1978) also emphasises that the use of relative values for rates of photosynthesis can induce completely erroneous changes in apparent photosynthetic efficiency. The data from Fig. 120 have been replotted on a relative scale and this form shows an apparently marked summer decline in photosynthetic efficiency (Fig. 121). Thus it is essential for PI curves to be presented in absolute terms to avoid these problems. However, in some algal populations there are seasonal changes in photosynthetic efficiency and Harris (1978) presents data for diatoms where P_{max} is correlated with photosynthetic efficiency, and I_k is therefore approximately constant (Fig. 122). Clearly different options are available to different algal populations, these options being heavily regulated by the seasonal environments of each population.

There are a number of alternative locations in the photosynthetic model system as to where discrete changes in the organisation and/or activity of the apparatus can be induced. In free-living algae the induction processes are regulated by specific environmental changes but are mediated by equally specific endogenous events. Prézelin (1981) presents an excellent summary of these endogenous events and suggests how the changes within the system will be expressed in the PI curve. These proposed changes in the system and hence in the PI curve are as follows:

6.2.1 *Change in size of photosynthetic units*

Since the chlorophyll *a* (chl *a*) cores of the photosynthetic reaction centres are assumed to be fairly uniform, changes in PSU size will be readily evident by changes in the size of the light-harvesting pigment component of the antennae. Other pigments may be involved in addition to chl *a* in some algal genera, and whole cell total pigment may change. In phycobilin systems, where no chl *a* is present at all, examination of potential endogenous events would require measurement of changes in one or more of the phycobilins. However, on a cellular or dry weight basis P_{max} would not change since only the *size* of the PSU is involved and not the density. Conversely, since the size of the light harvesting component has changed, a given amount of light energy will be utilised more efficiently and this will be reflected in a change in the initial slope of the PI curve. These events are summarised diagrammatically in Fig. 123 and have been reported for higher plants grown under lowered light conditions (Brown, Alberte and Thornber 1974; Alberte, Hesketh and Kirby 1976), diatoms exposed to lowered light levels (Perry *et al.* 1981), dinoflagellates (Prézelin and Sweeney 1978, 1979) and probably also blue-green algae (Jorgensen

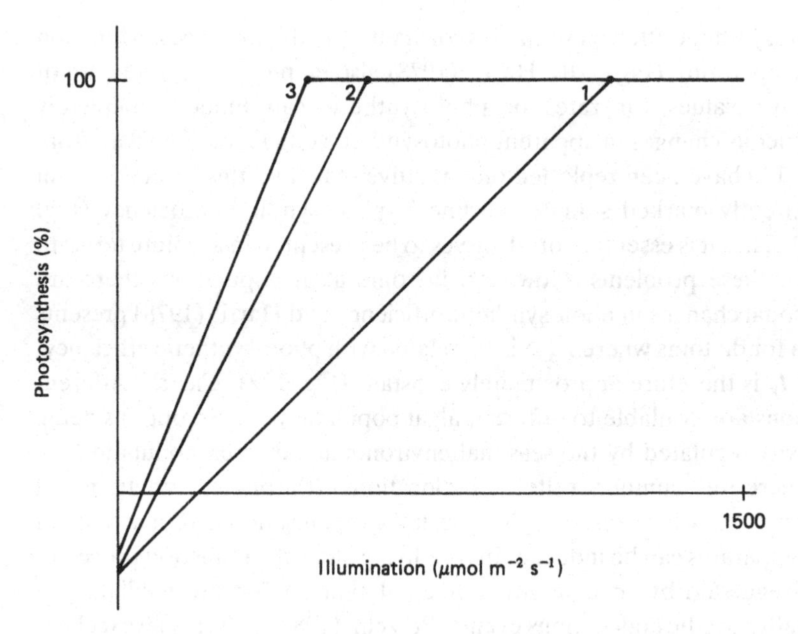

Fig. 121. The curves shown in Fig. 120 with photosynthesis expressed as a percentage of maximum photosynthesis. Note the apparent change in photosynthetic efficiency when photosynthesis is not expressed in absolute terms.

1969). The changes in dinoflagellates occur within a generation time and are always associated with major increases in pigmentation when brightly illuminated cultures are low-light adapted (Prézelin 1981). As yet, no evidence of comparable events in lichens is available.

6.2.2 *Changes in photosynthetic unit densities*

Since the proportion of total pigment when expressed on a cellular or dry weight basis changes as a direct function of the increase in PSU density there will be an equivalent increase in photosynthetic efficiency. Furthermore, since PSU density increases, P_{max} on a dry weight or cellular basis will also increase. However, on a chlorophyll basis, since PSU/chl *a* ratios would remain constant there should be no significant changes in the PI chl *a* curve. These relationships are summarised diagrammatically in Fig. 124.

Prézelin (1981) suggests that this strategy has not been documented very frequently, although examples from dinoflagellates, green algae, diatoms and higher plants have been reported (Jorgensen 1969, 1970; Patterson *et al.* 1977; Terri *et al.* 1977; Przélin and Sweeney 1979; Perry, Larsen and

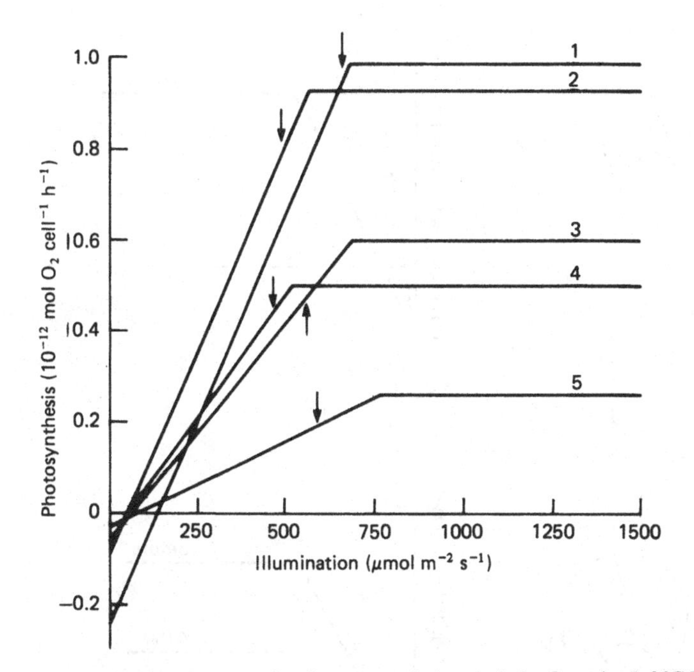

Fig. 122. PI curves for diatom populations in Lake Ontario. 1, 3 °C January; 2, 1-7 °C April; 3, 5.5 °C November; 4, 3 °C May; 5, 7.5 °C October. Mean photosynthetic efficiencies and P_{max} values are correlated; I_k (arrows) is therefore constant. The highest P_{max} values occur in populations from cold water in conditions of deep mixing. (From Harris 1978.)

Alberte 1981). Vierling and Alberte (1980) report a more complex series of events for blue-green algae where there appears to be a simultaneous increase in density of PS1 units and the light-harvesting antennae of PS2 at low light levels. So far, potential changes in PSU density have only been confirmed for the lichen *Peltigera praetextata* (Kershaw and MacFarlane 1980; Kershaw and Webber 1984) (see Section 6.3 below).

6.2.3 *The coupling or uncoupling of energy transduction between PS1 and PS2*

The energy flow through existing PSUs can be interrupted without any alteration of their size or their physical density and thus without any change in total pigmentation. Their functional density, however, is altered to a considerable extent and thus both light-limited and light-saturated rates of photosynthesis are dramatically changed, reflecting the different electron flow between a discrete number of total photosystems. These relationships are expressed diagrammatically in Fig. 125. The coupling or uncoupling of PSUs is completely reversible and is achieved rapidly. The

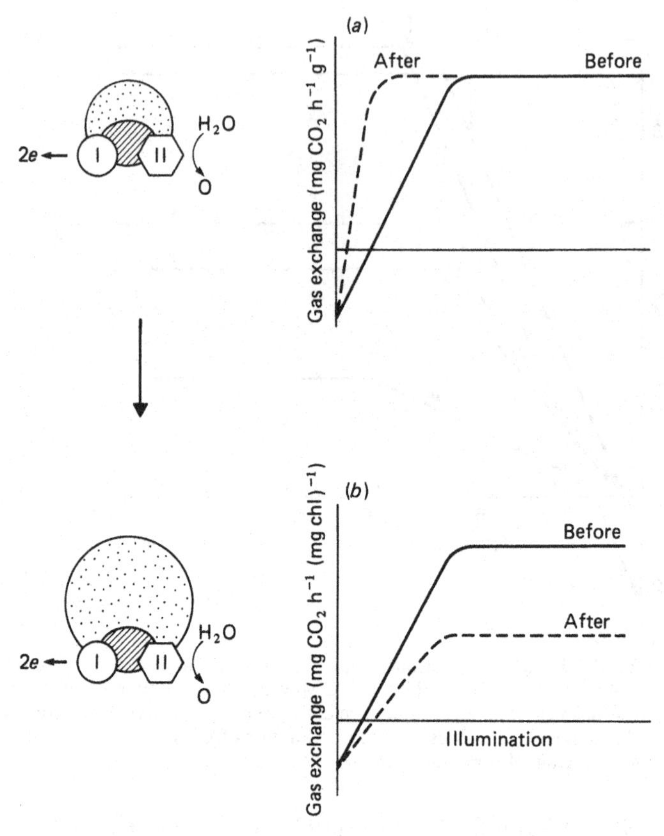

Fig. 123. A schematic representation of the relationship between altered photosynthetic unit size and changes in the PI curves, expressed either on a cellular basis (*a*) or a chl *a* basis (*b*).

resultant capacity changes are characteristic, for example, of the daily photosynthetic periodicity in dinoflagellates which are often additionally regulated by a biological clock (Prézelin, Meeson and Sweeney 1977; Harding *et al.* 1981), but they are equally related to thermal environmental changes in higher plants (Björkman, 1981). The mechanism of these rapid switches is still unknown. These capacity changes have now been proposed for a number of lichens (Kershaw *et al.* 1983; MacFarlane, Kershaw and Webber 1983) and are probably widespread.

6.2.4 *Alteration of 'downstream' enzymatic rates*

'Downstream' events which are external to PSUs include alteration of electron transport rates and hence the rate of carbon dioxide reduction as well as chemical events which follow in the Calvin cycle. These

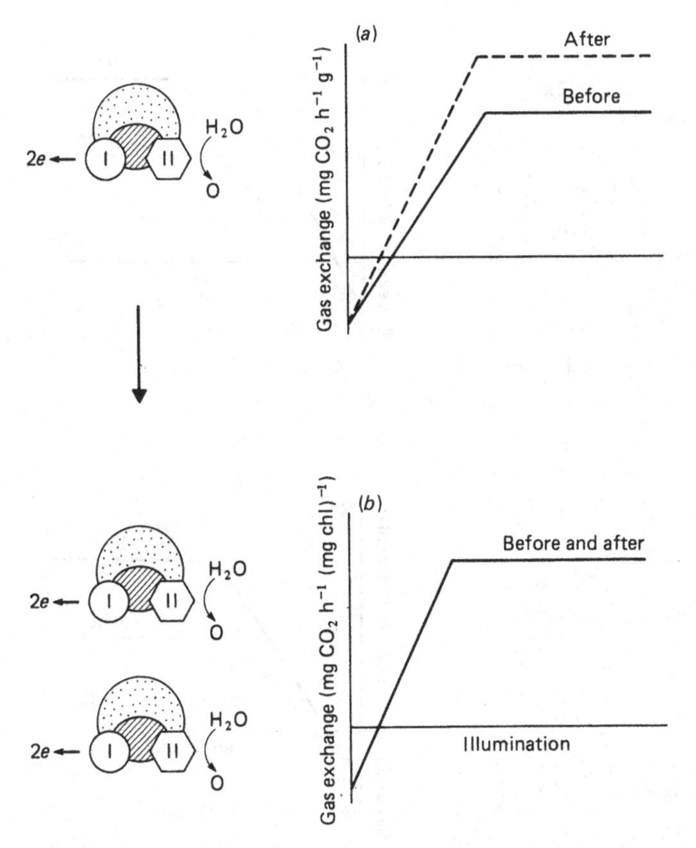

Fig. 124. A schematic representation of the relationship between altered photosynthetic unit density and changes in the PI curves, expressed either on a cellular basis (*a*) or a chl *a* basis (*b*).

latter processes are mediated by enzymatic activity and are completely independent of pigmentation, PSU size or PSU density. Light-limited rates of photosynthesis accordingly remain constant whilst P_{max} alters dramatically, presumably with corresponding endogenous activity changes in enzymes, cofactors or isozymes, although direct evidence of this is still unavailable for lichens (see below). These endogenous changes and the corresponding changes in the PI curve are summarised in Fig. 126.

Thus there appear to be four major locations in the algal photosynthetic pathway where endogenous changes can be regulated by environmental events, modifying the PI curve in a fairly predictable way. It is also likely that some algal genera may combine two or more of these strategies at particular seasons of the year. The final analysis of these interrelationships may therefore be considerably more complex than current

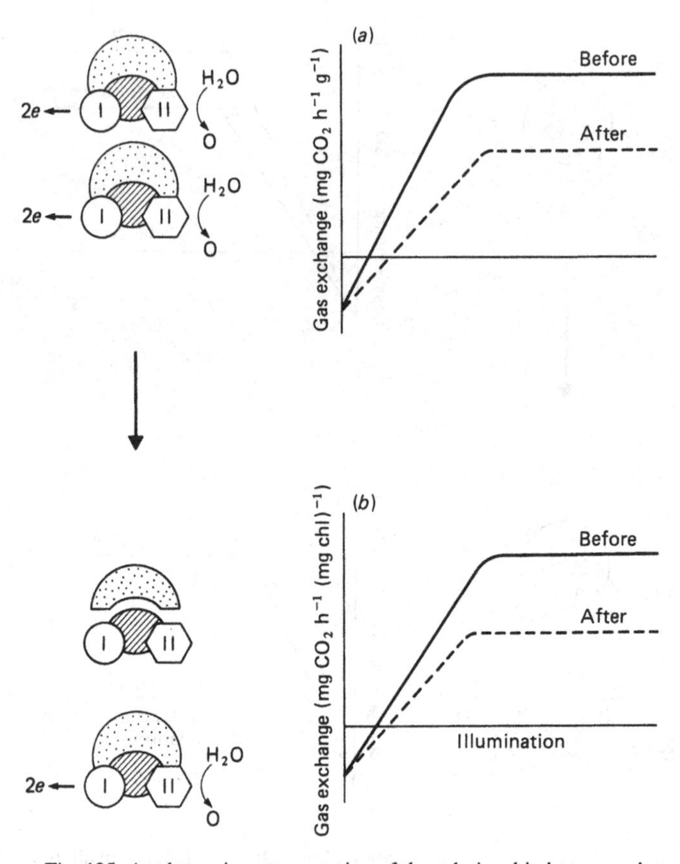

Fig. 125. A schematic representation of the relationship between altered photosynthetic energy transduction and changes in the PI curves, expressed either on a cellular basis (*a*) or a chl *a* basis (*b*).

ideas suggest. The data of Vierling and Alberte (1980) would appear to present one such second-order interaction. Similar capacity changes in lichens are probably widespread (Kershaw 1975*b*; Larson and Kershaw 1975*c*,*d*,*e*) but confirmation is largely still required.

6.3 Comparable events in lichens

It is evident then that the photosynthetic seasonal events that have been documented so far in lichens can be provisionally re-interpreted mechanistically by examining the seasonal PI curves in terms of both mg CO_2 h^{-1} g^{-1} and mg CO_2 h^{-1} (mg chl *a*)$^{-1}$. MacFarlane *et al.* (1983) and Kershaw *et al.* (1983) have proposed uncoupling of energy transduction from the antennae and PS1 and PS2 systems, resulting in marked changes in photosynthetic capacity in *Cladonia rangiferina*, *C. stellaris* and *Pel-*

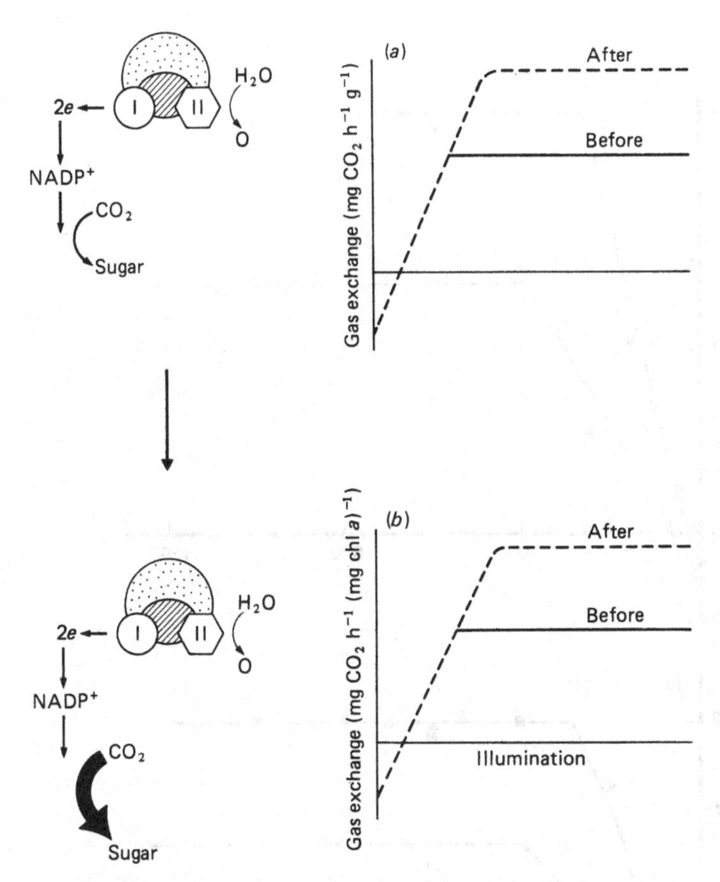

Fig. 126. A schematic representation of the relationship between altered photosynthetic enzymatic reactions and changes in the PI curves, expressed either on a cellular basis (*a*) or a chl *a* basis (*b*).

tigera praetextata respectively. In addition the increase in summer capacity in *P. praetextata* previously reported by Kershaw and MacFarlane (1980) reflects the potential synthesis of additional PSUs during conditions of full woodland canopy (Kershaw and Webber 1984), whilst seasonal changes of photosynthetic capacity in *P. rufescens* involve enzymatic rate changes in the Calvin cycle (Brown and Kershaw 1984).

The seasonal changes reported in *Cladonia rangiferina* are given in Fig. 127 for the shade ecotype (MacFarlane *et al.* 1983). In mid-summer P_{max} at light saturation is *c.* 9.25 mg CO_2 h^{-1} g^{-1} declining to *c.* 5.5 mg CO_2 h^{-1} g^{-1} in winter. There is no change in respiration rate but a significant change in the initial slope of the PI curve (α) between winter and summer. Equally light saturation (I_k) changes from *c.* 250 μmol m^{-2} s^{-1} in summer to *c.* 400

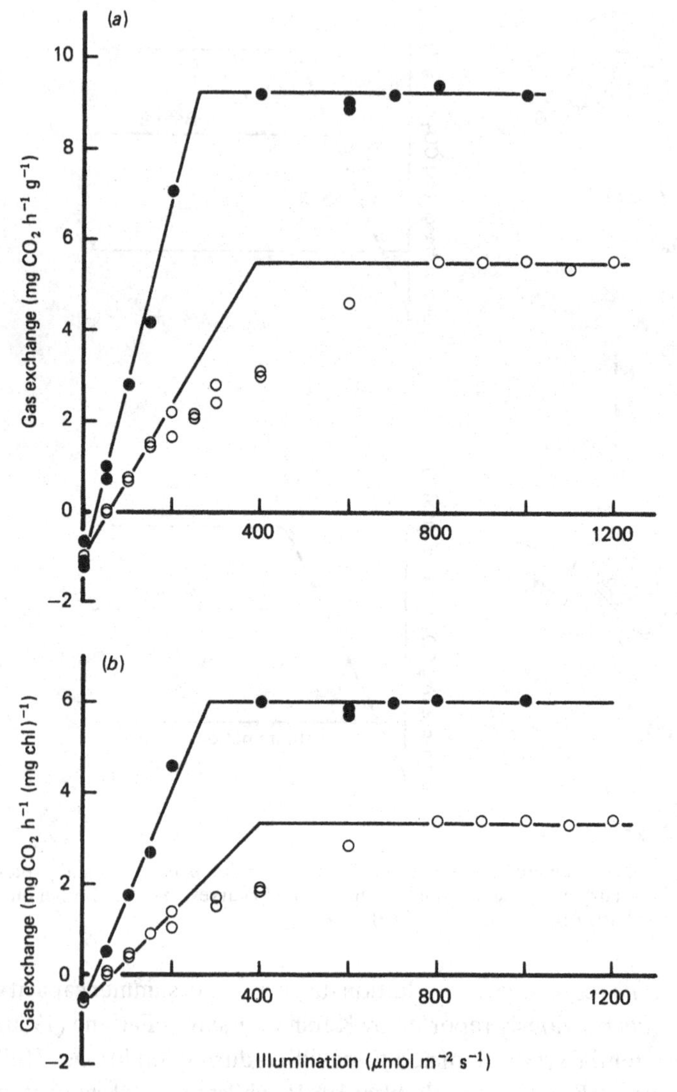

Fig. 127. Summer (●) and winter (○) PI curves for *Cladonia rangiferina* showing uncoupling of energy transduction, expressed either on a gram dry weight basis (*a*) or a chl *a* basis (*b*). (From MacFarlane *et al.* 1983.)

$\mu mol\ m^{-2} s^{-1}$ in winter. When expressed as $mg\ CO_2\ h^{-1}$ $(mg\ chl\ a)^{-1}$ (Fig. 127*b*) there is essentially no change in the pattern of the winter and summer PI curves and accordingly the capacity change is interpreted as uncoupling of energy transduction. Kershaw *et al.* (1983) report an almost identical response in *C. stellaris* and one which is also controlled seasonally not only

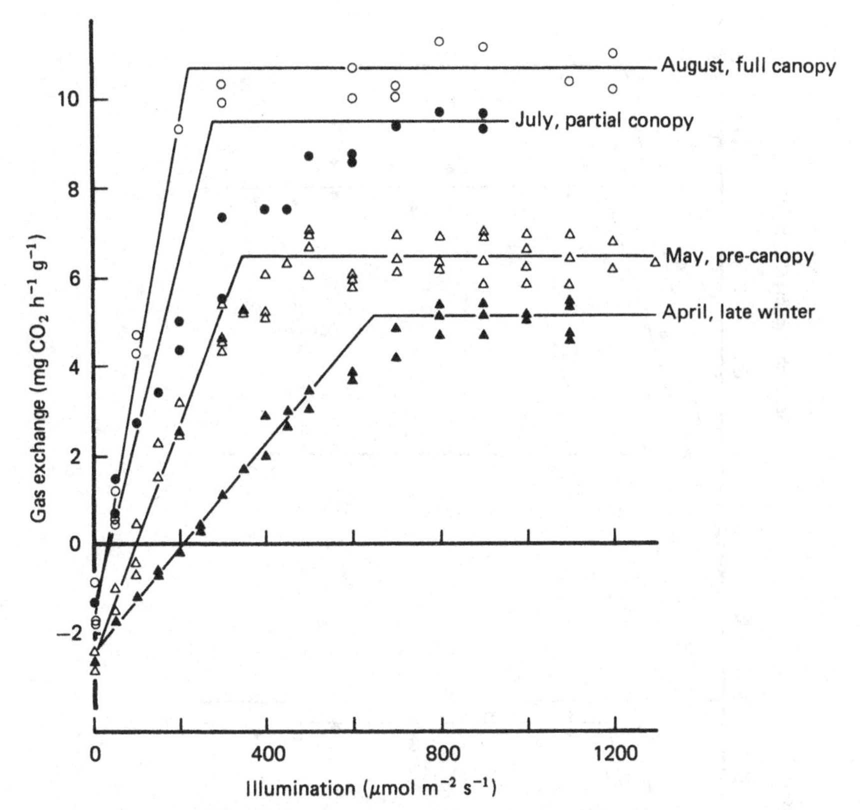

Fig. 128. Spring and summer PI curves for *Peltigera praetextata* showing uncoupling of energy transduction in the winter and the synthesis of additional PSUs in mid-summer. (From Kershaw and Webber 1984.)

by temperature but also by daylength (MacFarlane *et al.* 1983).

The seasonal events in *Peltigera praetextata* are considerably more complex and the PI curves at optimum thallus water content are given in Fig. 128 (Kershaw and Webber 1984). In winter, P_{max} at light saturation is *c.* 5.0 mg CO_2 h^{-1} g^{-1} with light compensation at 200 μmol m^{-2} s^{-1} and I_k at *c.* 650 μmol m^{-2} s^{-1}. In early summer, following coupling of energy transduction in additional PSUs, P_{max} increases to *c.* 6.5 mg CO_2 h^{-1} g^{-1}, light compensation decreases to 100 μmol m^{-2} s^{-1} and I_k to *c.* 350 μmol m^{-2} s^{-1}. After the development of the tree canopy there is a marked increase in the number of PSUs and a significant increase in α, with light compensation declining still further to *c.* 25 μmol m^{-2} s^{-1} and I_k to *c.* 250 μmol m^{-2} s^{-1}. Under full canopy conditions dark respiration declines significantly from *c.* 2.75 mg CO_2 h^{-1} g^{-1} to *c.* 1.75 mg CO_2 h^{-1} g^{-1}. The interpretation of the response to full canopy conditions in terms of additional PSUs is confirmed

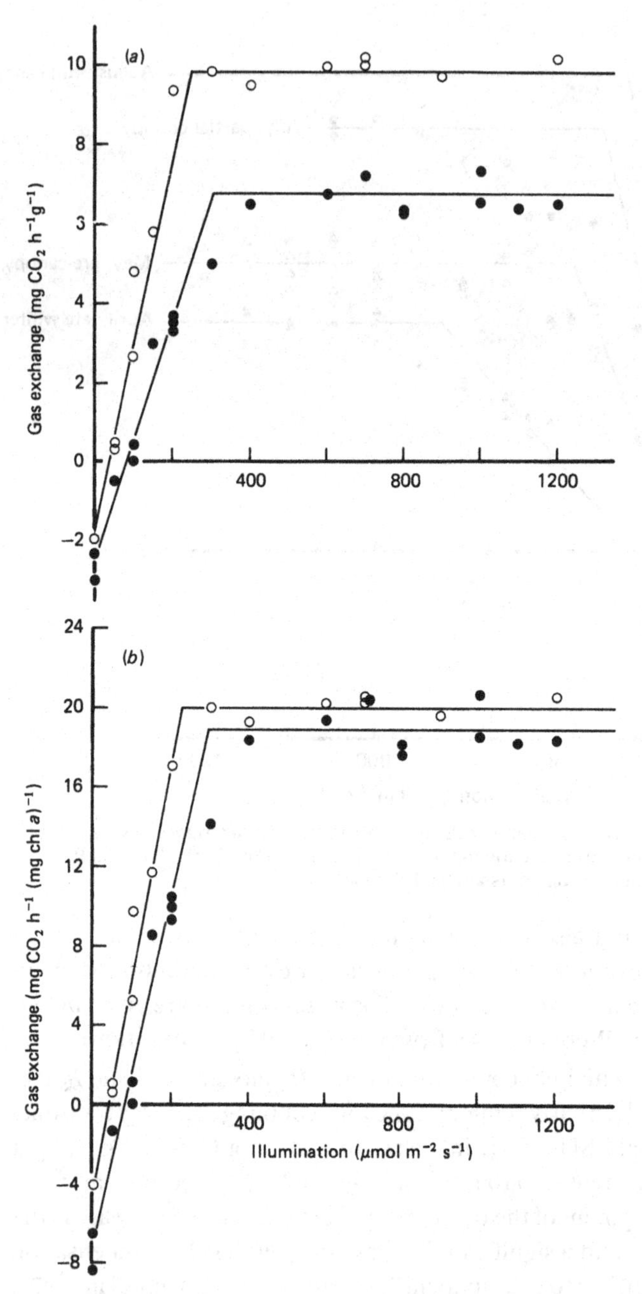

Fig. 129. PI curves for *Peltigera praetextata* in summer expressed on (*a*) a gram dry weight basis and (*b*) a chl *a* basis. ○, full canopy; ●, pre-canopy. (From Kershaw and Webber 1984.)

when pre-canopy and full canopy PI curves are expressed in micrograms of chl *a* (Fig. 129). Essentially the curves are identical with only the residual dark respiration differences evident.

These results agree fairly well with the data previously provided by Kershaw (1977*a*,*b*) and Kershaw and MacFarlane (1980) which, however, were expressed as a seasonal net photosynthetic response matrix. Although this approach did not allow any interpretation of the potential mechanisms of such seasonal changes, it is of considerable interest that the wide range of capacity changes documented above now fit logically together. The original data (Kershaw 1977*a*), however, did not use sufficiently high experimental light levels in winter and as a result the increase in net photosynthetic rate in early summer appeared to be considerably more substantial than it actually is. In addition, re-examination of the early summer and autumn response to temperature in terms of PI curves has failed to detect any seasonal adjustment of P_{max} to ambient temperature. In contrast, the net photosynthetic seasonal response of *Peltigera rufescens* to temperature does show a very clear cut response in summer (Brown and Kershaw 1984). The full seasonal response matrix is given in Fig. 169 (p. 254) and shows a good degree of capacity adjustment at 35 and 45 °C, but only above light saturation. The PI curves expressed both as per gram dry weight and per milligram chl *a* show the full extent of the adaptation to summer temperature, with rates increasing from *c*. 5.5 mg CO_2 h^{-1} g^{-1} to *c*. 9 mg CO_2 h^{-1} g^{-1} in summer while α remains constant (Figs. 130 and 131). Accordingly the change in P_{max} reflects an enzymatic rate change.

It is also interesting to re-examine Stålfelt's data (Fig. 115) in the light of these interpretations. Making allowances for the limited replication it seems likely that *Evernia prunastri* uncouples PSUs in winter whilst *Ramalina farinacea* perhaps acclimates, although the respiratory change complicates the interpretation. Re-examination of the data for *Parmelia disjuncta* (Kershaw and Watson 1983), where summer capacity change is evident only at high light levels, similarly suggests acclimation of an enzyme rate. Unfortunately the data for *Bryoria nitidula* and *Alectoria ochroleuca* (Kershaw 1975*b*; Larson and Kershaw 1975*d*,*e*) are too incomplete to substantiate fully acclimation as the mechanism of the summer change in P_{max}. It is important here to recognise that the net photosynthetic response at light values above saturation defines acclimation and where experimental procedures do not include illumination levels beyond *c*. 800 μmol m^{-2} s^{-1} photosynthetically active radiation it is usually impossible to confirm or eliminate acclimation as a photosynthetic strategy, particularly in species from open habitats.

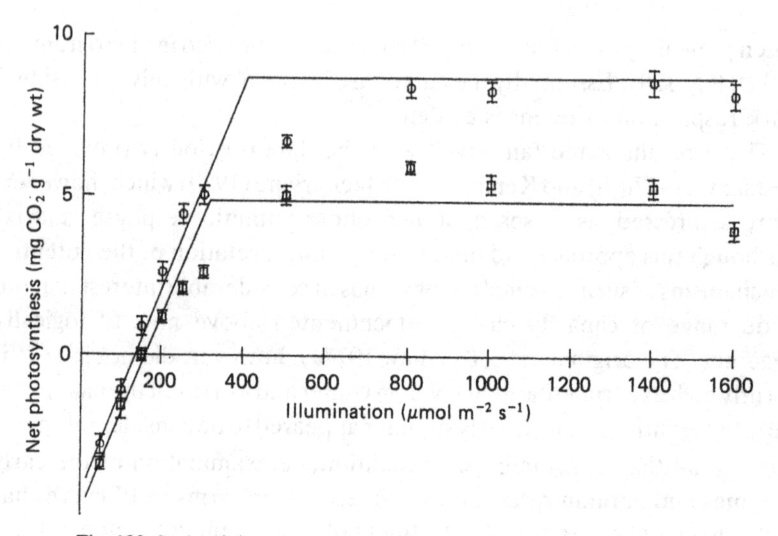

Fig. 130. Spring (□) and summer (○) PI curves for *Peltigera rufescens* expressed in terms of grams dry weight. Vertical lines show S.E.M. (From Brown and Kershaw 1984.)

Not all acclimation is to high summer temperatures. Coxson, and Coxson and Kershaw (1983*c*) (1983) have examined the seasonal net photosynthetic response to temperature in *Caloplaca trachyphylla* and have demonstrated low temperature acclimation during winter chinook snow-melt sequences. The full response matrix is given in Fig. 132, and shows seasonal constancy of net photosynthesis and respiration except at 7°C. The corresponding PI curves (Fig. 133) fully confirm the ability of *C. trachyphylla* to acclimate to winter snow-melt temperatures. Larson (1980) has similarly suggested that *Umbilicaria deusta* also responds during snow-melt conditions in the spring by acclimating to low temperature, but additional data are required to substantiate this.

Thus it now becomes provisionally a relatively simple matter to distinguish between the different mechanisms involved in the range of photosynthetic capacity changes so far identified in lichens. Equally the confusion over the use of the term 'acclimation' is eliminated. Capacity changes evident at *low* light only (change in α) represent a change of PSU size; capacity changes at both high and low light (change in α *and* P_{max}) reflect coupling or uncoupling of energy transduction; capacity changes evident only close to, or above light saturation, indicate acclimation or enzymatic rate change. It remains, however, to confirm these interpretations at the biochemical and enzymatic level.

It is evident, then, that lichens exhibit a range of mechanisms that

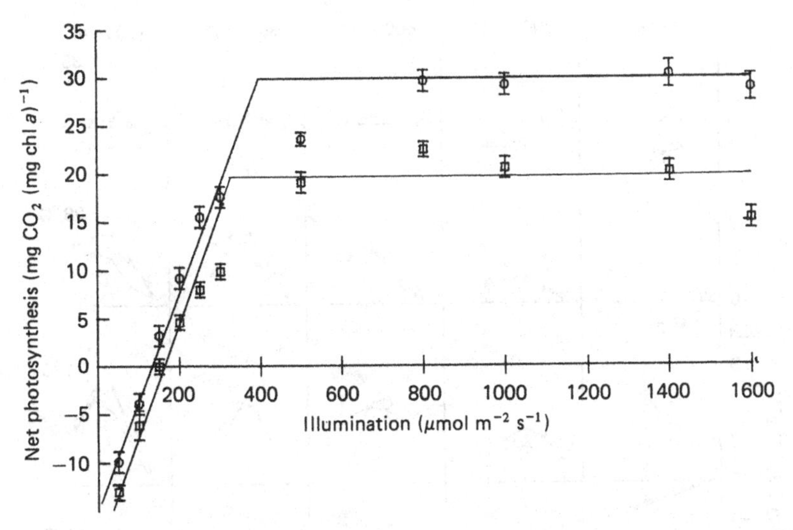

Fig. 131. Spring (□) and summer (○) PI curves for *Peltigera rufescens* expressed in terms of mg chl *a*. Photosynthetic efficiency remains more or less constant and the capacity change accordingly reflects acclimation of an enzyme mediated step. Vertical lines show S.E.M. (From Brown and Kershaw 1984.)

control photosynthetic capacity changes throughout the year. Some of the mechanisms clearly confer a photosynthetic advantage as light and temperature change seasonally. The significance of reversible coupling and uncoupling of energy transduction, however, remains obscure. The potential for combinations of seasonal photosynthetic events as differential strategies is discussed in a later context.

6.4 The induction of seasonal photosynthetic capacity changes

The first experimental induction of the low-light response of *Peltigera scabrosa* simply involved storing air-dry replicates (thallus moisture level 7-10% of the oven-dry weight) at 15 °C under 20 mol m^{-2} s^{-1} illumination with a 12 h day/night photoperiod. Under these low thallus moisture conditions it required 3 weeks of storage at low light to induce the required response (Fig. 118, p. 178). Conversely the effect of returning the low-light adapted replicates to high-light conditions was apparent within only 2 days. The data thus substantiate the striking levels of metabolic response in air-dry lichen thalli. Equally the winter/summer photosynthetic coupling and uncoupling in *Peltigera praetextata* can be readily induced in air-dry experimental replicates. Kershaw and MacFarlane (1980) collected thallus replicates locally in the field during April and these were stored both moist and dry at 0-2 °C with a 12 h day of 300 μmol m^{-2} s^{-1}

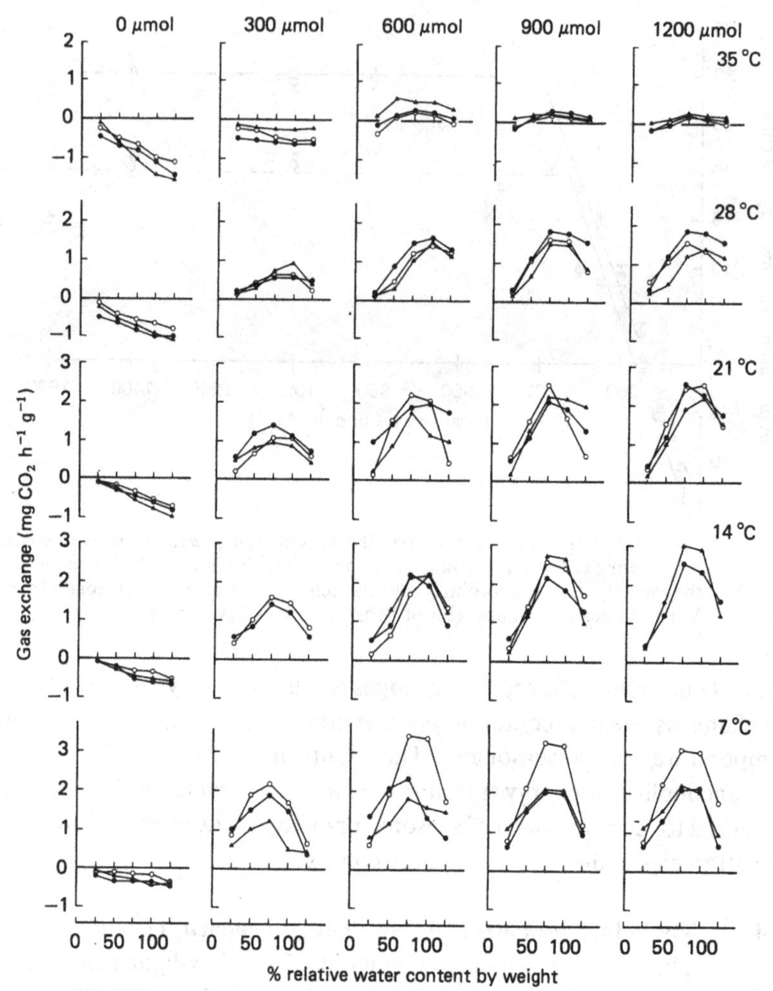

Fig. 132. The full gas exchange seasonal response matrix for *Caloplaca trachyphylla* showing the marked low-temperature acclimation in the winter replicates. Collection dates: ○, 8 January; ●, 8 May; ▲, 13 July. (From Coxson and Kershaw 1983*c*.)

illumination. At this period of the year the metabolic response to storage at 0 and 15 °C is completely plastic (see below). Experimental replicates were stored under similar light conditions at 15 °C for 7 days and on day 8 returned to 0 °C storage. Two sets of replicates were used throughout the experiment, air-dry replicates having a thallus moisture content of 7-12% of the oven-dry weight and thalli sprayed daily with a fine mist of distilled water and having a thallus moisture content of 200-400% of the oven dry weight. The hydrated replicates were soaked overnight at the storage

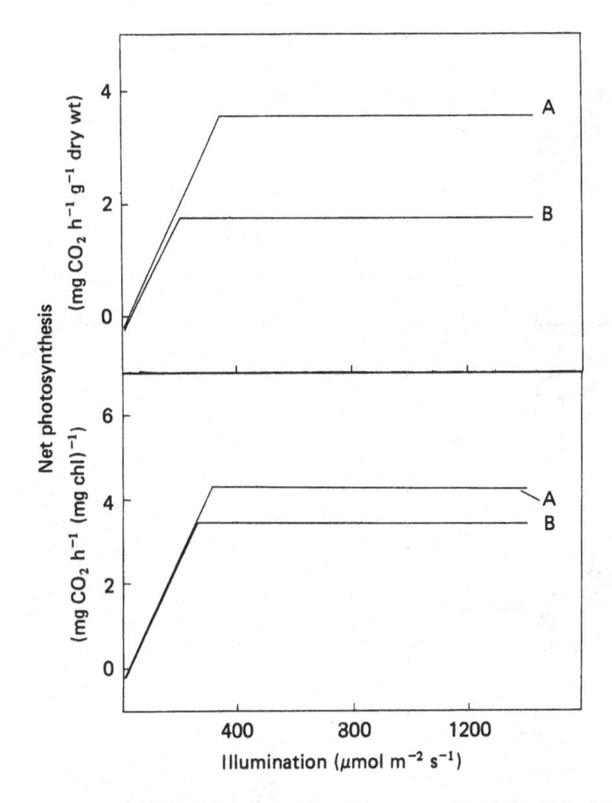

Fig. 133. The contrasting PI curves at 7 °C for *Caloplaca trachyphylla* stored at (A) low and (B) medium temperatures, showing low-temperature acclimation of P_{max}. The regressions are expressed in terms of grams dry weight and mg chl *a*. (From Coxson and Kershaw 1983*c*.)

temperature but the dry replicates were hydrated for only 2 h at the storage temperature, so that the net photosynthetic rates in the latter are slightly depressed compared with those of the replicates soaked overnight.

The results for the two groups are given in Fig. 134. Uncoupled rates in the hydrated replicates showed maximum values, at 200% water content by weight, of 3.5 mg CO_2 h^{-1} g^{-1}; this is in close agreement with the winter rates previously reported by Kershaw (1977*a*). Upon transference to 15 °C the replicates showed no response to the higher storage temperature after 1 day, but after 2 days a full level of photosynthetic adjustment had been achieved. This was then maintained during the subsequent 15 °C storage treatment. After being returned for 1 day to a storage temperature of 0-2 °C there was an immediate return of net photosynthetic rates to their previous uncoupled winter level. Concurrently there was no significant change in respiration rates. *Peltigera polydactyla* when air-dry or fully hydrated also

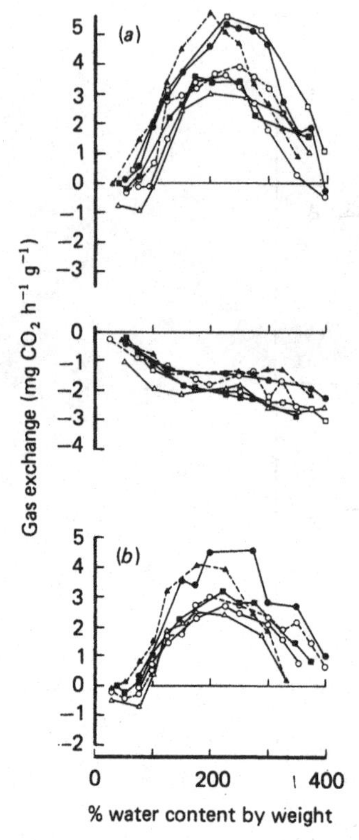

Fig. 134. The induction of coupling and uncoupling of energy transduction in PSUs of *Peltigera praetextata* in (*a*) moist and (*b*) air dry replicates. △, initial winter response; ○, after 1 day; ●, after 2 days; □, after 3 days; ▲, after 7 days storage at 15 °C. Replicates were then returned to low-temperature storage for 1 day (○) and re-examined after 3 days (■). Each line is the mean of four replicates with maximum standard errors of 0.4 mg CO_2 h^{-1} g^{-1}. (From Kershaw and MacFarlane 1980.)

shows a closely similar response to temperature treatment in the spring (Fig. 135). The winter net photosynthetic rates were approximately 2.5-3.0 mg CO_2 h^{-1} g^{-1}, somewhat lower than those of the hydrated replicates, presumably as a result of the limited pre-experimental soaking period. When moved to a 15 °C storage temperature for 2 days, maximum net photosynthetic rates of 4.0-4.5 mg CO_2 h^{-1} g^{-1} were generated. Again these are somewhat lower than the equivalent rates in the hydrated replicates but proportionately show the same increase over the winter rates. The ability of lichens to track continuously the environmental changes in the field is also evident in the data of Kershaw (1977*c*). Fresh material of *Peltigera praetextata* and *P. polydactyla* was collected from a woodland site near Waterdown, Ontario, on 1 March, when winter temperature condi-

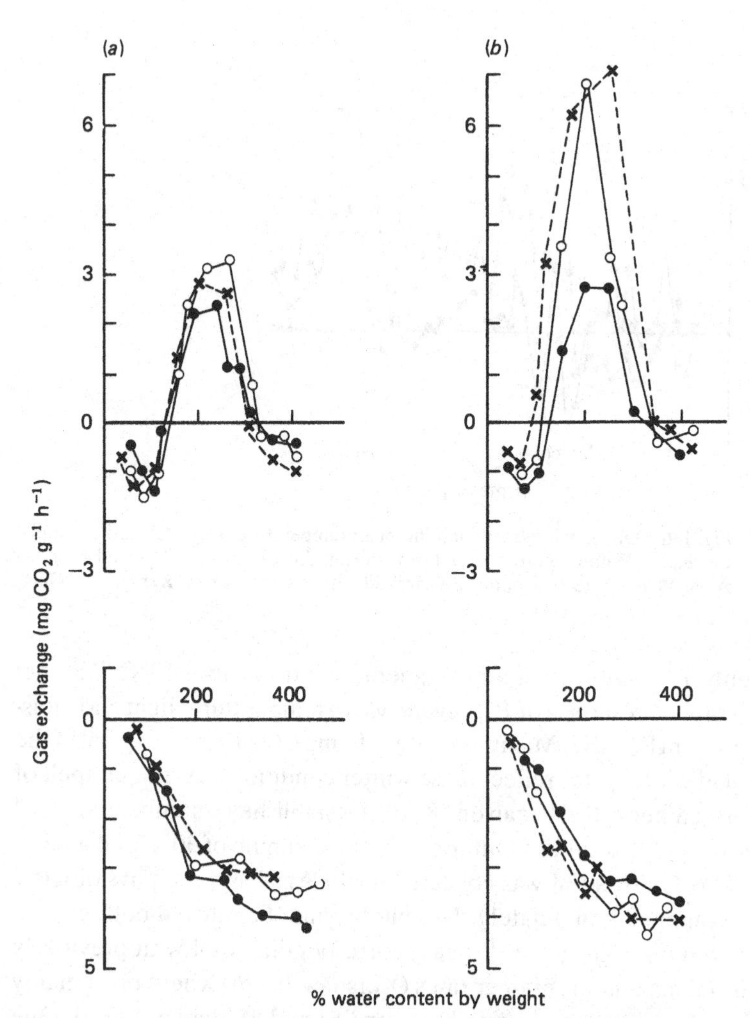

Fig. 135. Coupling of energy transduction in PSUs of *Peltigera polydactyla*. Net photosynthetic rate at 25 °C and 300 µmol m^{-2} s^{-1} and respiration rate at 25 °C and 0 µmol m^{-2} s^{-1} for control replicates in April (*a*) and replicates stored moist at 25°C (*b*). The control replicates and day 1 treatment (•) show a characteristic winter level of net photosynthesis which increases to its full summer level by day 2 (○). This level of response is still maintained by day 7 of the experiment (×). Each point is a class mean, with S.E.M. < 0.5 mg CO_2 g^{-1} h^{-1}. (From Kershaw 1977*c*.)

tions were still quite severe, on 19 April following a period of unusually warm weather; and again on 26 April after winter conditions had returned briefly. These fluctuations of ambient temperatures throughout the initial experimental period are given in Fig. 136.

Immediately before the first collection maximum temperatures had

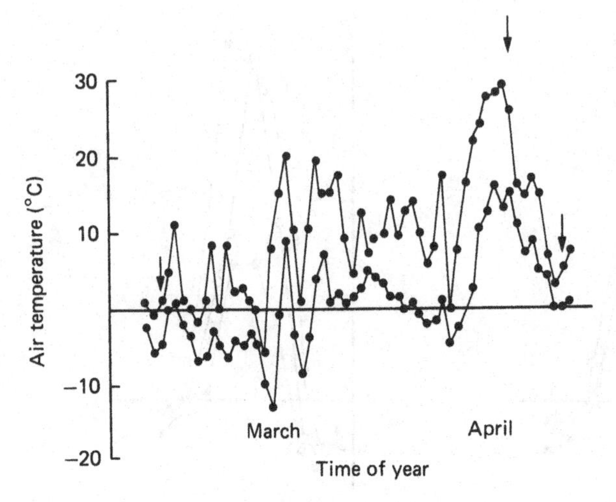

Fig. 136. Daily maximum and minimum air temperatures recorded at the Hamilton Royal Botanic Garden, Ontario, during the experimental period March-April 1976. Collection dates are marked with arrows. (From Kershaw 1977*c*.)

been only 1 °C, with minimum temperatures down to −10°C. The net photosynthetic response of *P. polydactyla* to temperature, light and moisture is given in Fig. 137. Maximum rates of 3 mg CO_2 g^{-1} h^{-1}, typical of the uncoupled winter rate, reflect these winter conditions. A unique spell of very hot weather which began on 18 April, established a number of record temperatures for southern Ontario, with a maximum of 30 °C and a minimum of 16 °C. Material was collected on 19 April and the physiological matrix examined immediately. Net photosynthetic rates in both species were now 6 mg CO_2 g^{-1} h^{-1}, again corresponding well with previously established maximum summer rates (Kershaw 1977*b*) where presumably coupling of energy transduction between PS1 and PS2 had occurred. One week later minimum field temperatures returned to 0°C and the net photosynthetic response at all temperatures and light levels again reflected the uncoupled winter level (Fig. 137).

Early experimental procedures (Kershaw 1977*b*) to adjust winter replicates to summer levels of net photosynthesis involved maintaining experimental lichen material in a moist condition during storage at an induction temperature of 25-30 °C. This procedure was adopted in the belief that such metabolic changes would require a concurrent level of photosynthetic and respiratory activity only achieved in hydrated thalli. The belief that air-dry lichens are completely inactive metabolically and therefore are not susceptible to any induction or stress parameters, is widespread. We have

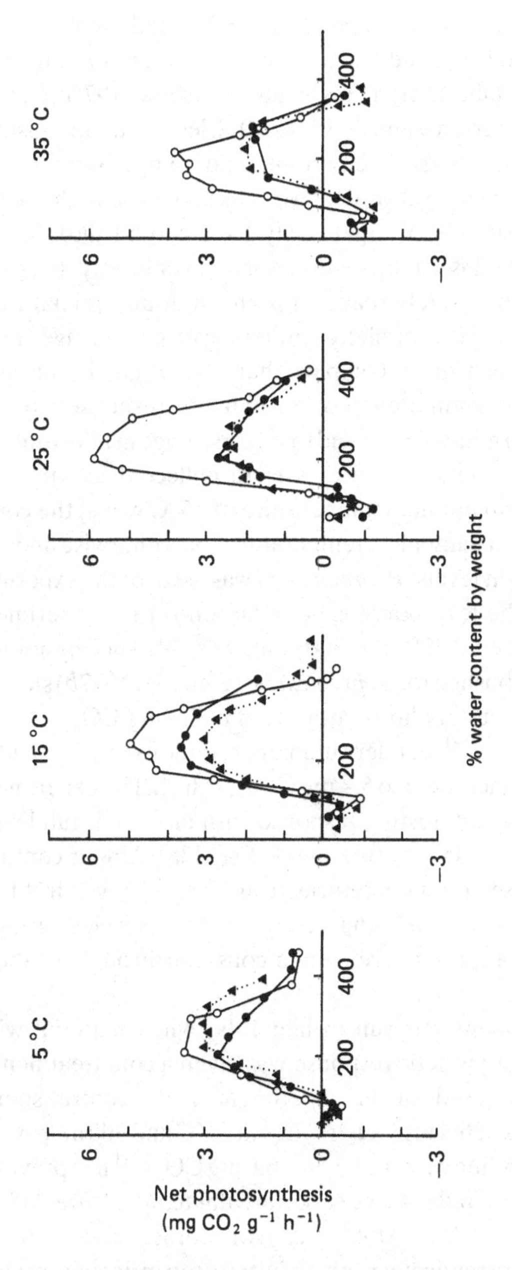

Fig. 137. The seasonal response matrix (at 450 μmol m^{-2} s^{-1}) for *Peltigera polydactyla* during the period of marked temperature fluctuation shown in Fig. 140. ●, 1 March, winter collection; ○, April collection after warm weather; ▲, April collection after return of cold weather. (From Kershaw 1977*c*.)

ourselves, for example, consistently emphasised the need to collect air-dry material (where possible) to reduce physiological acclimation in transit between the field and laboratory (Larson and Kershaw, 1975*c*,*d*,*e*; Kershaw 1975*a*,*b*, 1977*b*; Kershaw and Smith 1978). Clearly our optimism was misplaced. The ability to track environmental temperature or light changes continuously when wet or dry, would seem to be highly advantageous to an organism which largely only fixes carbon dioxide when hydrated. When the thallus is then hydrated, however briefly, the photosynthetic system is immediately ready to perform at an optimal rate.

The environmental control of photosynthetic capacity changes in *Peltigera* is, however, much more complex than the original data above suggest. Subsequent examination showed that the responses to lower temperature storage in mid-summer and to warm storage in mid-winter are quite different. Experimental replicates were collected in winter and stored at a summer ambient field temperature of 25 °C whilst the control replicates were stored at winter low temperatures but otherwise under the same daylength and light levels. *P. praetextata* was used for the experiment and the net photosynthetic response of both the control and experimental replicates were examined at 25 °C and 300 μmol m^{-2} s^{-1}. The physiological matrix previously established for *P. praetextata* (Kershaw 1977*b*) showed a maximum winter rate of net photosynthesis of *c*. 3.0 mg CO_2 g^{-1} h^{-1} at 15 °C and 300 μmol m^{-2} s^{-1}. Under summer conditions at 25 °C and 300 μmol m^{-2} s^{-1} the rate increased to 5.4 mg CO_2 g^{-1} h^{-1}. The experimental treatment shows a very interesting response with an initial, full level of warm acclimation after 2 days of treatment (Fig. 138). Under continued treatment this response gradually returns from 5 mg CO_2 g^{-1} h^{-1} to the control level of 3 mg CO_2 g^{-1} h^{-1} characteristic of this species during the winter period. Respiration rates remained constant throughout the experimental treatment.

The inverse experiment was run in late July, when material with a mid-summer net photosynthetic response was given a cold treatment. *P. polydactyla*, which was used for the experiment, had a control summer level of photosynthesis of 6.4 mg CO_2 g^{-1} h^{-1} at 25 °C and 300 mol m^{-2} s^{-1} illumination corresponding with the level of 6.0 mg CO_2 g^{-1} h^{-1} previously reported for this species under these experimental temperature and light conditions (Kershaw 1977*b*). After 1 day of storage at 5°C the experimental temperature replicates show an initial drop in net photosynthetic rate indicative of a winter response, but full net photosynthetic rates were restored by day 3 (Fig. 139). Again, respiration rates remained constant throughout the treatment.

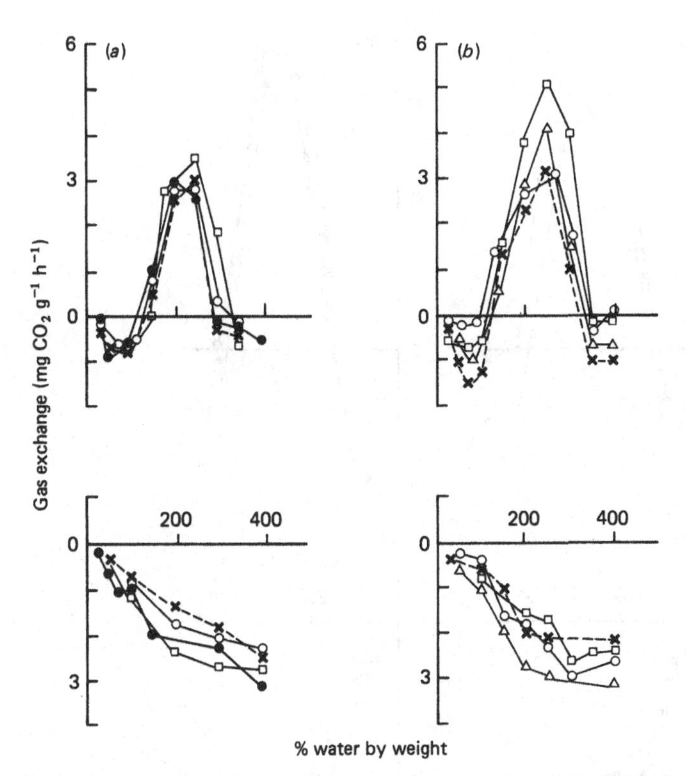

Fig. 138. The net photosynthetic rate at 25 °C and 300 μmol m^{-2} s^{-1}, and respiration rate at 25 °C and 0 μmol m^{-2} s^{-1} in *Peltigera praetextata* for control replicates in December (*a*) and replicates stored continuously moist at 30 °C (*b*). The control replicates and day 1 treatment (○) show a characteristic winter level of net photosynthesis. Net photosynthesis in the experimental replicates increases to its full summer level by day 2 (□). This is gradually reversed by day 3 (△) to the winter level by day 7 (×), despite continued warm storage. Each point is a class mean with S.E.M. < 0.5 mg CO_2 g^{-1} h^{-1}. (From Kershaw 1977*c*.)

It appears, then, that there is a 'window' in April when the typical pattern of summer/winter photosynthetic capacity in *Peltigera* is plastic, reflecting coupled or uncoupled energy transduction responding quickly and simply to changes in environmental ambient temperature. Subsequently, however, Kershaw and MacFarlane (1980) showed that a rapid daylength change *without any temperature change* also induced a similar marked photosynthetic capacity change. Experimental replicates were collected locally in early July, stored moist under a 12/12 h photoperiod and then moved to an 18/6 h photoperiod with a constant 15 °C temperature. The results are clear-cut (Fig. 140) and show a very marked capacity change on days 2 and 3 after the short-day treatment commenced, but then a return to a normal summer photosynthetic capacity by day 4. Kershaw

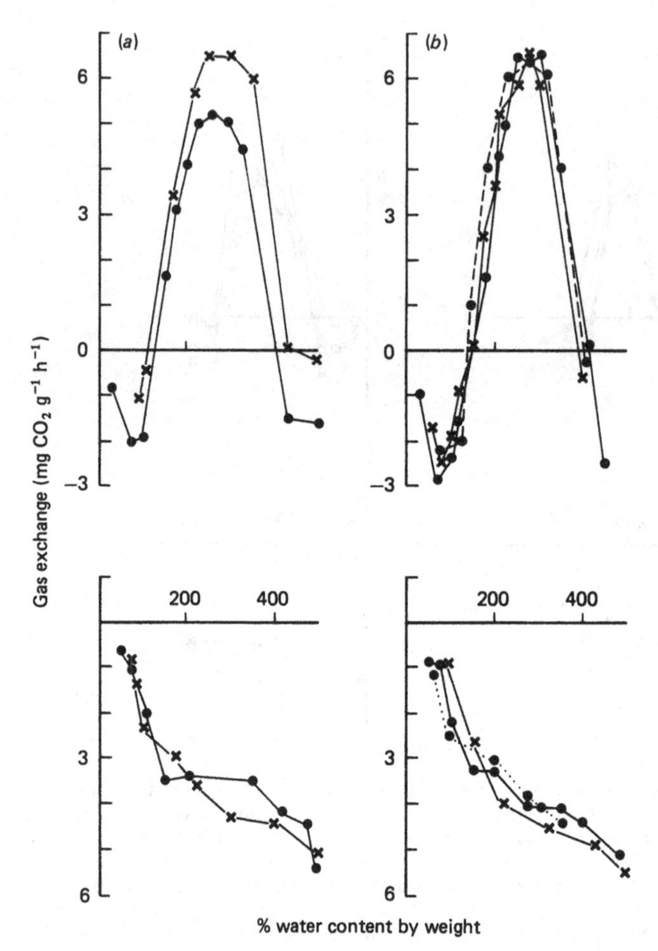

Fig. 139. The net photosynthetic rate at 25 °C and 300 μmol m^{-2} s^{-1} and respiration rate at 25 °C and 0 μmol m^{-2} s^{-1} in *Peltigera polydactyla* in summer control replicates (*b*) and replicates stored moist at 5 °C (*a*). There is an initial net photosynthetic response after 1 day of cold treatment (•) which is restored by day 3 (×). Each point is a class mean with S.E.M. < 0.5 mg CO_2 g^{-1} h^{-1}. (From Kershaw 1977*c*.)

and MacFarlane (1980) initially interpreted the 'crash' of photosynthetic capacity as a stress response with the lichen 'adapting' to the changed daylength after a few days.

MacFarlane *et al.* (1983), however, recognised that this response could in fact be part of the summer/winter capacity change as a result of uncoupling of energy transduction in some PSUs, and that not just temperature but also daylength was important. Thus replicates of two ecotypes of *Cladonia rangiferina* stored at 15 °C in April surprisingly showed very little

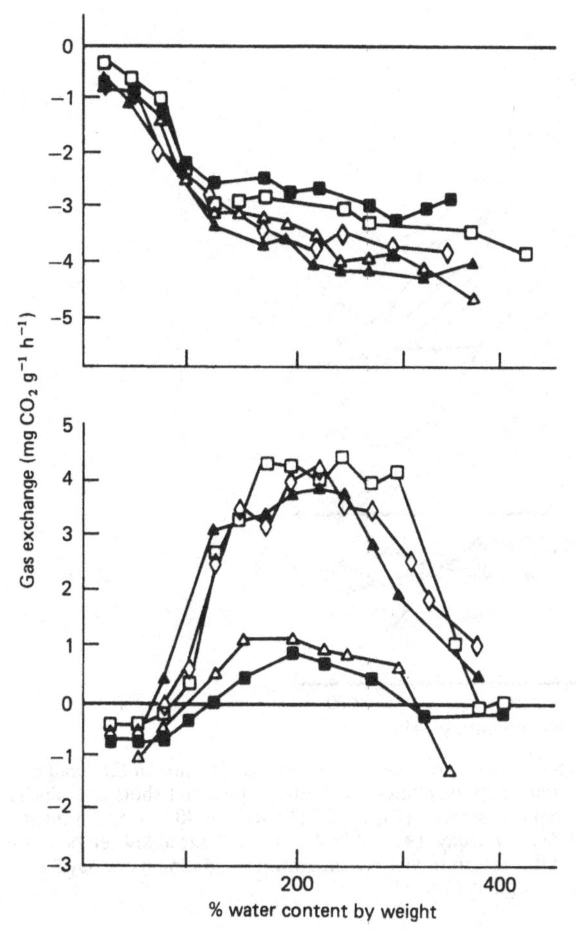

Fig. 140. The effect of an abrupt daylength change on the net photosynthetic and respiration rates in *Peltigera praetextata*. Control replicates under a 12/12 h light/dark photoperiod on day 1 (□) and day 4 (◇). Experimental replicates transferred to a 18/6 h light/dark photoperiod on day 2 (■), day 3 (△) and day 4 (▲). (From Kershaw and MacFarlane 1980.)

capacity response. The possibility that the control mechanism depended on simultaneous daylength and temperature changes was examined in late July by storing one group of replicates at winter temperatures with a summer daylength (16/8 h, day/night) and a second group at winter temperatures coupled with a short day (8/16 h day/night). Only the combined low-temperature and short day treatment produced a response; the results are given in Fig. 141. Kershaw (1977*b*) reported an initial response when *Peltigera* was stored at winter temperatures in July (see above) and

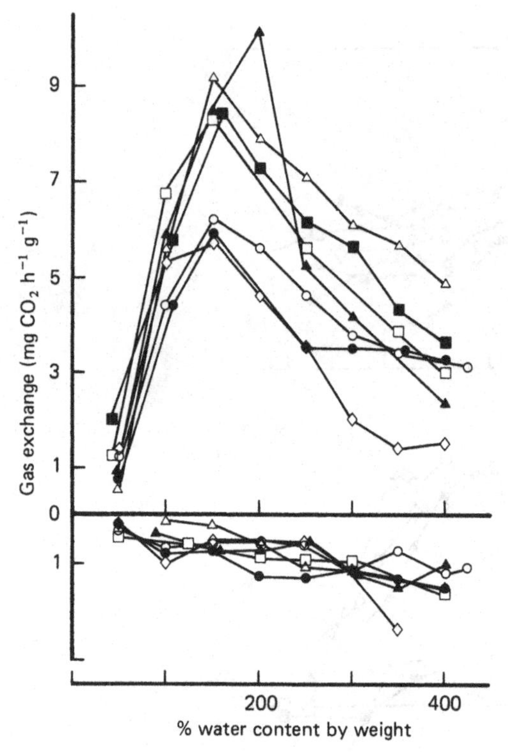

Fig. 141. The experimental uncoupling of energy transduction in *Cladonia rangiferina* in mid-summer by combined low temperature and short day air-dry storage. Control replicates: day 1 (□), day 22 (■) and day 40 (△). Experimental replicates after 1 day (○), 2 days (●) and 23 days (◇) storage at low temperature with short days. After return to warm temperatures and long days: day 3 (▲). (From MacFarlane *et al.* 1983.)

accordingly the combined daylength/temperature treatments were extended over a 6 week period. Maximum net photosynthetic rates in the control *Cladonia* replicates at the start of the experiment were *c*. 8.25 mg CO_2 h^{-1} g^{-1} and were still at this level on day 27. Replicates stored at 0°C, with a short day photoperiod, immediately exhibited winter rates of net photosynthesis (6.2 mg CO_2 h^{-1} g^{-1}) which were maintained throughout days 2, 3 and 4 and were still evident on day 23 (5.7 mg CO_2 h^{-1} g^{-1}, S.E.M. 0.3 mg CO_2 h^{-1} g^{-1}). These replicates were then returned to summer storage conditions (15/25 °C with a long day photoperiod), with the immediate restoration of high and stable photosynthetic capacities (days 26-41, Fig. 141). However, in January the alternative long-day, warm-temperature treatment failed to re-establish summer rates.

It is probably significant that the earlier *Peltigera* experiments reported by Kershaw (1977*c*) actually used a *neutral daylength* of 12 h that was typical, in fact, of the April natural seasonal 'window' for capacity changes. During the period of full summer when daylength is not neutral, two signals and not one are required to induce a complete winter capacity response in *Cladonia rangiferina*. However, under full winter conditions it has so far been impossible to induce a summer response experimentally and it appears that there is an additional endogenous sequence of events, over which we have as yet no control. Furthermore it does not necessarily follow that all lichens which exhibit uncoupling of energy transduction will respond identically to daylength and temperature signals and further work on a range of species is now required before we have a full and generalised level of understanding. Equally acclimation of photosynthetic optima to snow-melt ambient temperatures in *Caloplaca trachyphylla* can so far only be achieved experimentally in winter, corresponding with the natural timing in the field. Prézelin and co-workers have documented a number of examples of phytoplankton where daily periodicity of photosynthesis is regulated by a biological clock (Prézelin and Ley 1980). It appears probable that there is also a similar level of control of the seasonal timing of photosynthetic capacity changes present in the phycobionts of lichens.

7

Net photosynthetic optima and thermal limits

Since the last major review of the physiology and the ecology of lichens (Ahmadjian and Hale 1973), considerable advances have been made in our overall understanding of the response of lichens to their environment. This is particularly true of the information now available on the response of a lichen to its thermal environment. Since many lichens are characteristically found growing on apparently hostile surfaces and often in very extreme environments it has been tempting to believe that all lichens can therefore withstand severe stress. Several earlier documented examples apparently substantiated this belief, although it is now evident that, in common with all generalisations, there are many exceptions to the rule. However, the belief that lichens in all cases can withstand severe environmental stress also embraced the obvious corollary that lichens must, therefore, be physiologically adapted to these environments. In general, arctic lichens were thus assumed to have a low temperature optimum and tropical lichens a high temperature optimum for net photosynthesis. Equally the respiratory quotient (Q_{10}) should contrast markedly between these two climatically extreme geographical regions. Again these further generalisations are now seen to be true to an extent, but there are of course numerous exceptions. The specific thermal *operating* environment of a lichen is often very different from the ambient temperatures found in either arctic or in tropical situations, as has been emphasised in Chapter 1. The operating thallus temperature of *Parmelia disjuncta*, typical of boundary-layer rock surfaces in the low Arctic, is probably little different from that of a pendulous *Usnea* growing at 50 m in tropical forest. It would not be surprising, therefore, to find comparable metabolic responses to temperature in both species.

7.1 The potential fallacy in contrasting photosynthetic temperature optima in arctic and tropical lichens.

Ahmadjian (1970), Greene and Longton (1970), Lamb (1970), Farrar (1973), Kallio (1973) and Kappen (1973) have all commented on the low temperature optima of arctic lichens. This conclusion is based initially

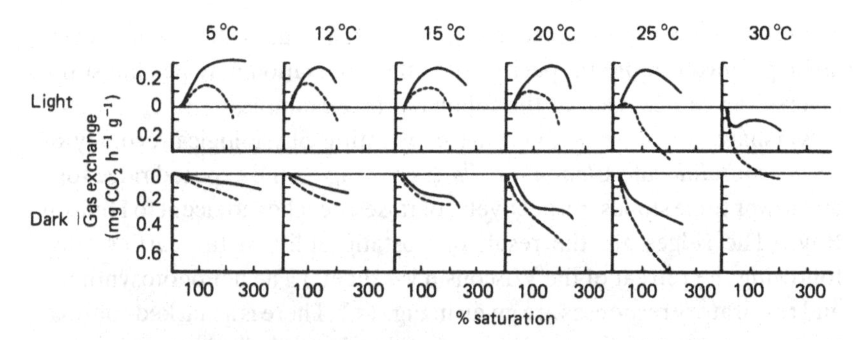

Fig. 142. The contrasting respiratory and net photosynthetic responses (at 150 μmol m^{-2} s^{-1}) to temperature in *Bryoria nitidula* (----) and *Cladonia stellaris* (—). (From Kershaw 1975*b*.)

on the work of Lange (1962, 1965), who contrasted the interaction between net photosynthesis and temperature in a number of tropical species with the interaction within arctic and alpine species. The results that he obtained were interpreted as showing temperature optima below 15°C and often approaching 0°C in arctic species as opposed to optima above 18°C for tropical species. However, Lange's results are based on net photosynthetic rates measured at thallus saturation only, rather than from the integral of the drying response curve. At thallus saturation, net photosynthesis can be heavily dominated by respiration, particularly if the internal diffusive resistance to carbon dioxide is also high. The resultant low net photosynthetic rates at full thallus hydration then become very much a function of the respiration rate, in addition to the diffusive resistance. Since respiration is heavily controlled by temperature, net photosynthesis also appears to be equally correlated with temperature. The use of the net photosynthetic rate at saturation as a typical level of response of the species can thus give a very erroneous impression. This is certainly true for those species that exhibit low net photosynthetic values at saturation that rise to higher rates as the thallus dries. *Alectoria ochroleuca*, *Bryoria nitidula*, *Cetraria nivalis* (Larson and Kershaw 1975*d*,*e*), ridge-top populations of *Cladonia stellaris* (Kershaw 1975*b*) as well as *Stereocaulon paschale* (Kershaw and Smith 1978), all have this type of net assimilation response to moisture. Conversely where diffusive resistance is not markedly affected by the level of thallus hydration, net photosynthetic rates at saturation can be used safely to express the overall response to temperature, and Lange's data will presumably include such cases.

Consequently, although in a general sense tropical and arctic lichens may have high and low temperature optima respectively for net photosynthesis, definitive evidence to support such a relationship completely is lacking. Furthermore in specific cases the correlation may be almost the inverse, as can be seen in the following two examples:

Kershaw (1975*b*) examined the contrasting physiological ecology of *Bryoria nitidula* and *Cladonia stellaris* growing on the exposed ridge tops and lower ridge slopes, respectively, of raised beaches adjacent to Hudson Bay. (The ridges are the result of isostatic uplift of the earth's crust following the retreat of the Wisconsin ice sheet.) The net photosynthetic and respiratory responses are given in Fig. 142. There is a marked contrast in the response of the two species to levels of thallus saturation but particularly, in the present context, to thallus temperature. *B. nitidula* never achieves photosynthetic compensation at full thallus hydration and at temperatures above 20°C is totally dominated by very rapid increases in respiration rate. The optimum temperature for net photosynthesis is at 12-15°C but a good level of activity is also evident at 5°C and the overall pattern is characteristic of a low-arctic species and certainly of its specific ridge-top niche. This environment is very exposed and characterised by low temperatures in both winter and summer. *C. stellaris*, on the other hand, has a very different response pattern with good rates of net photosynthesis up to 25°C without an excessive increase in respiration. Such a response pattern might perhaps be expected of a temperate species from a cool, moist, shaded environment. The particular niche it occupies along the Hudson Bay coast on the lower slopes of the beach ridges is indeed much less exposed than the ridge-top location of *B. nitidula*, is warmer, and the evaporative rates are significantly lower. Both species are widely distributed in harsh low-arctic regions but each of them is adapted to its *specific operating environment* and not *necessarily* to the *general environment*.

The same is true of *Peltigera rufescens*, a widespread species in Ontario and the northern United States. The response matrix is given in Fig. 169, (p. 254). *P. rufescens* is characteristic of very open habitats and is particularly abundant on glacial till surfaces adjacent to roads in central Ontario. MacFarlane and Kershaw (1978, 1980*a*) have documented the very high levels of illumination and thallus temperatures reached in this habitat where surface soil temperatures of 50°C and above are typical of summer conditions (see Fig. 3, p. 5). The closely adpressed habit of the lichen means that during thunderstorms the thallus can remain hydrated even under full radiation levels for several hours, and this is reflected in the temperature optimum of net photosynthesis. Thus maximum rates are generated at 35°C with very substantial rates maintained even at 45°C, light

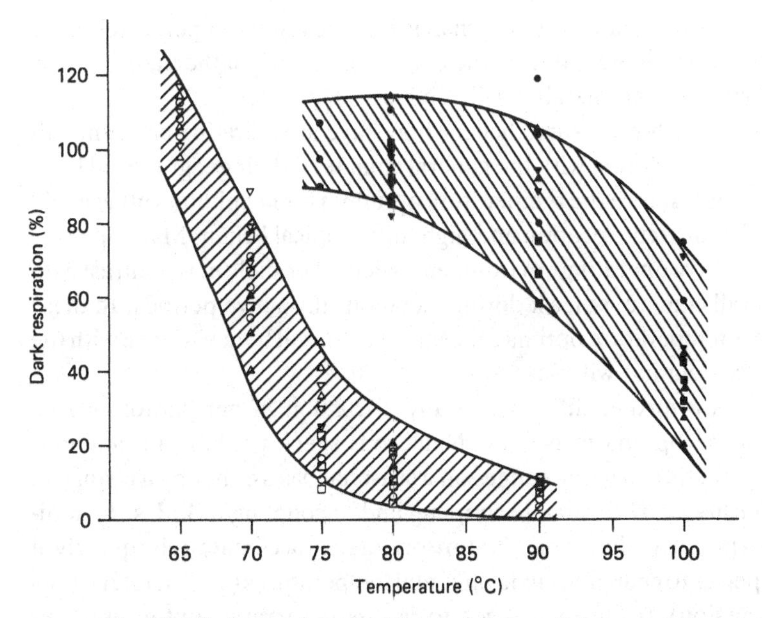

Fig. 143. The heat resistance of the species of two lichen associations, indicated by rates of carbon dioxide release (% of the controls) after heat treatment at different temperatures for 30 min. Usneetum barbatae: △, *Alectoria sarmentosa*; ▽, *Evernia prunastri*; ○, *Usnea dasypoga*; □, *Ramalina farinacea*. Fulgensietum continentale: ●, *Cladonia pyxidata* var. *pocillum*; ▲, *Cladonia rangiformis* var. *pungens* f. *foliosa*; ▼, *Cladonia rangiformis* var. *pungens*; ■, *Cladonia convoluta*. (From Lange 1953.)

saturation occurs above 1000 μmol m^{-2} s^{-1} (Fig. 131, p. 193). Again the physiological response of the lichen is adapted to its specific operating environment and not to the overall temperate-continental latitudinal climate.

The data of Lange (1969) for *Ramalina maciformis* also show the same adaptation of the lichen to its own operating environment rather than to the overall climate of a hot desert and offer a very good contrast with the adaptations outlined above for *Peltigera rufescens*. The net photosynthetic temperature optimum of *R. maciformis* ranges from 15°C under medium levels of illumination up to 20°C at the maximum experimental radiation levels employed. Lange discusses this apparent inconsistency for a hot-desert lichen in terms of the levels of metabolic activity induced by dewfall in the early morning when temperatures and light levels are quite low. Under full radiation conditions and with the prevailing high thallus temperatures characteristic of hot desert conditions, the thallus is completely dry and metabolically inactive. Accordingly, adaptation of the temperature optimum for gas exchange is to the lower temperatures of the operating environment during the early morning when the thallus is partially hydrated and metabolically active.

Thus in conclusion, some lichens may have an operating environment that correlates closely with the general climate of their geographical location. In this instance the temperature optimum for net photosynthesis will be low for an arctic location and high for a tropical lichen. Many species, however, have operating environments which thermally may contrast with the overall climate, at least during metabolically active periods. In these instances temperature optima for gas exchange will be at variance with the overall correlation with latitude.

There is a further difficulty in any discussion of net photosynthetic temperature optima in lichens. Numerous species exhibit pronounced photosynthetic capacity changes on a seasonal basis either by a change in PSU density or by reversible coupling and uncoupling of PSUs. In addition, net photosynthetic temperature optima can acclimate quite quickly in some species to changing environmental temperatures (see Chapter 6 for a full discussion). In these instances, to discuss temperature optima is rather pointless without including the degree of regulation that a species may or may not achieve.

7.2 The lichen response to high-temperature stress

There has been general acceptance of the concept of lichen resistance to high-temperature stress. This belief was based on the evidence presented by Lange (1953), who examined the heat resistance of a number of lichen species. The temperature accepted as an index of heat resistance in air-dry thalli was that which caused a 50% reduction in the normal respiration rate. Indices ranged from 70°C in *Alectoria sarmentosa* to 101°C in *Cladonia pyxidata*. Thalli which were fully hydrated and then heat-stressed differed little from other kinds of plant tissue, the limits of heat resistance ranging from 35 to 46°C. There is a considerable body of literature on the effects of heat stress on plants, including lichens and mosses (Lange 1953, 1955), which has been well summarised by Levitt (1980) and Kappen (1973), but although this includes actual examples of thermal stress affecting lichens, as a concept it has not been accepted ecologically. The problem arose initially over the choice of an exact criterion which could be used to measure the effects of heat stress. Further difficulties have also arisen over the method of application and duration of the heat-stress period. Typically, stress temperatures of 60-105°C for a short period have been used. As a result, although Lange (1953) showed a good correlation between the thermal resistances of lichen species from the exposed Fulgensietum continentale association, which contrasted with those of species from the less resistant arboreal Usneetum barbatae association, the actual stress temperatures used were all highly unrealistic ecologically. Thus to induce severe stress effects (reduction of respiration rate by 50%) in species belonging to the Usneetum barbatae association, a temperature of 75-80°C was necessary, and to stress species belonging to the Fulgensietum continentale association a temperature of 100°C was required (Fig. 143). These temperatures probably never occur naturally even under the most extreme hot-desert surface conditions and accordingly Kappen (1973) summarises the heat resistance of desiccated thalli as being 'extremely heat tolerant in the air-dried state'.

More recently Kershaw and Smith (1978) and MacFarlane and Kershaw (1978, 1980*a*) have re-examined the potential effects of high summer temperatures on air-dry lichens and have shown that some species are quite sensitive, even when dry, to temperatures which exceed those of their operating environment. The approach employed a more ecologically based treatment, with potential stress temperatures being applied over several days or weeks rather than as a brief single exposure to an extremely high temperature. Similarly, changes in net photosynthesis, and nitrogenase activity where applicable, were used in addition to respiration changes to assess the potential effects of the stress temperatures.

In all cases replicates were collected in the field and stored at 15°C in controlled-environment growth chambers for several days before experimental treatment. Control replicates of each species were stored at a constant day/night temperature of 15°C with a 12/12 h light/dark photoperiod and an illumination level of either 20 μmol m^{-2} s^{-1} (*Peltigera scabrosa* and *P. praetextata*) or 300 μmol m^{-2} s^{-1} (*P. rufescens*). Temperature stress was applied to air-dry replicates (water content by weight within 8% of the oven-dry weight) by storing them in growth chambers with identical light and daylength conditions but with day temperatures of 25, 35 or 45°C coupled with a 15°C night temperature. The temperatures were designed to simulate realistic levels of potential stress in the field and were applied for varying experimental periods up to a maximum of 40 days. Thus soil surface temperature under full radiation conditions in open habitats can reach 55-60°C, whereas surface temperatures in closed-canopy woodland rarely exceed 15°C. Accordingly 'stress' is used here in the sense of a temperature treatment which would be a normal temperature for an exposed surface but would greatly exceed the surface temperature in closed-canopy woodland and be potentially highly stressful to woodland

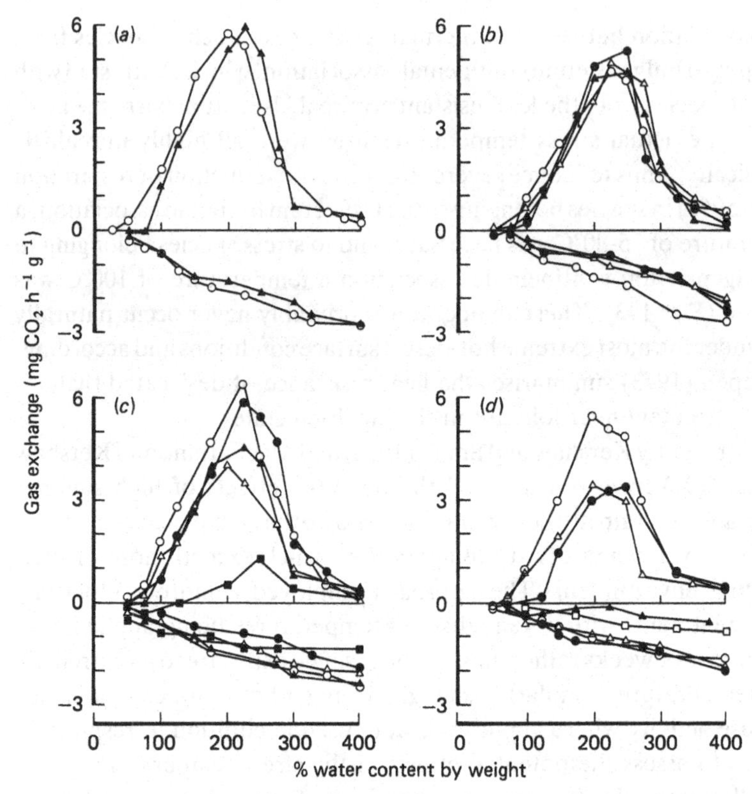

Fig. 144. Net photosynthetic and respiration rates in *Peltigera praetextata*. Thallus replicates were stored air-dry at (*a*) 15 °C, (*b*) 25 °C, (*c*) 35 °C and (*d*) 45 °C during a 12 h daylength period, and at 15 °C at night. ○, Day 1; ●, day 7; △, day 14; ▲, day 21; (35 °C only); ■, day 32; □, respiration day 21 (45 °C only). (From MacFarlane and Kershaw 1980*a*.)

species. The length of the stress periods was similarly chosen to duplicate summer stress periods typical of low-arctic and southern Ontario habitats respectively.

Thallus temperatures of *P. praetextata* and *P. rufescens* obtained by using embedded microthermocouples and measured in adjacent woodland and roadside environments respectively have been given previously (Fig. 3, p. 5). Lichen thallus temperatures in excess of 60°C were recorded during the summer solstice and at solar noon for *P. rufescens*, whilst simultaneously, thallus temperatures in *P. praetextata*, in adjacent woodland, only briefly exceeded 30°C during some sunfleck activity in the late afternoon. The responses of the two species to thermal stress reflect the contrasting natures of their operating environments. Thus in *P. praetextata* at 45°C there was a sharp decline in the net photosynthetic rate from a

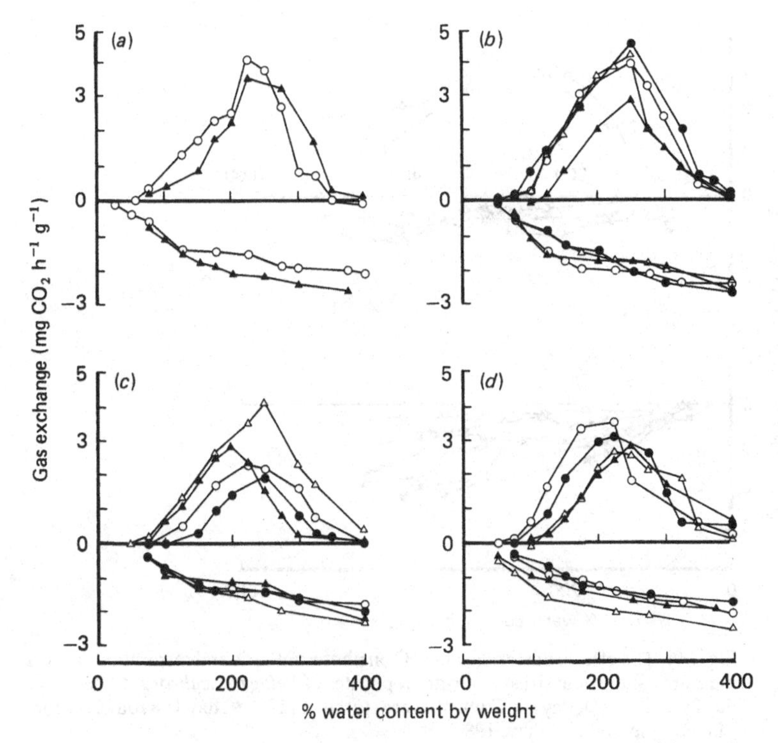

Fig. 145. Net photosynthetic and respiration rates as measures of thermal stress in *Peltigera rufescens*. Thallus replicates were stored air-dry at (*a*) 15 °C, (*b*) 25 °C, (*c*) 35 °C and (*d*) 45 °C during a 12 h daylength period, and at 15 °C at night. ○, day 1; ●, day 7; △, day 15; ▲, day 22. (From MacFarlane and Kershaw 1980*a*.)

normal maximum of 5.5-6 mg CO_2 h^{-1} g^{-1} (see also Kershaw 1977*c*) down to *c*. 3.5 mg CO_2 h^{-1} g^{-1} by day 7. By day 21 the replicates were apparently severely stressed and did not even reach light compensation under an illumination level of 150 μmol m^{-2} s^{-1}. Under a temperature stress of 35°C some decline in net photosynthesis was evident after 14 and 21 days followed by a further marked decline after an additional 11 days (Fig. 144). Conversely respiration was little affected by the temperature stress with the exception of the replicates stored at 45 °C, where after 21 days there was a significant decline from maximum values of *c*. 2.0 mg CO_2 h^{-1} g^{-1} to 0.8 mg CO_2 h^{-1} g^{-1}. As an extreme contrast a population of *P. rufescens* was collected from northern Ontario, from an area of burnt woodland where it is a characteristic species of the early successional recovery phases (Scotter 1964; Kershaw 1978). The equivalent 'stress' storage treatments at 35 and 45°C had no deleterious effects at all on gas exchange (Fig. 145).

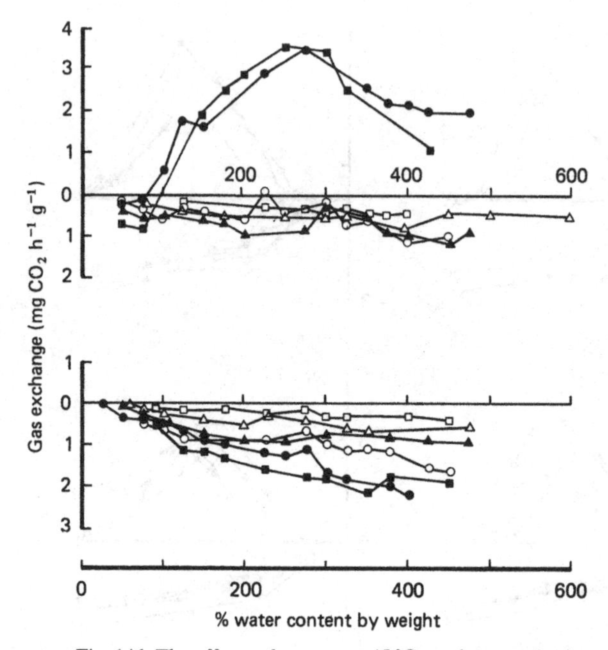

Fig. 146. The effects of storage at 45 °C on photosynthetic and respiration rates as measures of thermal stress in air-dry replicates of *Peltigera scabrosa*. ○, day 1; ▲, day 2; △, day 7; □, day 14. Controls were stored at 15 °C: ●, day 1; ■, day 7. (From MacFarlane and Kershaw 1980*a*.)

In contrast to the robust attributes of the northern population of *P. rufescens*, *P. scabrosa* collected from the same northern site but with a totally contrasting operating environment, shows an extreme level of thermal sensitivity. This species is restricted to deep moss mats under full spruce canopy shade and accordingly will rarely be exposed to temperatures even as high as 15°C. Net photosynthesis in the *P. scabrosa* replicates stressed at 45°C was almost eliminated after a single 12 h period at this temperature (Fig. 146) and respiration rates were similarly affected. With continued treatment, respiration rates progresssively declined and clearly photosynthetic activity had completely ceased by day 7. The pattern of response at 35°C (Fig. 147*a*) showed a progressive and significant decline in photosynthetic activity and finally its complete elimination by day 14. At 25°C there was again clear evidence of stress (Fig. 147*b*), maximum levels of net photosynthesis falling from *c*. 3.7 mg CO_2 h^{-1} g^{-1} to 2.0 mg CO_2 h^{-1} g^{-1} after 20 days of storage at this temperature. After an additional 7 days of stress treatment, rates were down to *c*. 1.3 mg CO_2 h^{-1} g^{-1}.

Stereocaulon paschale shows an equal level of thermal sensitivity with the virtual elimination of all photosynthetic activity after two stress periods

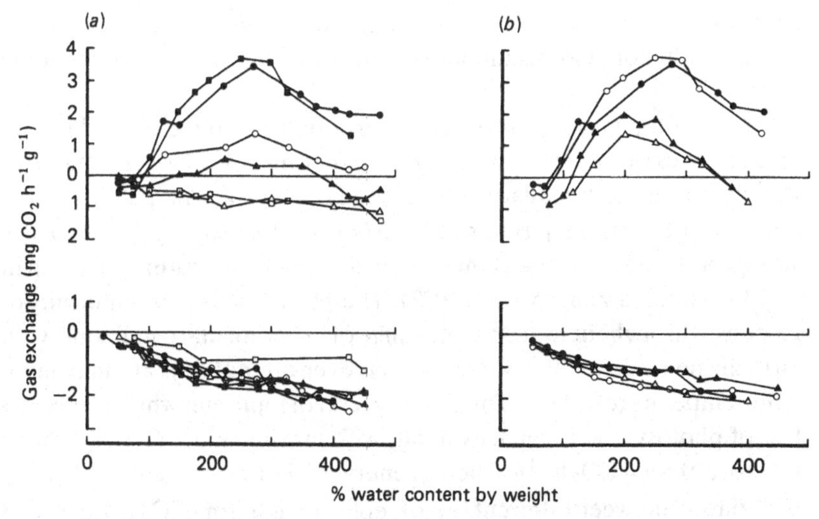

Fig. 147. (*a*) The effects of storage at 35 °C on photosynthetic and respiration rates as measures of thermal stress in air-dry replicates of *Peltigera scabrosa*. ○, day 1; ▲, day 2; △, day 7; and □, day 14. Control replicates stored at 15 °C: ●, day 1; ■, day 7. (From MacFarlane and Kershaw 1980*a*.) (*b*) The effects of 25 °C air-dry storage on photosynthetic and respiration rates as measures of thermal stress in air-dry replicates of *Peltigera scabrosa*. ▲, day 20; and △, day 28. Controls stored at 15 °C: ●, day 1, ○, day 31. (From MacFarlane and Kershaw 1980*a*.)

at 45°C, and at 35°C stress effects also are rapidly evident (Kershaw and Smith 1978). The successional sequence following fire in lichen woodland has been documented by Maikawa and Kershaw (1976) and it is apparent that there is a delay of approximately 60 years before *Stereocaulon* enters the succession. Kershaw (1977*a*) has suggested that the extreme surface microclimate of these burnt surfaces (Kershaw, Rouse and Bunting 1975; Kershaw and Rouse 1976; Rouse 1976) is largely responsible for this long delay in the development of the mature lichen surface of these low-arctic woodlands. Kershaw (1978) has recorded temperatures in excess of 45°C occurring for up to 5 h over *Biatora granulosa* colonising recently burnt surfaces (Chapter 1). Conversely, in very open woodland temperatures only briefly exceed 40°C and with a normal level of canopy shading are usually below 30°C. These microclimate observations correlate extremely well with the observations here on thermal stress response in *Stereocaulon*. A species that cannot maintain its photosynthetic capacity after even 12 h of stress at 45°C would not be able to compete successfully in a surface regime where temperatures in excess of 50°C have been recorded. Similarly, long-term stress effects, which are evident at 35°C, offer a reasonable explanation for the 60 year delay in the entry of *Stereocaulon* into the succession. After 60 years the tree density is sufficient to reduce direct surface radiation and maximum temperatures on average are less than 30°C.

The data for *Parmelia disjuncta* offer a final, extreme example of an arctic species with a remarkable degree of thermal resistance – essential for its survival in the boundary-layer niche it occupies. The pattern of net photosynthesis and respiration in *P. disjuncta*, after varying periods of dry storage at 15, 25, 35 and 45°C maximum daytime temperatures, is given in Fig. 148 (Kershaw and Watson 1983). There is a remarkable uniformity of response throughout the complete range of experimental conditions, with virtually no evidence of any stress effects even after 21 days at a maximum daily temperature of 45°C. This is the only set of replicates which shows any loss of photosynthetic capacity at all, with maximum net photosynthetic rates of *c*. 0.8 mg CO_2 h^{-1} g^{-1} being generated. However, the overall range of variation between different sets of replicates is 0.2 mg CO_2 h^{-1} g^{-1}, and this slight apparent loss of photosynthetic capacity could either be interpreted simply as part of the normal degree of between-experimental variation, or could equally reflect a slight stress response after 3 weeks of storage at 45°C daytime temperature.

The thermal environment of *P. disjuncta* is dominated by boundary-layer conditions coupled with the effects of thallus colour (Kershaw 1975*a*), and this has been discussed in Chapter 1 (see Fig.6, p. 8). However, the advantages of a dark thallus in winter can be offset by correspondingly high thallus temperatures in the summer, and this is clearly seen in the temperature data. In this sense it is not surprising that *P. disjuncta* is adapted to tolerate levels of thermal stress that are a normal component of its summer environment but which are remarkable at first sight for a low-arctic species.

The effects of thermal stress are not restricted to reduction of net photosynthetic capacity or respiratory rates, and the effects of thermal stress on nitrogenase activity have been briefly discussed previously in relation to the marked summer decline in rates reported by MacFarlane and Kershaw (1977) (Figs. 78 and 79, p. 112). This decline in nitrogenase activity was interpreted as the result of high summer temperatures in the field. MacFarlane and Kershaw (1980*a*) also examined experimentally the effects of thermal stress on nitrogenase activity in the contrasting species *Peltigera praetextata* and *P. rufescens*. The effect of up to 40 days of stress at 25, 35 and 45°C on nitrogenase activity in both species is given in Fig. 149. The control replicates which were stored at 25°C for 40 days showed rates which compare very favourably with the normal levels of nitrogenase

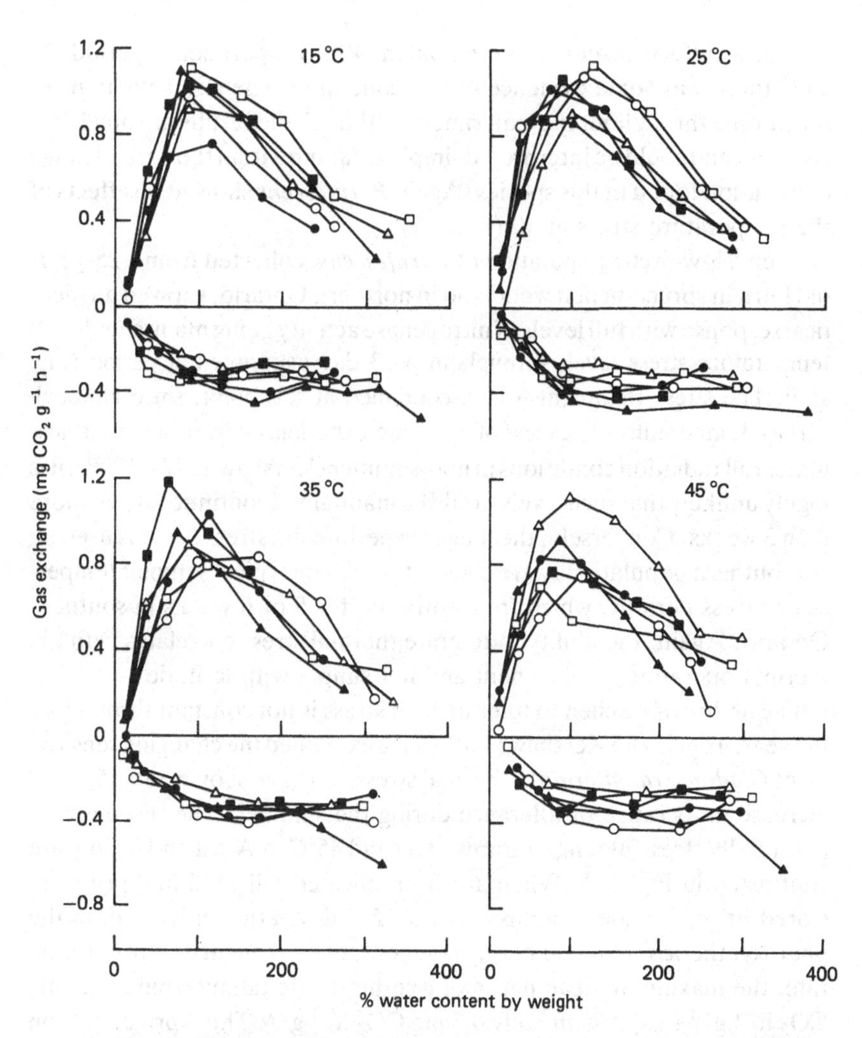

Fig. 148. The pattern of net photosynthesis and respiration in *Parmelia disjuncta* after varying periods of air-dry storage at 15 °C (controls) and at 25, 35 and 45 °C maximum daytime temperatures. ● day 1; ○ day 2; ■ day 4; ▲ day 7; △ day 14; □ day 22. Each point is a mean value for each 50% moisture class and maximum standard error bars are less than 0.2 mg CO_2 g^{-1} h^{-1}. (From Kershaw and Watson 1983.)

activity in *P. praetextata* and *P. rufescens*, i.e. 10-15 nmol C_2H_4 h^{-1} mg^{-1} and 5-6 nmol C_2H_4 h^{-1} mg^{-1} respectively (MacFarlane and Kershaw 1977). The activity of the *P. praetextata* replicates stored at 45°C, however, fell rapidly from 12 nmol C_2H_4 h^{-1} mg^{-1} on day 1 to 1.5 nmol C_2H_4 h^{-1} mg^{-1} by day 27, with only a trace of activity by day 34 and no activity at all by day 40. In marked contrast the equivalent replicates of *P. rufescens* showed a uniform response throughout the 40 day experimental period. At 35°C there was some evidence of a decline in nitrogenase activity in *P. praetextata* throughout the experiment, although the results are much less clear-cut and could be interpreted simply as falling within the normal range of variation found in this species. Again *P. rufescens* showed no effects of the temperature stress at all (Fig. 149).

Even a low-arctic population of *P. rufescens*, collected from a 25-year-old burn in spruce-lichen woodland in northern Ontario, showed an identical response with full levels of nitrogenase activity being maintained at all temperature stress levels throughout a 23 day experimental period (Fig. 150). The stress treatment was discontinued at this point, since although surface temperatures in excess of 50 °C are experienced over burnt surfaces under full radiation conditions in mid-summer (Kershaw 1977*a*, 1978), it is highly unlikely that such levels would be maintained continuously for more than 3 weeks. Conversely, the longer experimental stress periods used on the southern populations were chosen to simulate typical summer temperature stress periods, which frequently run for 5 or 6 weeks in southern Ontario. Again, the ability to tolerate thermal stress correlates with the thermal operating environment and not simply with latitude.

The ability of a lichen to tolerate heat stress is not constant throughout the year. Tegler and Kershaw (1981) have examined the changing sensitivity of *Cladonia rangiferina* to thermal stress and have shown a 10-15 deg C increase in its range of tolerance during the summer. The responses to potentially stressful temperatures of 35 and 45°C in April and in July are contrasted in Fig. 151. When air-dry replicates collected in April were stored at 35°C under conditions of a 12 h day, after only 14 days the photosynthetic rate could barely compensate the concurrent respiration rate, the maximum mean net photosynthetic rate falling from *c.* 2.3 mg CO_2 h^{-1} g^{-1} to approximately 0.3 mg CO_2 h^{-1} g^{-1}. The April collection effectively represents the response in early spring, immediately following snow-melt, which can be sometimes delayed in the Muskoka region on Ontario until late April. By mid-summer the July replicates stressed at 35°C under conditions of a 12 h day developed a maximum net photosynthetic rate of *c.* 2.2 mg CO_2 h^{-1} g^{-1} after 14 days which, although below the initial summer maximum mean value of *c.* 3.0 mg CO_2 h^{-1} g^{-1}, could fall within the degree of variation often found in *C. rangiferina*.

There is a similar development of thermal tolerance in the replicates which were stored at 45°C under conditions of a 12 h experimental day. In the April collection, photosynthesis was virtually eliminated after 7 days of stress treatment. Furthermore, after an additional period of 7 days heat

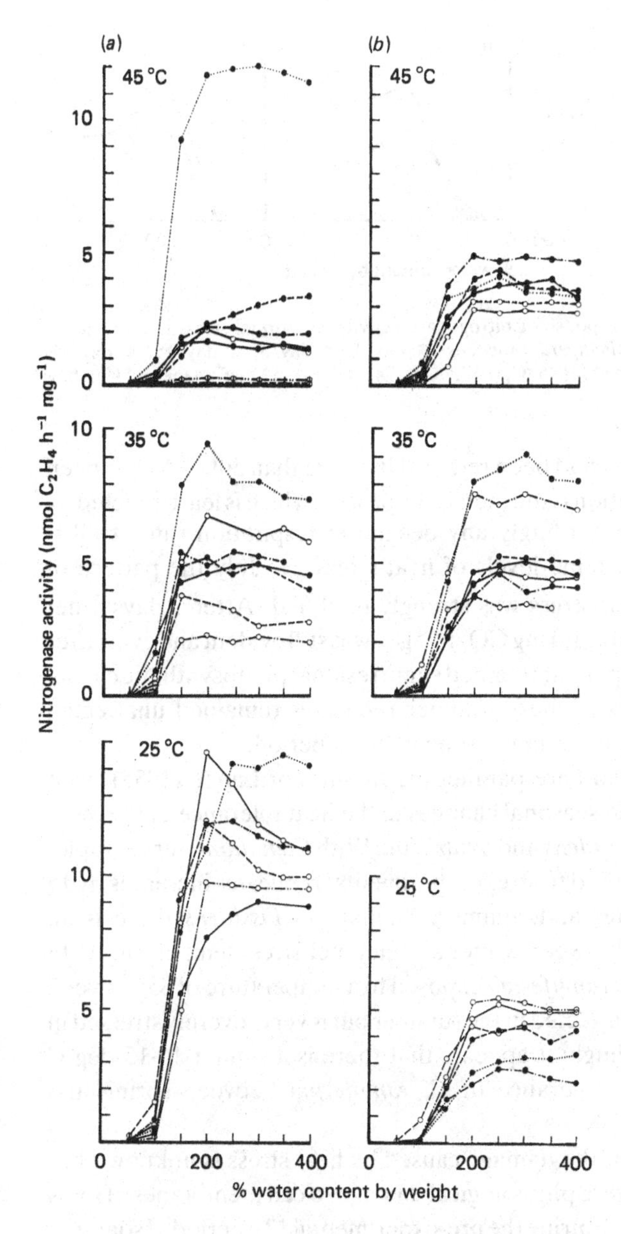

Fig. 149. The effects of thermal stress on the nitrogenase activity of *Peltigera praetextata* (*a*) contrasted with the effects on *P. rufescens* (*b*). Replicate thalli were stored air-dry at 45, 35 and 25 °C during a 12 h daytime period and at 15 °C at night. ●····● day 1; ●----● day 6; ●–● day 12; ●–·–● day 21; ○–○ day 27; ○–·–○ day 34; ○····○ day 40. (From MacFarlane and Kershaw 1980*a*.)

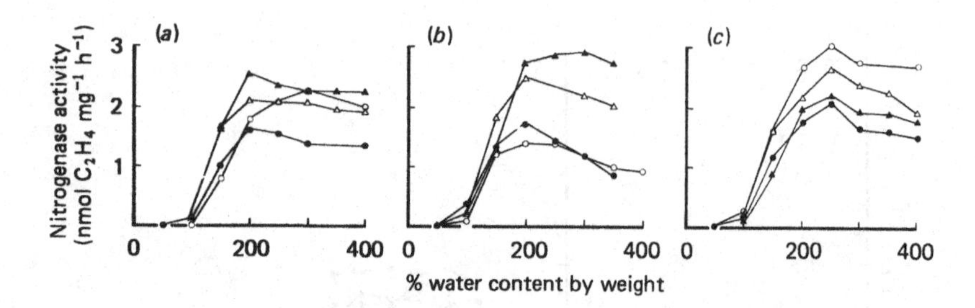

Fig. 150. The response of nitrogenase activity to thermal stress in a low-arctic population of *Peltigera rufescens*. ○ day 1; ● day 8; △ day 15; ▲ day 23. Experiment run at (*a*) 25 °C; (*b*) 35 °C; (*c*) 45 °C. (From MacFarlane and Kershaw 1980*a*).

stress even respiration had been reduced by more than 50%. As has been shown above, respiration is the metabolic process which is least affected by thermal stress and accordingly any decline in respiration rate at all is indicative of very extreme levels of heat stress. By July the pattern of response to 45°C heat stress was strongly modified. After 7 days a net photosynthetic rate of *c*. 1.4 mg CO_2 h^{-1} g^{-1} was still evident and even after 21 days the algal component retained some residual photosynthetic capacity. Respiration rates in these summer replicates remained unaffected throughout the 45°C experimental heat stress period.

The results reported here parallel the findings of Lange (1955), who demonstrated marked seasonal changes in the heat tolerance of the moss genera *Ctenidium*, *Fissidens* and *Syntrichia*. With *Ctenidium*, for example, the temperature required to stress experimental replicates increases by 15 deg C between winter and summer. Similary in *Fissidens* there is an increase of 10 deg C between winter and summer stress temperatures. In comparison *Cladonia rangiferina* exposed to a temperature of 35°C over 7 days is completely unaffected in the summer but is very severely stressed in early spring. Accordingly it appears that there is a similar 10-15 deg C increase in the heat tolerance of *C. rangiferina* between spring and summer.

The exact nature of the damage caused by heat stress is unknown but certainly includes severe physiological damage to cell membranes. This is evident in all replicates during the pre-experimental 12 h period of soaking, when there is both a marked coloration of the water and a characteristic odour. Such an extreme level of physical damage is almost certainly irreversible. However, this cannot as yet be verified, since effective long-term growth of lichen material has still to be achieved experimentally.

From the data so far available it appears that the most sensitive indicator

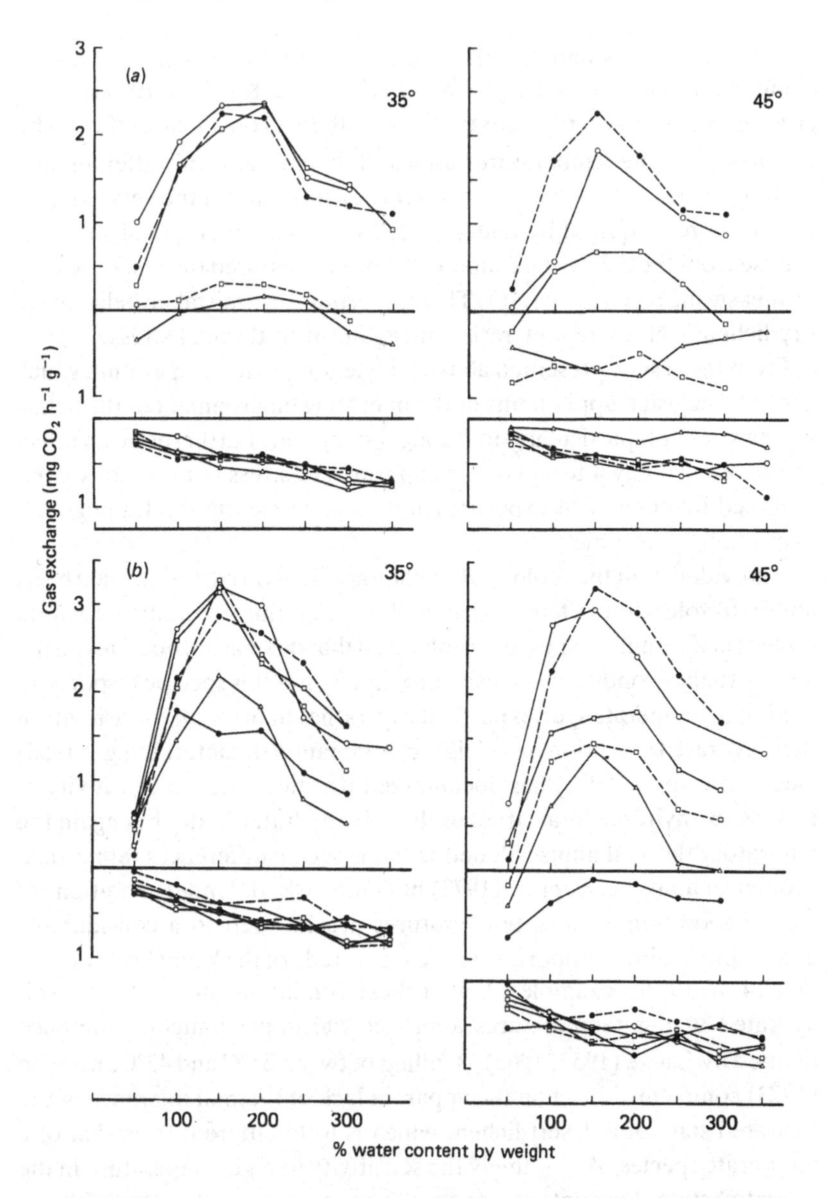

Fig. 151. Net photosynthetic and respiration rates as measures of thermal stress in *Cladonia rangiferina* collected (*a*) shortly after snow-melt in April, and (*b*) in July. Thallus replicates were stored air-dry at 35 °C or 45 °C. The response to these storage conditions as a potential level of stress was monitored as gas exchange rates after day 1 (●--●), day 2 (○–○), day 4 (□–□), day 7 (□- -□), day 14 (△–△) and day 21 (●–●). (From Tegler and Kershaw 1981.)

of thermal stress is photosynthesis, the capacity for nitrogenase activity being maintained much longer (MacFarlane and Kershaw 1980*a*). Respiration is even more resistant to stress, with rates only being affected by the most extreme temperatures used and then usually only after an integrated period of 2 to 3 weeks. This certainly accounts for the very extreme temperatures required by Lange (1953) to obtain any thermal stress response from the lichens he examined, since he assessed the response only by measuring respiration rates. This has resulted in a mistaken belief that a dry lichen is an extremely resistant organism to thermal stress.

From the results presented above it is clear that any temperature which exceeds the usual upper limits of the operating environment of the lichen can be stressful, particularly to the algal symbiont. Furthermore an effect can be induced by a few extreme exposures to stress but is readily compounded by number of exposures at a more moderate level, integrated over a period of time.

It is evident that the ecology of a lichen species is in part controlled by its ability to tolerate the temperatures of its operating environment within quite specific limits. It must be emphasised that this control operates on the air-dry thallus condition and in all probability it is this specific response in arctic and temperate species particularly, rather than thermal sensitivity in their hydrated condition, which is the dominant parameter. Lange (1953) (see also Kappen 1973) has documented the much greater sensitivity of lichens to environmental stress in the fully hydrated state, but again the laboratory thermal limits will define interspecific differences rather than ecological limits, as Kappen (1973) has emphasised. Under conditions of free evaporation, thallus temperatures are lowered to a considerable extent and in direct proportion to the magnitude of the latent heat flux (cf. Fig. 14, p. 19, for example). Under these conditions rarely, if ever, will hydrated thallus temperatures approach the upper limits of tolerance defined by Lange (1953, 1965) as falling between 35°C and 43°C. Kappen (1973) comments in fact on the apparent lack of thermal tolerance of the hydrated state of a desert lichen, which is little different from that of a temperate species. Accordingly the sensitivity to high temperature in the hydrated state does not appear to interact to any great extent with the ecology of a lichen in most instances. However, in tropical regions, where continuously high levels of humidity are a regular feature of the environment, exceptions to this generalisation may well occur. Equally the response matrix of *Peltigera rufescens* (Fig. 169, p. 254) also points to some exceptional situations found in temperate regions.

7.3 The lichen response to low-temperature stress

Kappen (1973) has summarised the early information available on the level of resistance to low temperatures found in lichens. Kappen and Lange (1972) showed that *Ramalina maciformis*, *Umbilicaria vellea* and *Roccella fucoides*, which colonise very contrasting environments (hot-desert, low-arctic and maritime habitats respectively), all withstood cooling of their desiccated thalli to −196°C and showed normal carbon dioxide uptake almost immediately following rehydration. Kappen (1973) summarises other documented examples of the marked tolerance of lichens, in the dry state, to extremely low temperatures, and it perhaps seems reasonable to accept that *most* lichens when dry can tolerate low temperatures. However, a similar general interpretation was also made of the resistance to high-temperature stress and perhaps a note of caution should be expressed, since as yet there is still only a very limited data base available on low-temperature stress effects.

The problem is a complex one since any experimental manipulation has to control for the rate of freezing, the level of stress to be applied, the length of the stress period, the rate of thawing and finally a realistic estimate of physical and physiological damage. In addition most manipulations have utilised freezing in the dark and since such experiments are extended over a time period there is a serious lack of a natural photoperiod. Thus, the natural combination of low temperature and short days in the field rapidly induces the uncoupling of photosynthetic units (PSUs) in several lichens (MacFarlane *et al.* 1983; Kershaw *et al.* 1983). This is evident as a marked decline in photosynthetic capacity during the winter. The same response can be extended well into the natural summer by coupling a short-day treatment with a low temperature, and it is important to distinguish between this response and a potential freezing stress response where capacity is irreversibly affected by low temperature itself rather than continuous darkness as a 'short' day.

Larson (1978) examined the effects of 3½ years of air-dry storage on *Alectoria ochroleuca* at −60°C. The data show that there is no significant change in the pattern of either net photosynthesis or respiration except at 21°C and 150 μmol m^{-2} s^{-1} (Fig. 152). Since *A. ochroleuca* has been shown to be capable of photosynthetic acclimation to temperature (Larson and Kershaw 1975*c*,*d*) this discrepancy at 21°C could be explained simply as a further response at this particular temperature (cf. Fig. 116, p. 174). These results suggest that, at least for *A. ochroleuca*, plants can be collected in the field and held frozen in the dried state for extended periods of time.

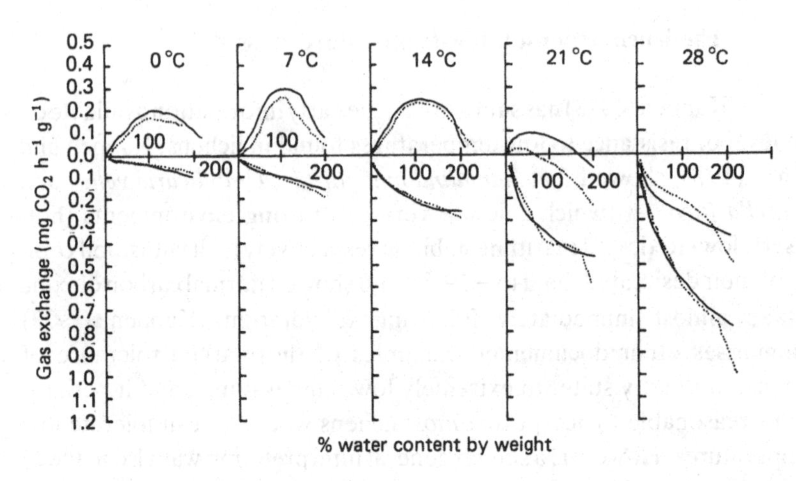

Fig. 152. The response of net photosynthesis (at 150 μmol m^{-2} s^{-1}, upper series of curves) and respiration (lower curves) to thallus temperature and moisture content for lichens freshly collected (continuous lines) in October 1974 (for photosynthesis) and April 1974 (for respiration) and for lichens collected in September 1973 and then stored for 3½ years at −60 °C (dotted lines). The 95% confidence interval amounts to 0.075 mg CO_2 h^{-1} g^{-1} and is shown in the 7 °C panel. Each curve is the mean of five replications. (From Larson 1978.)

Implicit in this statement, however, is the tenet that the photosynthetic and respiratory response subsequently examined may indeed reflect the winter metabolic condition. This is also seen very clearly in the extensive comparative data for a range of *Umbilicaria* species examined by Larson (1983*a*).

Replicates were collected in both (late) summer (July-August) and in mid-winter and frozen either dry or fully hydrated for 9 months at −25°C. Subsequently replicates were returned to laboratory temperatures and then soaked for 1 h in water adjusted to the requisite experimental temperature. Freezing damage was assessed by examining net photosynthetic and respiratory responses at a combination of four light and three temperature levels and comparing data with the normal response matrix for the species (Larson 1980). Subsequently the permanence of the stress effects were examined over a recovery period of 28 days. The results for *Umbilicaria vellea* are particularly interesting. The normal summer/winter respiratory comparison showed a marked decrease in respiration rate in the winter. This has been discussed previously above. The same pattern is evident in the summer replicates frozen dry, where respiration rates declined significantly and apparently the treatment at low temperature in the dark is sufficient to induce the normal winter response. The winter replicates, frozen dry, showed typical winter rates which are not significantly different from those of the winter controls. In marked contrast the replicates frozen when fully hydrated, irrespective of when they were collected, exhibited a massive increase in respiration rate and Larson (1983*a*) shows that in fact this represents severe irreversible damage with coloured pigments actually eluted from the experimental material. This confirms the previous data of Kappen and Lange (1972) who showed that European populations of *U. vellea* were also rather more sensitive to freezing stress than other lichen species.

Larson (1983*a*) also shows that the normal increase in respiratory activity in the winter, together with the correspondingly small decrease in net photosynthetic rate characteristic of *U. deusta*, could be induced in the summer replicates by the dry freezing treatment. *U. deusta* is seriously stressed, however, when frozen in the hydrated state in summer. The winter replicates, on the other hand, showed no significant change after 9 months of storage frozen either in an air-dry condition or when fully hydrated. If anything, there was a slight increase in net photosynthetic rates at low temperatures, which further emphasises the potential which some lichens have for low-temperature capacity adjustments.

The careful and detailed data set given by Larson (1983*a*) shows that in fact *U. vellea* and *U. deusta* are two extremes of a continuous range of response seen in the other species of *Umbilicaria* he examined. It seems clear, from both the data for *Alectoria* and *Umbilicaria* and the general summary provided by Kappen (1973), that freezing air-dry thalli even to very low temperatures and for extended periods of time does not materially damage either the phycobiont or mycobiont. What is important, however, is that if the species exhibits any characteristic winter capacity changes, then cold storage will probably induce these responses, certainly if the material is collected in mid- to later summer. The dark experimental conditions will presumably suffice as a short-day treatment which, when coupled with low temperatures, has now been shown to be quite effective in forcing an out-of-season response in a number of other species (see discussion above). Such responses should, however, be reversible with a correct storage temperature coupled with the requisite daylength, and accordingly freezing in an air-dry state perhaps offers the easiest long term method of storing experimental material, provided the necessary experimental manipulation of any seasonal capacity changes has been previously defined.

The response of a fully hydrated lichen to freezing temperatures is quite different and it seems clear that there is a continous range of response correlating with the ecological preferences of a species and the specific nature of its operating environment (as characterised by *Umbilicaria vellea* and *U. deusta* above: Larson, 1983*a*). It is quite probable that other species may be shown to be even more sensitive than *U. vellea* and perhaps damaged by even 'chilling' rather than freezing temperatures, particularly if such an event occurs in the early summer. Certainly the data for *U. deusta* emphasise the considerable difference in sensitivity to freezing between winter and summer replicates. This, when added to the documented examples of high-temperature stress and to the seasonal capacity changes seen either as gas exchange response curves or as PI curves, further emphasises the potentially ubiquitous occurrence in lichens of internal rhythms triggered by environmental signals.

8

Respiration and growth

Despite the essential nature of respiration for the growth and maturation of a plant, in virtually all the ecophysiological studies of the past 20 years, respiration rates have been only arbitrarily and routinely documented. We still only have a very poor understanding of the significance of respiration in the ecology of a species. Respiration has largely been viewed as a negative process, partly as a result of the rather naive energy-flow models popular some 10 years ago. Carbon loss via respiration was virtually regarded as a drainage of essential units pouring in an uncontrolled way into an infinite sink. Clearly, the balance between respiration and photosynthesis is indeed important to the well-being of a plant but many vital metabolic processes also function through the provision of energy via respiration.

In the limited number of ecophysiological studies on lichens that are available, an equally cursory examination of respiration has usually been made. Lange (1953) appears to have recognised the fundamental importance of respiration and used it as a measure of stress response to high temperatures. Otherwise respiration has been recognised as an important component of net photosynthesis at high levels of thallus hydration, an important source of energy during dark fixation of nitrogen, but again only important ecologically as a carbon loss. Smith and his co-workers have critically documented the range of carbohydrates appearing as photosynthetic products in lichens and the wide-ranging importance of mannitol as a fungal respiratory substrate; this has been well reviewed by Richardson (1973) and Smith (1980). As yet, though, we have no specific information as to the proportion of respiration attributable to each of the bionts and, as Farrar (1973) points out, neither do we have any idea of the specific respiratory pathways in lichens or the range of substrates, other than mannitol, which may be involved. The available data base is very restricted and the focus here will be to examine the environmental control of rates of respiration and particularly those situations where respiratory losses have been implicated as being significant to the ecology of a specific group of lichens.

8.1 Resaturation respiration

Ried (1960*b*) demonstrated a pronounced increase in the respiration rate immediately following thallus rehydration in a number of lichen species. He also showed an apparent correlation between the size of this respiratory burst and the corresponding habitat characteristic of different lichen species. The respiratory burst was apparently more pronounced in lichens from mesic habitats and this relationship has since been reported by Smith and Molesworth (1973) and Farrar and Smith (1976). Dilks and Proctor (1974) and Krochko, Winner and Bewley (1979) have also used the magnitude of the respiration burst as a measure of drought stress. Smith and Molesworth (1973) define three distinct stages that are present in the complete sequence of events in *Peltigera polydactyla*: The first minute is characterised by a large and rapid, non-metabolic release of carbon dioxide. This is followed by an extensive period of resaturation respiration (the respiration burst of Ried 1960*a*). They report that this phase initially represents a five-fold stimulation of respiration which declines gradually over a period of 10 h. The final phase of the whole sequence, basal respiration, then remains constant with time. Farrar (1973) suggests that the wetting burst is probably physical in origin and he subsequently has confirmed this, showing that dead thalli and dried filter paper exhibit the same burst of carbon dioxide. Brown, MacFarlane and Kershaw (1983) have further examined the wetting burst by drying experimental replicates of *Peltigera polydactyla* in helium or carbon-dioxide-free air. The rapid burst of carbon dioxide production in *P. polydactyla* was restricted to the first 2 min following imbibition of water (Fig. 153). The rate of emission of carbon dioxide from the thalli that had been dried in air was 4.35 μmol CO_2 g^{-1} min^{-1} in the first 30 s, declining very steeply over the next half minute. In the replicates dried previously in a stream of commercial-grade compressed air (which contains only a residual trace of carbon dioxide) the immediate carbon dioxide emission following hydration was almost entirely eliminated, and drying in helium resulted in the complete absence of any carbon dioxide burst in the first 30 s. During the burst, the start of dark respiration as measured by oxygen uptake was immediately apparent in the replicates dried in carbon-dioxide-free air, and was also evident in the pattern of carbon dioxide production.

After the physical release of carbon dioxide following the initial 30 s of imbibition, rates of carbon dioxide production reached 2.39 μmol CO_2 g^{-1} min^{-1} within 5 min, declined rapidly over the first 2 h to 0.85 μmol CO_2 g^{-1} min^{-1} and subsequently declined very slightly over the next 20 h (Fig. 154). Oxygen uptake closely parallels the pattern of carbon dioxide production,

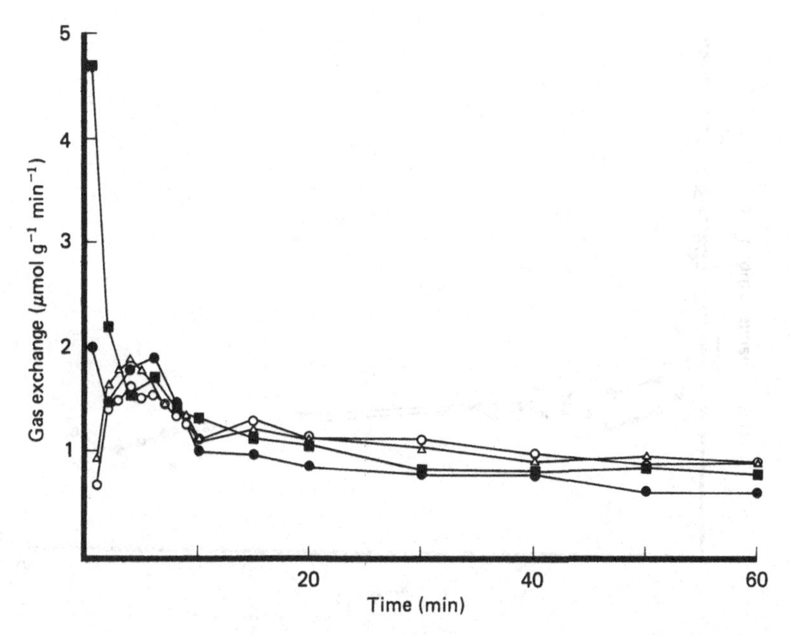

Fig. 153. The immediate effects on the wetting burst and resaturation respiration of drying *Peltigera polydactyla* in carbon dioxide-free air (○, oxygen uptake; ●, carbon dioxide evolution) or in air (△, oxygen uptake; ■, carbon dioxide evolution). (From Coxson *et al.* 1983*a*.)

with initial rates of 1.82 μmol O_2 g^{-1} min^{-1} being developed within 2 min. Again these declined after 2 h (to 0.6 μmol O_2 g^{-1} min^{-1}) and then subsequently fell only slightly over the following 24 h experimental period. Smith and Molesworth (1973), however, report an apparently greater level of resaturation respiration, also using *Peltigera polydactyla*. The discrepancy between their results and the data presented by Brown *et al.* (1983) probably lies partly in the choice of an experimental control and partly in the greater sensitivity and flexibility of an oxygen electrode. Thus Brown *et al.* (1983) contrast resaturation respiration with the basal rate after 1 h (cf. Fig. 153 above) rather than a rate taken after 20 h (Smith and Molesworth 1973), when there is continued decline of respiration, possibly as carbon reserves are slowly depleted. The resultant comparison then yields an apparently more substantial level of resaturation respiration. Considerable emphasis has been placed on the potentially serious losses of carbon as a function of resaturation respiration in environments where alternating wet and dry periods occur frequently (Farrar 1976*a*,*b*; Farrar and Smith 1976).

Resaturation respiration, however, presents a rather varied picture and

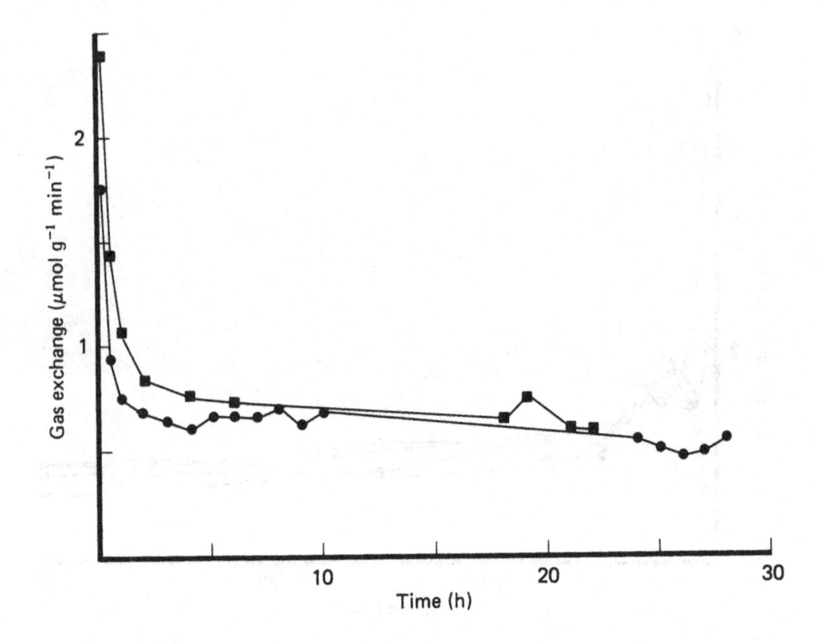

Fig. 154. Carbon dioxide evolution (■) and oxygen uptake (•) in *Peltigera polydactyla* over a 24 h period following resaturation. (From Coxson *et al.* 1983*a*.)

different lichen species appear to show different rates of resaturation respiration which last for different lengths of time. *Chondropsis* follows the *Peltigera* pattern, with resaturation respiration peaking at 3 min and basal respiration established by 40 min (Rogers 1971). Species of *Umbilicaria* as well as *Hypogymnia physodes* also complete the resaturation phase in *c.* 40 min (Farrar and Smith 1976; Larson 1979). Increased respiration may last for up to 2 h in *Xanthoria* (Smith and Molesworth, 1973), 3 h in *Cladonia stellaris* and *C. evansii* (Lechowicz 1978) and 5 h in *Cetraria cucullata* (Lechowicz 1981). It becomes difficult with such a range of resaturation values to estimate the impact of frequent wetting and drying cycles. This was particularly emphasised by Brown *et al.* (1983) who demonstrated that there is also a very marked seasonal component to the size of the resaturation peak in *Peltigera polydactyla*. Resaturation respiration was seen repeatedly in *P. polydactyla* when experimental material collected frozen or saturated from under snow in December and February and during snow-melt in April was subsequently being dried under storage conditions in a growth chamber. However, in the experimental material collected in May, following a more extensive dry period, the peak of resaturation respiration was completely absent. Oxygen uptake simply

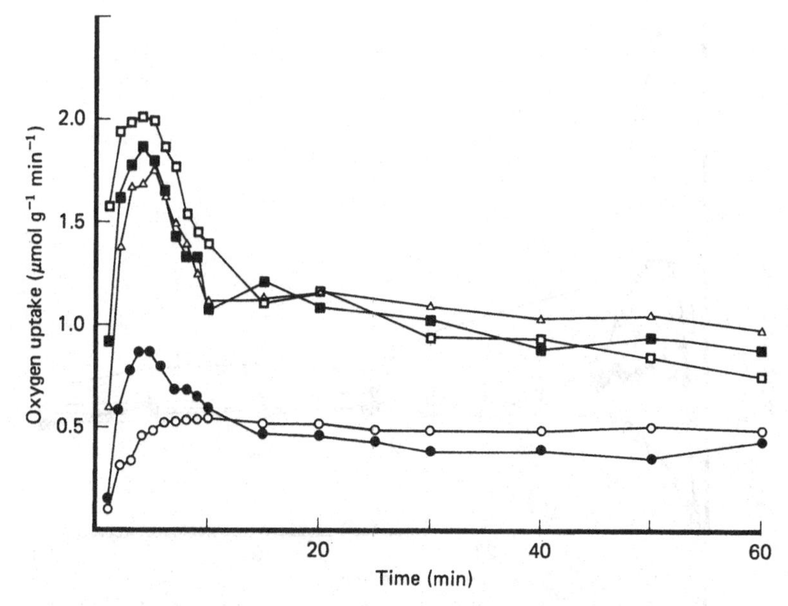

Fig. 155. The seasonal pattern of oxygen uptake following resaturation in *Peltigera polydactyla*. □, December; ■, February; △, April; ○, May; ●, June. (From Coxson *et al.* 1983*a*.)

rose gradually to 0.5 μmol g^{-1} min^{-1} over the first 5 min of hydration and then remained constant for the subsequent experimental period of 1 h. However, by June, material collected following a 2 week period of wet weather showed a partial recovery of resaturation respiration activity (Fig. 155).

Two major environmental interactions were potentially involved in this winter-spring sequence: the length of time the lichen thalli were saturated in the field and the differential rates of drying to which they were subsequently exposed. The experimental combination of length of thallus saturation with drying rate is given in Fig. 156. Extension of the period of full thallus hydration from 3 h to 3 days, coupled with subsequent fast drying, increases the resaturation respiration rate from 0.88 to 2.43 μmol O_2 g^{-1} min^{-1}. This is somewhat higher than the rates found in material collected during snow-melt and allowed to dry under storage conditions in the growth chamber. Different combinations of soaking and rate of drying produce intermediate results. A 3 day soak followed by an intermediate drying time of 5 h induces a resaturation respiration rate which is equivalent to a short 3 h soak coupled with fast drying. The 7 day presoak coupled with an intermediate drying rate generates a resaturation respiration peak of 1.32 μmol O_2 g^{-1} min^{-1}, and this suggests that the length of the

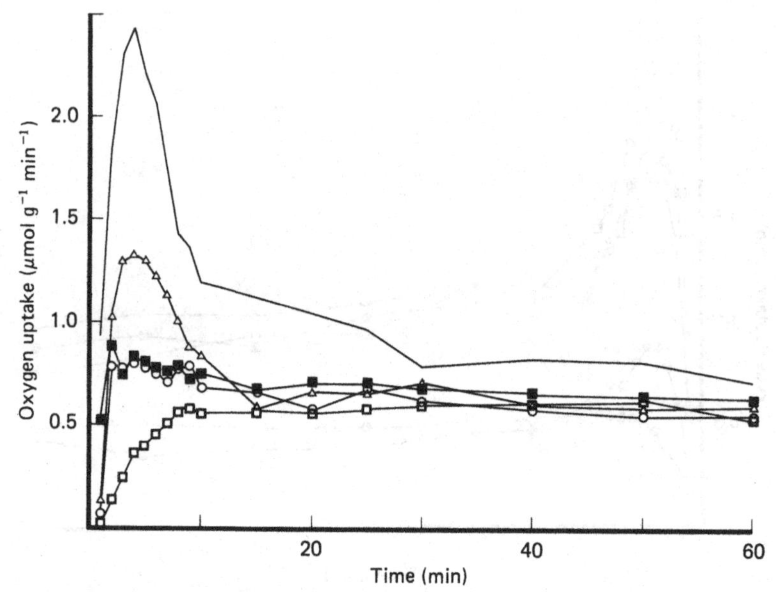

Fig. 156. Oxygen uptake following resaturation in *Peltigera polydactyla*. Replicate thalli were maintained fully saturated for either short or long periods and then dried either rapidly or slowly. ■–■, 3 h wet/1 h drying; □–□, 3 h wet/5 h drying; —, 3 days wet/1 h drying; ○–○, 3 days wet/5 h drying; △–△, 7 days wet/5 h drying. (From Coxson *et al.* 1983*a*.)

hydration period is more important than the speed of drying. Light, and hence the provision of an immediate supply of respiratory substrate, does not appear to be involved, since neither set of replicates held at thallus saturation for 3 h under either 250 or 850 μmol m^{-2} s^{-1} illumination and then dried at an intermediate rate over a 5 h period showed any resaturation respiratory peak on rehydration.

The dependence of resaturation respiration on prior events in the field is of considerable interest and significance in any consideration of carbon balances, but mechanistically is extremely puzzling. A fast drying rate produces a higher resaturation respiration rate than does slow drying, a pattern which has also been reported in *Tortula ruralis* (Krochko *et al.* 1979). However, the length of time over which the lichen remains fully saturated more profoundly influences resaturation respiration and is largely responsible for the very marked seasonal effects reported by Brown *et al.* (1983) for *Peltigera polydactyla* collected in southern Ontario. Significantly, though, throughout the year there is usually a large stochastic weather component in many temperate regions and to predict the extent of carbon loss from limited laboratory studies is impossible. Accordingly, the deleterious effect of alternating wet and dry periods on the overall carbon balance of a lichen must be questioned. Certainly the resaturation losses in *Peltigera praetextata* growing in woodland in southern Ontario are not likely to be significant in comparison with the high rates of photosynthesis generated in spring, summer and autumn, irrespective of canopy conditions (Kershaw 1977*a*,*b*; Kershaw and MacFarlane 1980). Conversely other species may be more susceptible, but a careful analysis of the carbon balance of the lichen as well as the thermal, light and moisture operating environment would be required on a seasonal basis.

The potential significance of resaturation respiration and the resultant increased carbon demand on the phycobiont leading to a poor seasonal carbon balance has also been exaggerated in relation to the reported losses of photosynthate from the phycobiont. Thus, Drew and Smith (1967) and Chambers, Morris and Smith (1976) report that 62% of the fixed ^{14}C in *P. polydactyla* appears in mannitol after 45 min of fixation, and 72% after 17 h. Farrar (1978) estimates that 90% of all the carbon fixed in photosynthesis passes to the fungus in lichen containing either blue-green algae or *Trebouxia*, and Smith (1980) concludes that photosynthate efflux from lichen algae is of massive proportions. These estimates, however, are derived from glucose and ribitol transfer rates at full thallus hydration, since the $^{14}CO_2$ label was derived from [^{14}C]bicarbonate solution. Certainly in *P. polydactyla* and *P. praetextata* 'massive' effluxes of glucose only occur at maximum levels of thallus hydration and during a full drying curve much of the photosynthate is retained by the phycobiont (Tysiaczny and Kershaw 1979; MacFarlane and Kershaw 1982). A drying cycle accordingly is an essential component of the carbon balance between mycobiont and phycobiont. Assuming similar environmental control of ribitol transfer from *Trebouxia*, the efflux at full thallus hydration becomes a modest proportion of the retained photosynthate at lower levels of thallus moisture and similarly resaturation respiration losses become a much less significant amount of this retained photosynthate. At the moment the overall evidence points to alternating wetting and drying periods as essential rather than ecologically stressful events for most lichens. The exceptions may occur in semi-aquatic species which are wet for rather more extensive time periods and, as Ried (1960*a*,*b*) and Brown *et al*. (1983) have shown, will thus have a fairly substantial resaturation burst on a continuous basis.

8.2 Respiration and thallus hydration

The pattern of respiratory response to the degree of thallus hydration is quite variable. Snelgar, Brown and Green (1980) examined a range of *Peltigera*, *Pseudocyphellaria* and *Sticta* species and demonstrated a full and remarkable range of response to thallus hydration. The basic response pattern most frequently observed by Lange (1969) in *Ramalina maciformis* shows a constant respiration rate over a wide range of thallus hydration levels, and this has prompted him to question the different patterns reported by other workers for a wide range of species (Lange 1980).

The shape of the response curve will reflect several interactive parameters: the specific level of thallus moisture at which a restriction on the biochemical and enzymatic events occurs, the interaction of this event with temperature, and the possible interaction of the thallus internal resistance with either oxygen diffusion in or carbon dioxide diffusion out. A sequence of respiratory/thallus moisture response curves taken from a range of species and sources is given in Fig. 157. *Peltigera dolichorhiza* (Snelgar *et al.* 1980), *P. aphthosa* (Kershaw and MacFarlane 1980) and *Ramalina maciformis* (Lange 1980) show a fairly constant respiration rate with decreasing thallus moisture, that then declines linearly at low levels of hydration. In *P. dolichorhiza* the final decline in respiration is not until quite low levels of thallus hydration have been reached, whereas the decline in photosynthesis seems to be more rapid. As a result, net photosynthetic rates at low levels of thallus moisture are dominated by the still active thallus respiration (see also Fig. 102). In *Pseudocyphellaria billardieri* and *Sticta caperata* (Snelgar *et al.* 1980) there is an initial constant rate of respiration which then falls in parallel with the declining net photosynthetic rates.

In all these cases the species grow in very mesic environments, and have thalli with low internal diffusive resistances. In contrast, however, Lange (1980) reports an identical pattern of response in *Ramalina maciformis*, a desert species, which has a high internal diffusive resistance (see above). Equally paradoxical is the linear decline in respiration rates as the thallus dries out which is often found in those species that have high internal diffusive resistances and markedly depressed net photosynthetic rates at full thallus saturation. Typical examples are reported for *Parmelia caperata*, *P. sulcata* and *P. physodes* (Harris 1971); for *Cladonia mitis*, *C. rangiferina* and *C. uncialis* (Lechowicz and Adams 1974), for *Alectoria ochroleuca* (Larson and Kershaw 1975*d*); for *Bryoria nitidula* and *Cladonia stellaris* (Kershaw 1975*b*); and for *Usnea* and *Pseudocyphellaria colensoi* (Snelgar *et al.* 1980). Several of these examples are included in Fig. 158. However, the pattern of respiratory response to thallus hydration can also be profoundly influenced by temperature and this is very evident in the data for *Peltigera polydactyla* and *P. praetextata* given by Kershaw

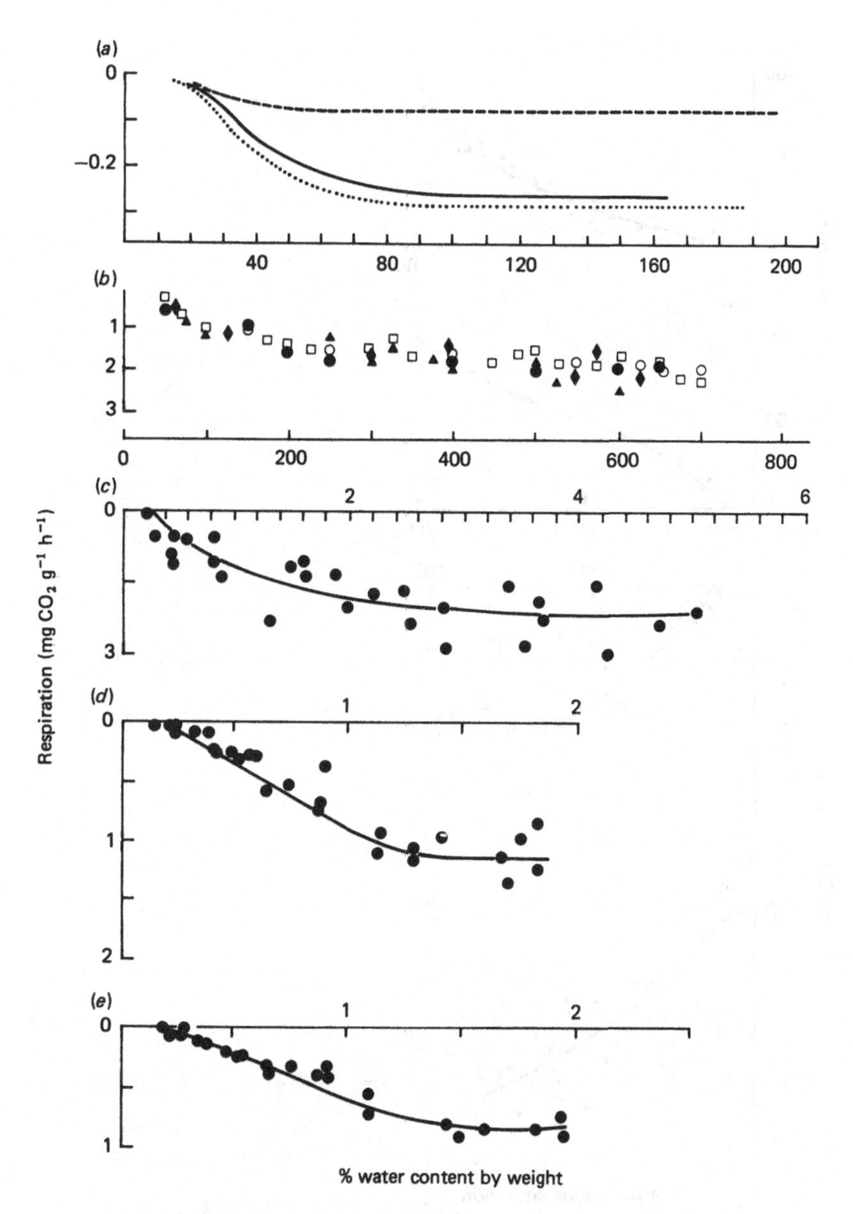

Fig. 157. Respiration rate as a function of thallus hydration, where the rate remains constant over the initial dehydration period. (*a*) *Ramalina maciformis*, (*b*) *Peltigera aphthosa*, (*c*) *Peltigera dolichorhiza*, (*d*) *Pseudocyphellaria billardieri*, (*e*) *Sticta caperata*. The lines and symbols in (*a*) and (*b*) represent replicates at different temperatures. (From (*a*) Lange 1969; (*b*) Kershaw and MacFarlane 1980; and (*c*,*d* and *e*) Snelgar *et al.* 1980, respectively.)

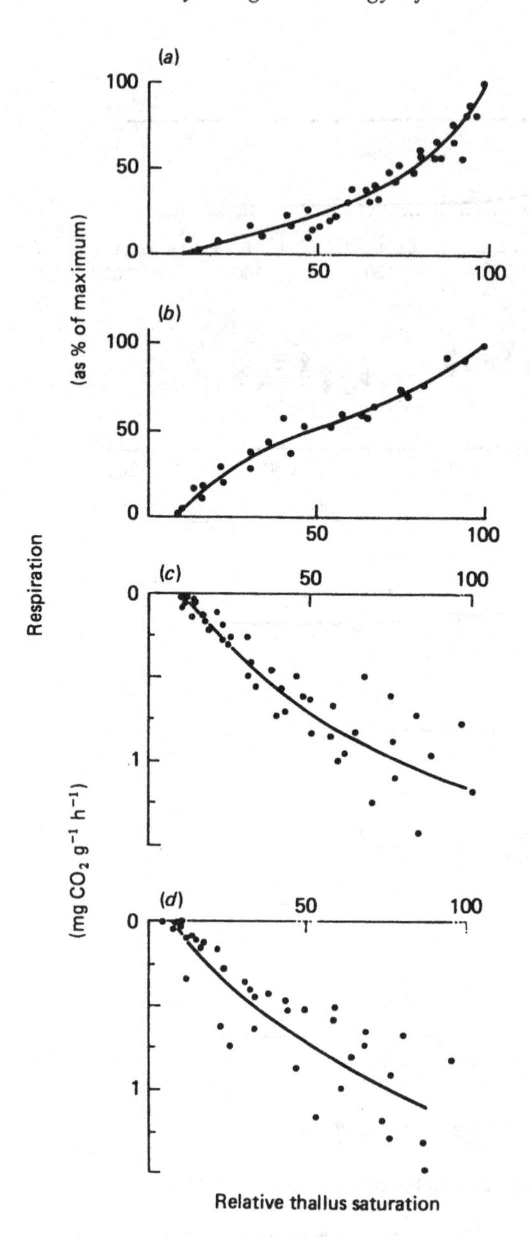

Fig. 158. Respiration in relation to the degree of thallus hydration in a number of species where the rate falls almost linearly as the thallus dehydrates: (*a*) *Parmelia caperata*; (*b*) *P. sulcata*; (*c*) *Pseudocyphellaria colensoi*; (*d*) *Usnea* sp. (From Harris, 1971; and Snelgar *et al*., 1980.)

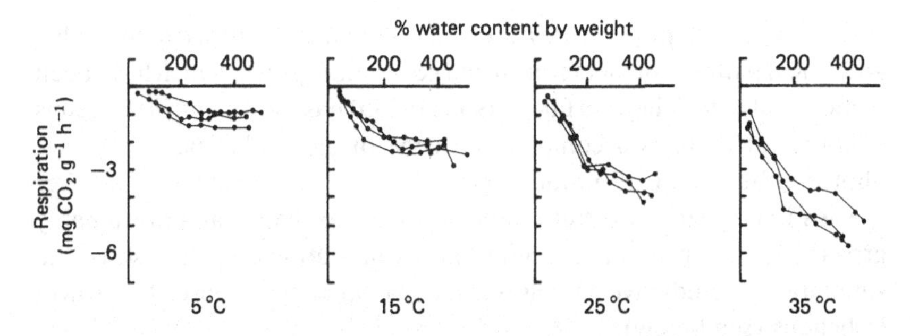

Fig. 159. The pattern of respiration in *Peltigera praetextata* where degree of thallus hydration interacts with temperature. Each line represents a different replicate. (From Kershaw 1977*a*.)

(1977*a*,*b*, Fig. 159). There is an almost constant respiration rate with decreasing thallus moisture at 5 °C, an intermediate response at 15 °C, but a steep linear decline at 35 °C.

At present an exact interpretation of these variable response patterns is uncertain and equally there does not appear to be any consistent correlation between the specific pattern of respiration in a species and its ecological preferences.

8.3 The interaction between respiration, thallus age and temperature

Kershaw (1977*a*) gives comparative net photosynthetic rates for young marginal lobes of *Peltigera polydactyla* which were compared with those for the older central parts of the thallus. The maximum mean rate of net photosynthesis in the young lobes is 4.2 mg CO_2 g^{-1} h^{-1} compared with 2.6 mg CO_2 g^{-1} h^{-1} for the older material (Fig. 102). The respiration rates show a similar correspondence ratio, with the young and old replicates generating 5.6 and 3 mg CO_2 g^{-1} h^{-1} respectively. Kershaw (1977*a*) explains the differences in terms of the anatomy of *P. polydactyla*, which has a much thicker medulla and more abundant dark hyphae on the underside of the older portions of the thallus. The dry weight of these older replicates is greater than younger replicates of an equivalent size and the gas exchange rates, when expressed on a dry weight basis, are accordingly less. These differences are large and can induce considerable scatter in net photosynthetic measurements unless careful control is maintained over the selection of replicates for each experiment. Nash *et al.* (1980) have examined an equivalent gradient of respiration rates in *Cladonia stellaris* and *C. rangiferina* where the very distinct vertical zonation of the phycobiont (Kärenlampi 1970; Kershaw and Harris 1971*c*) induces a corresponding gradient in photosynthesis but also an increase in respiration rate with increasing depth into the lichen mat. In marked contrast to the radial and apical patterns of metabolism and correlated growth which have been demonstrated in lichens so far, Larson (1983*b*) has shown that in the genus *Umbilicaria* there is a complete absence of any radial pattern in net photosynthesis and respiration rates. The potential existence of other species having similar diffuse patterns of gas exchange and presumably growth, is important. It certainly places limitations on the use of the simulation or predictive models that have so far been proposed for growth in lichens (see below).

The interaction of temperature with rate of respiration has been reviewed by Quispel (1960), Smith (1962) and Farrar (1973). Subsequent data have extended the information base but have not really added any new concepts. High values are consistently reported for the Peltigeraceae and Stictaceae (cf. Figs. 161 and 162) and there appears to be little correlation with latitude (Scholander *et al.* 1952; Scholander and Kanwisher 1959). However, no attempt was made by Scholander to relate the thermal operating environment to latitude and the possible existence of a correlation may have been obscured. For example, the comparison of metabolism in *Cladonia stellaris* and *Bryoria nitidula* (Fig. 142, p. 207) shows very different respiration responses to temperature, with the two species also occupying very contrasting thermal operating environments.

8.4 Seasonal respiratory changes

There is now abundant evidence of marked respiration changes in lichens on a regular seasonal basis. Kershaw and MacFarlane (1980) examined the increased density of photosynthetic units (PSUs) in *Peltigera praetextata* and *P. scabrosa* when stored under low light conditions, and this has been discussed above (Chapter 6). During the changes in PSU density brought about by low light, respiration rates in *P. scabrosa* increased significantly from 2.2 mg CO_2 h^{-1} g^{-1} to 3.5 mg CO_2 h^{-1} g^{-1} after storage at 20 μmol m^{-2} s^{-1}, but in contrast respiration in *P. praetextata* *decreased* from 3 mg CO_2 h^{-1} g^{-1} down to 1.8 mg CO_2 h^{-1} g^{-1}. Larson (1980) has also documented marked respiratory changes in *Umbilicaria* which are evident either as an increase or as a decrease in respiration rate depending on the species. Although Prézelin (1981) and Harris (1978) present a detailed review of photosynthetic events, neither has attempted to relate respiratory rates to the concurrent rate of photosynthesis. For example, a decrease in the light environment of *Peltigera scabrosa*, a low-arctic species, automatically signals a corresponding decrease in the thallus operating temperature. An increase in respiration rate may in part compensate for the otherwise reduced capacity for metabolism. Conversely the change of light level in *P. praetextata* in a temperate environment, at the reduced level of thallus operating temperature, would not limit metabolism to any great extent, and the reduction in respiration rate during the peak summer months could represent a similar levelling of metabolic capacity over the growing season. However, such an interpretation is highly speculative and the pattern of change in *Umbilicaria* species (which all grow in temperate Ontario) does not support such a concept, unless there is a very marked differential between the operating environments of each group of species.

Thus although such respiratory changes are potentially widespread in lichens and appear to be induced simultaneously either with changes in PSU density or by reversible coupling or uncoupling of PSUs their ecological and biochemical significance at the moment remain obscure. This is analogous to the situation in free-living algae where diurnal changes in both carboyhydrates and dark respiration rates of natural populations have been recorded. Ganf (1974) and Gibson (1975) have reported such changes in blue-green algae related directly to surface irradiance conditions. Similarly Harris (1973) reported diurnal, although smaller, changes in the rate of dark respiration of populations of green algae and diatoms taken from surface waters. Foy, Gibson and Smith (1976) have shown that daylength has a profound effect on the growth of a range of species of planktonic blue-green algae, and Harris (1978) concludes that internal regulation of carbohydrate metabolism and growth in algae requires further work. The respiratory changes in lichens could involve the phycobiont or the mycobiont or both partners, although the regularity of occurrence of change with photosynthetic daylength and temperature events strongly implicates the algal biont.

The situation in higher plants, however, is somewhat more encouraging. Simon (1979*a*), working with *Lathyrus japonicus*, has shown that the energy of activation (E_a) of the enzyme NAD malate dehydrogenase as well as its substrate-binding properties (K_m) (Simon 1979*b*) can experimentally be changed very rapidly. The changes in E_a in two cold-adapted clones of *Lathyrus* acclimated to 15/7 °C were contrasted with those in replicates acclimated to 30/22 °C. On day 16 the storage conditions of the two experimental sets of replicates were interchanged, causing an immediate response: E_a adjusted within 48 h and K_m yielded even more rapid results (Simon 1979*b*). Thus at least one of the major respiratory enzymes is capable of extremely rapid adjustment and certainly offers a sound explanation of summer/winter respiratory capacity changes.

Whether a similar set of enzyme responses is present in lichens remains to be seen.

8.5 Growth

Hale (1973) presents an excellent summary of the data available on lichen growth rates which has been more recently augmented by Armstrong (1976). Hale (1973) suggests that the growth pattern in lichens is restricted to centrifugal, marginal and apical growth; intercalary growth contributing to lobe elongation is minimal and is restricted to lobes 1-2 years old (Hale 1970). Growth in the older parts of the thallus is reflected in thallus thickness and occasionally in the production of vegetative propagules or apothecia. Fahselt (1976), for example, has shown that transplanting sections of *Parmelia cumberlandia* thalli to a more southerly location resulted in, in addition to an enhanced growth rate, an expansion of existing apothecia as well as the development of new ones. However, Larson (1983*c*) working with *Ramalina menziesii* has demonstrated a considerable potential in this species for unlimited growth, with intercalary growth playing a central role in the continuous expansion of the thallus nets. In the genus *Umbilicaria* there is also no defined centripetal growth and net photosynthetic and respiratory rates are uniform over the entire thallus. Dry weight increments are also uniform over the laminar portions of the thallus. In addition, however, there are substantial increases associated solely with the umbilicus, indicating marked centripetal growth.

Armstrong (1976) has summarised the changes in growth of a thallus throughout its life in terms of the different measures that have been employed to quantify centrifugal growth (Fig. 160). Two phases are evident, the first of which is an early non-linear phase, distinguished by a logarithmic increase in thallus radius, or by a decreasing radial growth rate (mm yr^{-1}) or by even more marked changes in (cm^2 cm^{-2} yr^{-1}). The second phase is linear and leads to fragmentation of the thallus in some species. It is distinguished by a constant increase in radius and hence a constant radial growth rate. When expressed in relative terms, however, a linear increase in diameter translates into a negative exponential decline in relative growth rate. Thus there are fundamentally different growth patterns in fruticose and foliose lichens. In some foliose species it also appears that there are a number of alternative intermediate growth patterns which can be characteristic for particular lichen species.

Kärenlampi (1971) examined the relative growth rates of a number of fruticose lichens in terms of a regression of daily rainfall and dry weight changes. His data for *Cladonia stellaris* show a highly significant regression

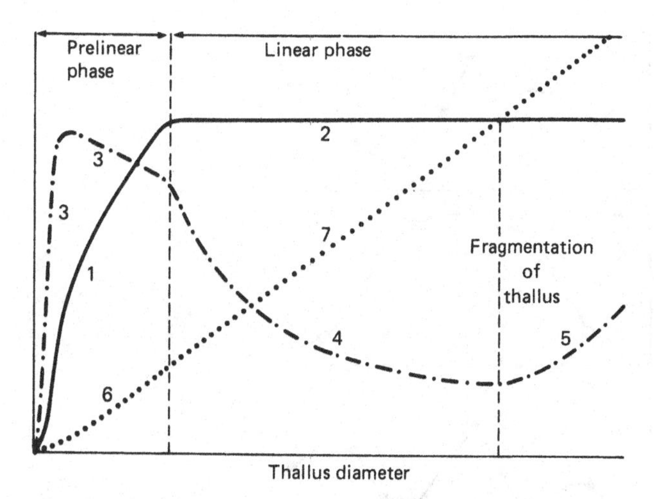

Fig. 160. A summary of changes in growth rate of a thallus through life and the corresponding growth phases: —, radial growth rate (mm yr^{-1}); —·—, relative growth rate ($cm^2\ cm^{-2}\ yr^{-1}$); ····, thallus radius (cm). 1, increasing radial growth rate in prelinear phase; 2, constant radial growth rate in linear phase; 3, changes in relative growth rate in prelinear phase; 4, negative exponential decline in relative growth rate in linear phase; 5, increasing relative growth rate after fragmentation of the thallus; 6, logarithmic increase in radius in prelinear phase; 7, constant increase in radius in linear phase. (From Armstrong 1976.)

for overall dry weight gain, with very marked apical growth when examined as sequential podetial lengths. There has been general agreement that rainfall exerts overall control over the growth of a lichen, reflecting the poikilohydric nature of lichen metabolism. As a result, since rainfall is often seasonally distributed, there is a corresponding distribution of growth rates, probably modified by the seasonal distribution of temperature. Hale (1970), for example, shows a very marked growth period in *Parmelia caperata* during May, June and July which probably represents the combination of rainfall distribution and warm temperatures. However, the most outstanding gap in our knowledge of lichens is reflected in our inability to grow replicates under controlled conditions, and as yet only very rudimentary information is available on the interaction of other environmental variables with moisture.

Armstrong (1973, 1977, 1982) in a sequence of elegant transplant experiments has confirmed, though, the integrated effects of rock aspect on the growth of several lichen species. The correlation between short-wave radiation and rainfall on the growth of *Physcia orbicularis*, *Parmelia conspersa*, *P. glabratula* subsp. *fuliginosa* and *P. saxatilis* is given in Fig. 161 and calls for little comment. Thin slates bearing colonies of these four

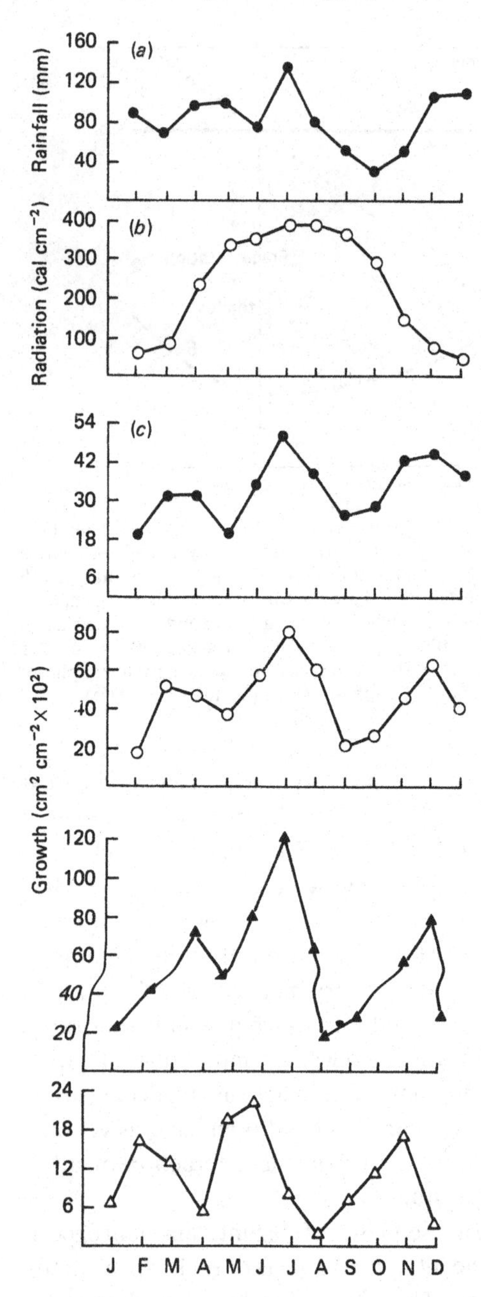

Fig. 161. The seasonal distribution of (*a*) rainfall and (*b*) solar radiation with (*c*) the corresponding growth patterns of *Physcia orbicularis* (●), *Parmelia conspersa* (○), *P. glabratula* subsp. *fuliginosa* (▲), and *P. saxatilis* (△). (From Armstrong 1973.)

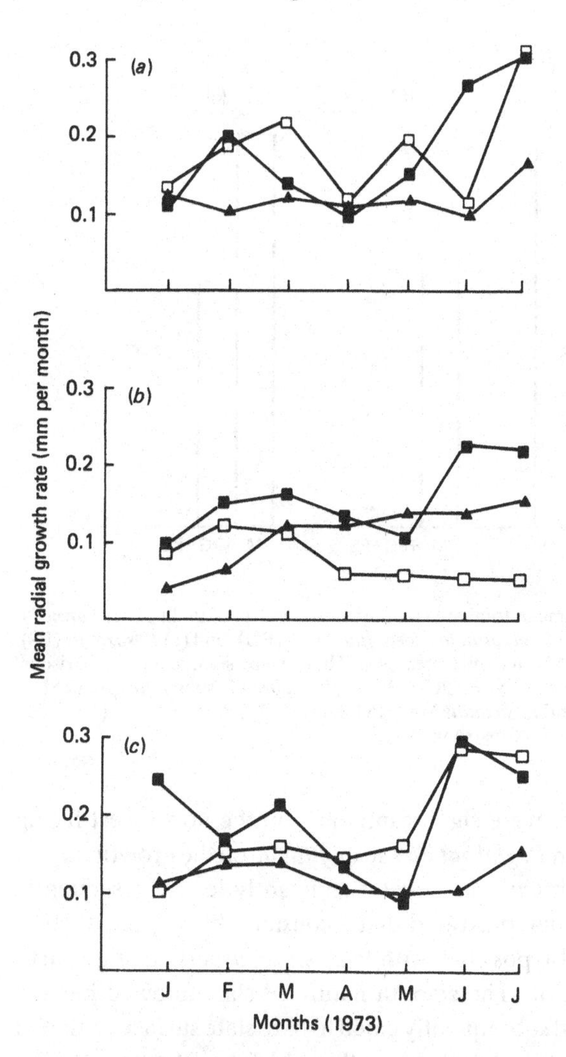

Fig. 162. The radial growth rate of thalli placed on horizontal boards (■) and after transplantation to south-east facing rock surfaces (□) and north-west facing rock surfaces (▲). (*a*) *Parmelia conspersa*, (*b*) *P. glabratula* subsp. *fuliginosa*, (*c*) *Physcia orbicularis*. (From Armstrong 1973.)

lichens were placed on horizontal boards under full natural radiation and 'grown' for 1 year before transplantation onto vertical rock faces with contrasting south-easterly and north-westerly aspects. Replicate thalli left on horizontal boards served as a reference growth controls. The transplant growth data for *Physcia orbicularis*, *Parmelia glabratula* subsp. *fuliginosa* and *P. conspersa* are given in Fig. 162. Growth rates of *Physcia orbicularis*

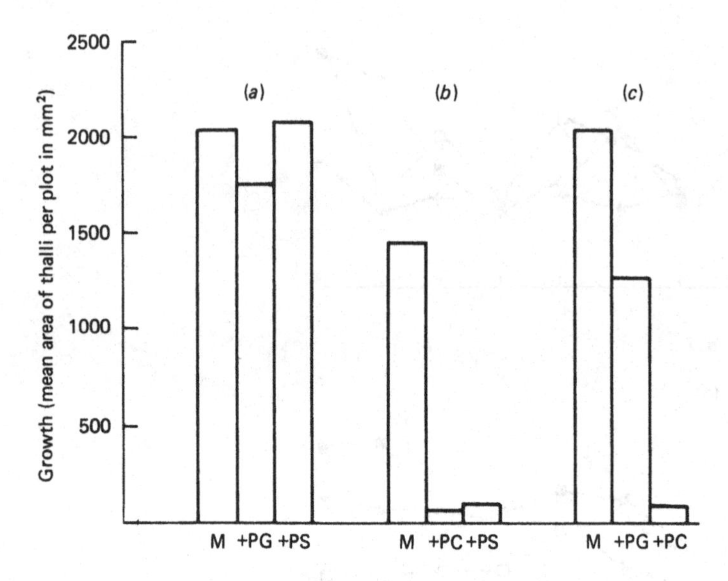

Fig. 163. Growth (mean total area in mm^2 after 3 years) of thalli of (*a*) *Parmelia conspersa* (PC), (*b*) *P. glabratula* subsp. *fuliginosa* (PG), and (*c*) *P. saxatilis* (PS) in monocultures (M) and mixtures (+). There were significant comparisons between treatments: PG M v. (PC + PS)/2*; P. *glabratula* subsp. *fuliginosa* M v. (PC + PS)/2***; and *P. saxatilis* M v. (PG + PC)/2***, PG v. PC***. (**P*, 0.05, *** *P*, 0.001.) (From Armstrong 1982.)

and *Parmelia conspersa* were significantly less on the north-west facing rock surfaces whereas in *P. glabratula* subsp. *fuliginosa* the growth rate on the south-easterly rock exposures was significantly less. Subsequently Armstrong (1982) has demonstrated that a considerable degree of competitive interaction is also possible, with *Parmelia conspersa* being a particularly strong competitor. The smooth nature of slate allowed healthy thalli to be removed and subsequently glued to bare slate surfaces either in monoculture or paired in 1 : 1 mixtures with each of the other two species used in the experiment. The outcome over 3 years is given in Fig. 163; a highly significant competitive sequence was found, with *Parmelia conspersa* > *P. saxatilis* > *P. glabratula* subsp. *fuliginosa*.

It now becomes very evident that the simulation model proposed by Harris (1969) and Kershaw and Harris (1971*a*,*b*) is essentially naive and contributes little to the current level of understanding that we have subsequently achieved. It is important to recognise, for example, that the driving function of a simulation model must be derived from the seasonal operating environment of the lichen, interacting with the seasonal capacities of photosynthesis and respiration (and in some cases nitrogenase activity) which in turn are mediated by the drying capabilities of the environment and lichen thallus. Further control may be exerted by the ionic environment, which is as yet largely unspecified, although potassium, phosphorus and nitrogen are clearly essential components. Finally an appropriate growth model is essential, and this may have to include competitive interactions with other associated species. So far, all computer simulations or regression models have assumed that photosynthetic carbon uptake is directly proportional to growth (net assimilation). The former process, however, represents the first in a complex series steps, mediated by a number of environmental parameters, that ultimately yield an increase in biomass. Any computer simulation has to convert arbitrary carbon units into dry weight production and it is this component of the model that experimentally still evades a solution.

9

Phenotypic plasticity and differential strategies

Weber (1962) in his eloquent dissertation on the wide range of morphological variation found in many species of crustaceous lichens, continually emphasised the profound interaction between the lichen and its environment: 'The bewildering array of thallus forms found within a single species over a richly diversified area is the result of subtly interacting factors of insolation, available moisture, etc., not the least important of which is the physical nature of the substrate itself.' There is almost a focus of attention at the actual level of the operating environment: 'This apparent paradox [erosion of old but not young thalli] simply emphasizes the dynamic nature of the interplay between climatic, edaphic, and biotic factors'. Again: 'a series connecting normal thalli with extremely vestigial ones indicates that there must be subtle variations in the microclimate. Here microrelief plays a significant role.' Weber concludes that a very large proportion of the variability he observed, particularly in the genus *Acarospora*, reflects environmental variation with no counterpart in the genotype. Now, 20 years further on we are faced with a slow accumulation of examples of apparent physiological plasticity with an accompanying question as to the genotypic or environmental control of this physiological variation.

The alternative question as to the level of advantage provided by, for example, a particular photosynthetic modification within a specific environment, also poses some severe problems. There appears to have been widespread acceptance that maximisation of carbon gain is a preferred strategy, often without due recognition that there is an associated metabolic cost of maintaining the adopted strategy. This cost represents a finite proportion of the benefit derived from the strategy and furthermore the cost-benefit ratio will be constrained by the physical operating environment. There is an additional difficulty involved in using a single criterion, carbon gain, as a measure of success in plants. Carbon fixation is biochemically quite remote from the end-products of photosynthesis – growth and reproduction. In other organisms and in higher plants there is considerable appeal in using reproductive success in addition to biomass production as an alternative and perhaps more satisfactory approach, since it effectively integrates the complex of biochemical events that follow the fixation of carbon dioxide. Unfortunately the problem is not so readily resolved in lichens since neither reproductive success nor growth have more than rarely been examined. At present we can only assume that there is a relationship between 'success' and net photosynthesis. 'Respiration' and its contribution to growth and reproduction is largely ignored. It must also be recognised that in many cases the ultimate understanding of the ecology of a plant will invoke escape from competition. There are numerous examples of the more successful growth of a species in a less stressful experimental environment and thus the conclusion that its normal field range reflects its inability to compete with other species. The results of Armstrong (1982) are thus centrally important in that they document in lichens, for the first time, the process of competition which is implicit in many ecological situations.

The few examples of physiological variation discussed below accordingly take the simplistic assumption that carbon gain, when optimised, is indeed fundamentally advantageous and correlated in some equally simple way with success. It is also assumed that competitive avoidance is implied in the development of differential strategies, a model of which is tentatively outlined below.

9.1 Phenotypic plasticity in the net photosynthetic-thallus hydration response

Harris (1971) showed that the net photosynthetic response to the level of thallus hydration was quite habitat specific for replicates of *Parmelia caperata* and *P. sulcata* growing at different heights in a tree. His data for *P. caperata* show that the optimum level of thallus hydration for maximum net photosynthesis is 50-55% and 70-80% relative water content in tree-top and tree-bottom ecotypes respectively. *P. sulcata* has a lower optimum relative water content for photosynthesis (at 45-50%) which is constant for tree-top and middle-zone ecotypes, but there is a pronounced decline in net photosynthetic rates at full thallus hydration with only 55-60% of maximum photosynthetic rates being generated in tree-top ecotypes compared with 85-90% in the middle-zone ecotypes. Harris concludes that the ability of some *Parmelia* species to exist in a variety of physiological states appears to have distinct ecological advantages. The data presented more recently by Coxson *et al.* (1983*a*), Green and Snelgar (1981*a,b*) and Lange and Tenhunen (1981) suggest that the difference in response could, however, be very simply explained by correlated differences in internal diffusive resistances of the thallus. The tree-top ecotypes, growing under higher levels of illumination, perhaps require a thicker upper cortex to prevent light stress to the phycobiont (see discussion below). This in turn results in an increase in diffusive resistance with its attendant problems of gas exchange at higher levels of thallus hydration. There is a reduced requirement at mid-tree height for screening of the phycobiont from excessive illumination, leading to reduced diffusive resistance and enhanced gas exchange at maximum thallus hydration. Whether such morphological changes are genotypic is unknown but the evidence presented below for *Cladonia rangiferina* would suggest not.

Kershaw and MacFarlane (1980) show an even more marked contrast between two ecotypes of *Peltigera aphthosa* which has been discussed briefly in Chapter 5 but is given here again in detail. The experimental material was collected from an area of spruce-lichen woodland in northern Ontario that had been partly burnt 25 years before. Two very distinct morphotypes are typically found in contrasting microclimates. Patches of old unburnt, closed-canopy woodland, with a dense ground cover of moss, support large colonies of the normal woodland ecotype. In contrast, on the very open and exposed soil surface of the adjacent burnt woodland a second ecotype is present which enters the successional sequence within the first 12 years. The contrasting response patterns to light stress and the very different water-holding capacities of the two ecotypes are given in Fig. 164(*a*).

The effects of high-light stress on the woodland population of *P. aphthosa* are clear cut and show a remarkable level of sensitivity to a relatively moderate light level. After 7 days of air-dry storage experimental replicates had net photosynthetic and respiratory rates of 2.8 mg CO_2 h^{-1} g^{-1}, but by day 14 there had been a dramatic decline in photosynthesis so that compensation point was not achieved. After 20 days of light stress net photosynthetic rates declined still further and there is little evidence that the rates differed significantly from the dark respiration rates. Dark respiration, which is probably largely fungal, is constant throughout the experimental period. The control replicates stored at 20 μmol m^{-2} s^{-1} maintain their level of metabolic activity throughout.

The results for the *P. aphthosa* population collected from the woodland burnt 25 years previously offer a remarkable contrast. Maximum rates of activity compared closely with the control replicates of the woodland population, except that maximum levels of thallus saturation did not exceed 425% water content by weight. This contrasted markedly with thallus moisture values of up to 1000% of the oven-dry weight found in fully saturated specimens of the woodland population. In addition to this very

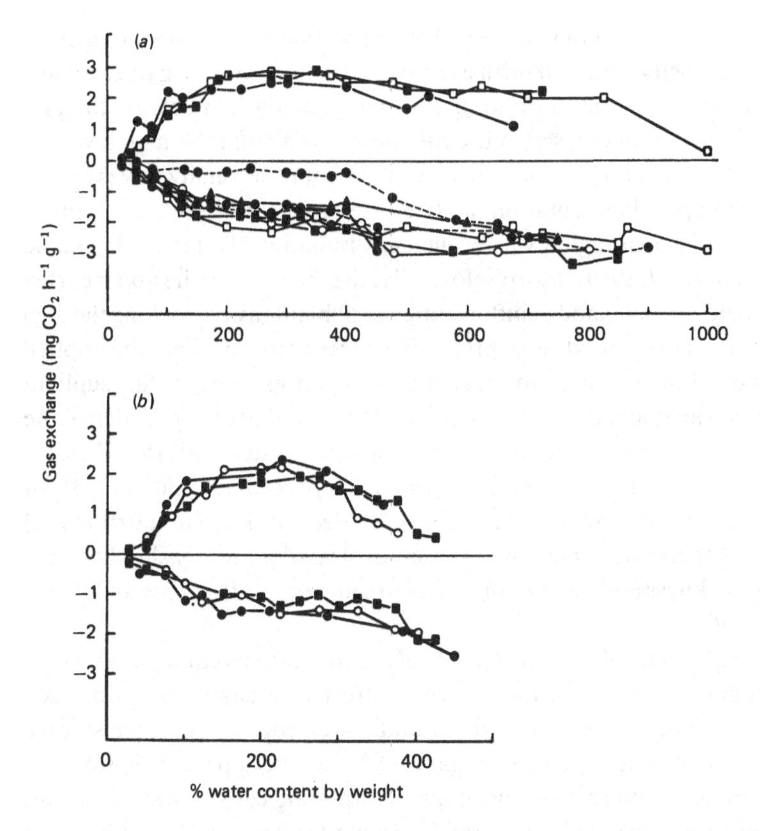

Fig. 164. The pattern of net photosynthesis and respiration with the contrasting thallus water-holding capacities in (*a*) the normal shade ecotype of *Peltigera aphthosa* and (*b*) the sun ecotype. The shade ecotype is sensitive to 350 μmol m^{-2} s^{-1} illumination, with net photosynthetic rates after 14 days (●----●) completely below compensation. ●–● and ■–■, control replicates; □–□ and ○–○, gas exchange after 7 and 20 days respectively. (From Kershaw and MacFarlane 1980.)

pronounced modification of water-holding capacity found in the population from the burnt woodland, there is no evidence of light stress at all, with both net photosynthetic and respiration rates being maintained throughout the experimental period (Fig. 164*b*).

The detrimental effects of growing higher plants from shaded habitats in strong light, including the destruction of chlorophyll and changes in chloroplast morphology, have been well established (Montfort 1950; Munding 1952; Gauhl 1969, 1970). However, in these cases full radiation conditions were required for a stress response to develop. The effects of quite moderate levels of illumination on the woodland population of *P. aphthosa* were initially surprising but as has been discussed previously (Chapter 5) the evolution of a lichen thallus appears to be a trade-off between moisture-holding capacity, internal diffusive resistance and screening of excessive illumination. The sun morphotype is extremely dark in colour, the pigmentation presumably screening the phycobiont from the high levels of solar radiation (a function which the spruce canopy normally serves for the shade ecotype.) Pigmentation has been frequently interpreted as being a protection against high levels of incident radiation (Bitter 1901; Galloe 1908; Laudi *et al.* 1969). Morphologically, the thickness of the upper cortex has also been implicated in influencing the light intensity reaching the alga and hence is regarded as a protective mechanism against chlorophyll oxidation. For instance, the fact that a lichen growing in full sunlight developed an upper cortex about twice as thick as that of a specimen of the same species from a shaded habitat was shown by Bitter (1901) for *Hypogymnia physodes*, Tobler (1925) for *Xanthoria parietina*, Galun (1963) for *Buellia canescens* and Looman (1964) for *Lecanora reptans*. Ertl (1951) extended these investigations and demonstrated a relationship between cortex thickness and habitat for *Peltigera praetextata*, *P. canina* and *Solorina saccata*.

Accordingly the shade ecotype of *Peltigera aphthosa* requires only a very thin upper cortex, has a low internal diffusive resistance and can, as a result, develop a massive medulla and lower cortex for water storage without significant impedance to gas exchange. The apparent physiological differences reflect anatomical and morphological variation and thus parallel those reported for *Parmelia* above (Harris, 1971). Although a similar explanation of the net photosynthetic differences between ridge-top and ridge-bottom morphotypes of *Cladonia stellaris* (Kershaw 1975*b*) may be involved, the response of net photosynthesis to temperature cannot be rationalised so easily. The response of net photosynthesis to the degree of thallus hydration and temperature in both ecotypes is given in Fig. 165. The ridge-top ecotype has marked morphological and possibly anatomical differences which result in an increased capacity to hold water. There appears to be a significant difference, however, in the response to temperature, with the ecotype from the ridge-top (a more exposed and colder habitat) showing higher rates at lower temperatures. This is in contrast to the lower-slope ecotype which has good levels of net photosynthesis still evident at 25 °C. Thus although variation in the response of net photosynthesis to thallus hydration, induced by differential diffusive resistances, appears to be quite widespread in lichens, there are also some instances where an additional explanation may be required.

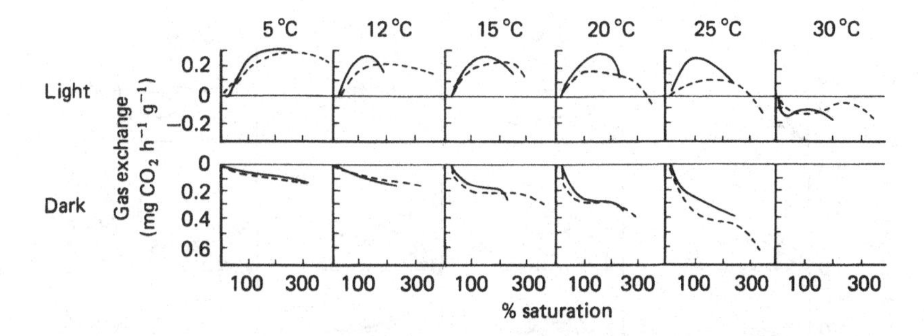

Fig. 165. Comparative net photosynthetic rates in *Cladonia stellaris* collected from the lower slope (—) and the summit of the ridge (----) of a raised beach in northwestern Ontario. Each line is the mean of four replicates, with a S.E.M. of less than 0.03 mg CO_2 h^{-1} g^{-1}. Light, 150 μmol m^{-2} s^{-1}; dark, 0 μmol m^{-2} s^{-1}. (From Kershaw 1975*b*.)

9.2 Phenotypic plasticity of photosynthetic capacity

MacFarlane *et al.* (1983) and Kershaw *et al.* (1983) have examined the seasonal, net photosynthetic response in sun and shade morphotypes of *Cladonia rangiferina* and *C. stellaris* respectively.

The seasonal response matrices for sun and shade forms of *C. stellaris* are given in Figs. 166 and 167. Although the patterns of net photosynthetic and respiratory response to both temperature and the level of thallus hydration are closely similar in sun and shade forms, there is a considerable difference between their maximum photosynthetic capacities (P_{max}). The P_{max} sun/shade ecotypic differences, however, are completely eliminated if the gas exchange rates are expressed in terms of milligrams chlorophyll *a* (chl *a*). For example, replicates of both ecotypes of *C. stellaris* examined at 25 °C under 300 μmol m^{-2} s^{-1} illumination essentially have identical net photosynthetic rates when expressed in terms of milligrams chl *a* rather than grams dry weight (Fig. 168). Despite the very contrasting light environment, and what at first sight appears to be an equally contrasting net photosynthetic seasonal response matrix, Kershaw *et al.* (1983) conclude that the differences are more illusory than real. They suggest that the shade environment could provide a positive feedback of surplus metabolites to provide maintenance energy for a larger algal population, and that it is this that establishes the marked capacity contrast rather than genotypic control of P_{max}. This conclusion is supported by the absence of any marked isoenzyme differences between the two ecotypes (Kershaw *et al.* 1983).

The pattern of net photosynthetic response and the capacity difference between the sun and shade ecotypes of *Cladonia rangiferina* is closely similar (MacFarlane *et al.* 1983). Again there is little evidence to suggest a

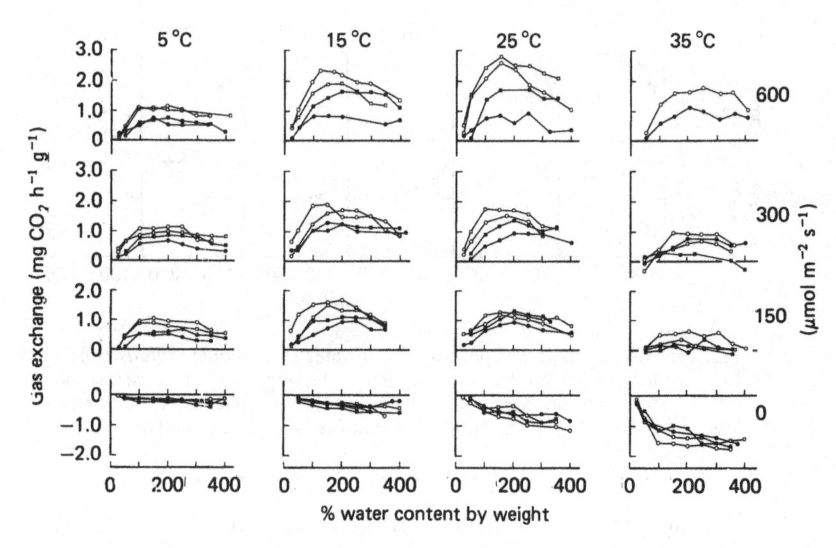

Fig. 166. The seasonal gas exchange response matrix for the sun ecotype of *Cladonia stellaris*. ●, April; ■, June; □, August; ○, September.

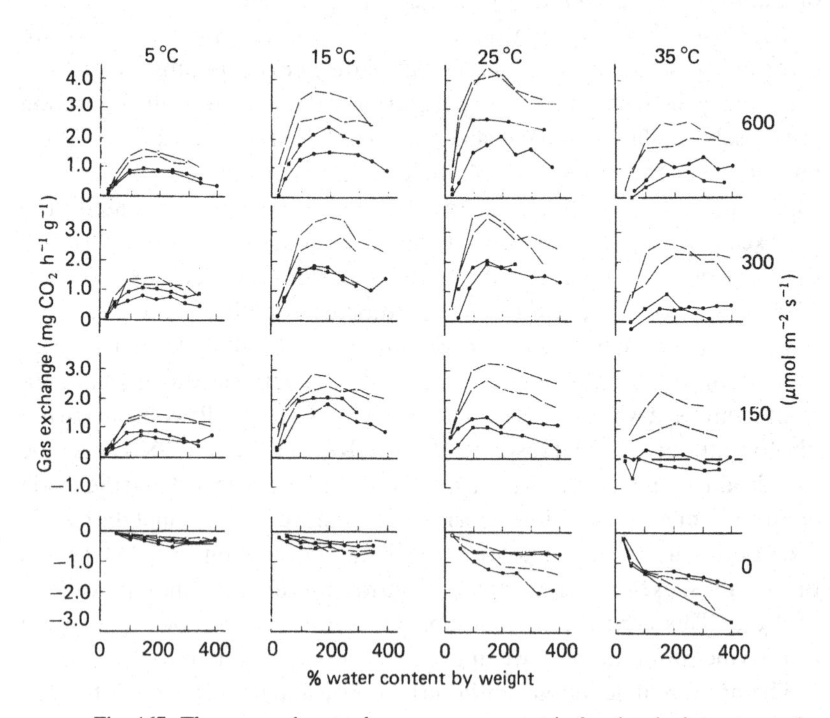

Fig. 167. The seasonal gas exhange response matrix for the shade ecotype of *Cladonia stellaris*. Graph symbols as in Fig. 166.

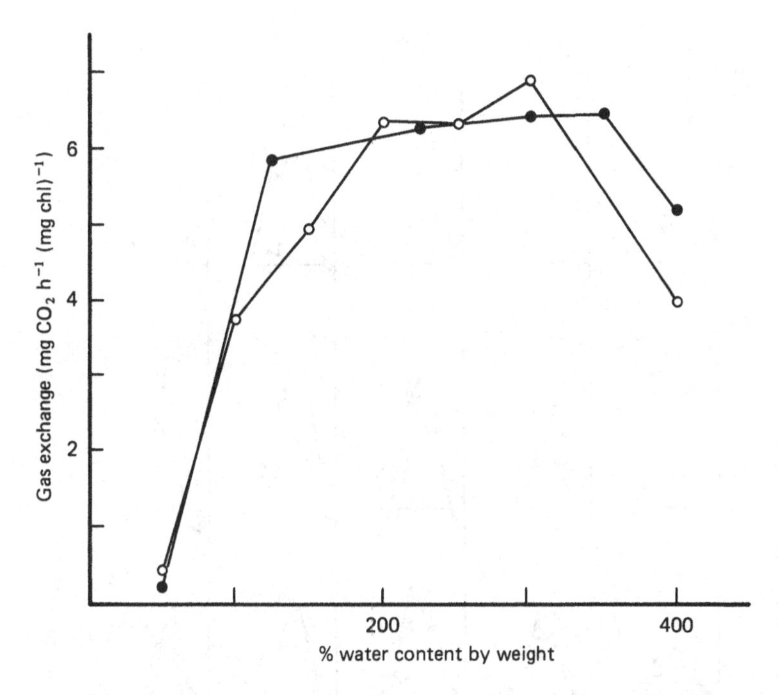

Fig. 168. The net photosynthetic rates at 25 °C and 300 μmol m^{-2} s^{-1} illumination for the sun ecotype (○) and the shade ecotype (●) of *Cladonia stellaris*, expressed on the basis of chlorophyll content.

genetic basis for the P_{max} contrast, and from the limited evidence to date it appears that considerable modification of the basic pattern of net photosynthesis can occur in a species simply as a result of morphological changes induced by different microclimates. Rapid uptake or conservation of thallus moisture by contrasting thallus morphologies, for example, also results in very different diffusion rates for carbon dioxide, particularly at thallus saturation.

The differences between two populations of *Peltigera rufescens* – one from an arctic environment and the other from a temperate habitat – suggest a very different situation (Brown and Kershaw 1984). The seasonal net photosynthetic-temperature response matrices for the two populations are given in Figs. 169 and 170. The photosynthetic- illumination (PI) curve for the southern population has been discussed previously; it was shown that there is a pronounced increase in photosynthetic capacity at 35 °C in the summer, without any change in photosynthetic efficiency when expressed in terms of either dry weight or milligrams chl *a* (see Fig. 131, p. 193). There is thus clear evidence of photosynthetic temperature acclima-

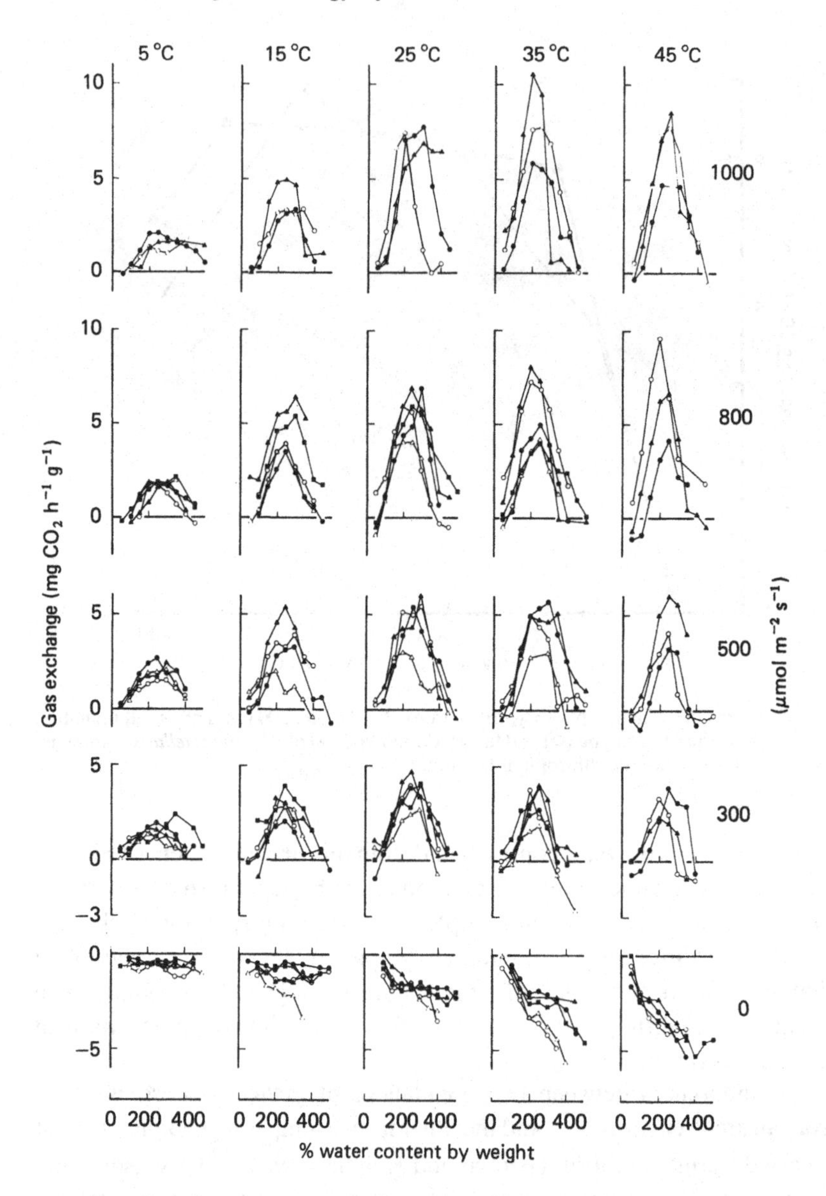

Fig. 169. The seasonal gas exchange response matrix for a southern population of *Peltigera rufescens*. There is a pronounced capacity change at high temperatures in the summer replicates. ■, January; △, April; ▲, June; ○, September; ●, December. (From Brown and Kershaw 1984.)

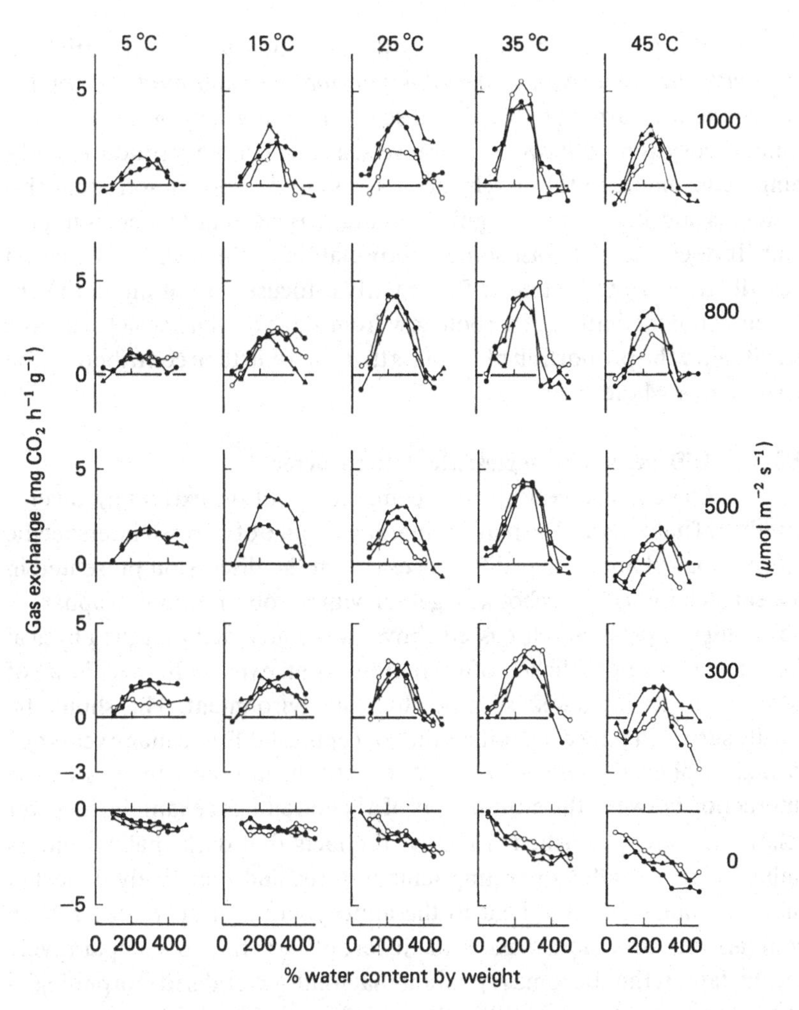

Fig. 170. The seasonal gas exchange response matrix for an arctic population of *Peltigera rufescens*. Photosynthetic capacity is constant throughout the year. ▲, July; ○, September; ●, November. (From Brown and Kershaw 1984.)

tion in the southern population of *P. rufescens*. In contrast, however, the northern population maintains a constant photosynthetic capacity throughout the year (Fig. 170).

It is difficult to see how enzymatic events in the Calvin cycle, potentially changes in fructose-1,6-bisphosphatase and/or ribulose bisphosphate carboxylase activity (Berry and Björkman, 1980; Badger, Björkman and Armond 1982), can be explained by simple morphological or anatomical modifications. As a result it seems reasonable to conclude that lichens as a group of plants have a well-developed capability for both genotypic and phenotypic photosynthetic plasticity, the latter source of variability *strongly correlated with morphology or anatomy and mediated by the* environment. Farrar (1976*a*) has discussed the concept of a lichen as a simple ecosystem with two trophic levels, an algal primary producer and a fungal consumer, and although Farrar's proposed carbon flow through this system is now known to be highly inaccurate, the actual concept is important. It does allow at least some interpretation of the levels of variation described above to be made. Particularly it focuses attention on the environmental mediation of carbon flow from alga to fungus which will, as a result, alter the proportion of biomass that each will then contribute to the system as a whole.

9.3 Differential strategies: a tentative scheme

Lichens are conspicuous components of an extremely diverse number of habitats and display a remarkable range of intra- or interspecific morphological and colour differences. From the discussion presented in the early chapters of this book, together with the observations of apparent physiological plasticity discussed above, it is concluded that the physical form and colour of a lichen often represents an extremely specific set of adaptations to an equally specific physical environment. This should be hardly surprising since it parallels the concepts established many years ago in higher plant ecology – yet only recently have such concepts as the interaction between the morphology of a lichen and, for example, its water relations been established. Equally, the facts that dark thallus colours induce higher thallus operating temperatures and that finely dissected thalli exchange sensible heat to the atmosphere at a facilitated rate in comparison with a more massive thallus are observations of comparatively recent date. It thus becomes apparent that some generalised morphological strategies are present within the wide range of forms documented in lichens as a whole.

Grime (1979) has presented a thoughtful and stimulating summary of how different strategies of higher plants interact with environmental and community processes to control the actual structure and composition of vegetation. Of the three broad categories Grime establishes – stress-tolerators, ruderals and competitors – lichens are included totally within the stress-tolerant class. Grime (1979) concludes, however, that in stress-tolerators the most important response to environmental variation is physiological rather than morphogenetic. This deduction is rather at variance with the conclusions outlined here, where both morphological changes and physiological adaptations are thought to be present. The examination made here in fact suggests that the energetic maintenance cost of an adapted physiological mechanism must be commensurate with any benefits derived from it. Furthermore, physiological plasticity will frequently be correlated with morphology. It must also be correlated with predictable environmental changes. Finally the existence of competition between lichens also suggests that lichens are not exclusively stress-tolerators.

A provisional scheme of the adaptations that have so far been documented in lichens is given in Fig. 171 together with their extreme and generalised environmental correlations. The scheme is certainly incomplete, since it is dominated by temperate and arctic examples and totally ignores the ionic environment. It may perhaps be sufficiently general to apply to at least some desert and tropical situations and serves here as a synthesis of some of the ideas that have been presented previously.

The scheme is derived from the basic energy balance equation:

$$Q^* = LE + H + G + P, \quad (8)$$

where the net radiation Q^* over a vegetated surface provides energy for evaporation of water (LE), heating of the surface (H), heating of the soil column (G) and photosynthesis (P). These individual energy components are then further modified by specific morphological, anatomical or edaphic constraints, within the overall limits of a latitudinal zone. This allows several characteristic net photosynthetic patterns to develop that will maximise carbon gain as a successful strategy either throughout the year or at quite specific times of the year, in response to particular seasonal combinations of environmental variables.

The ecology of a lichen is primarily controlled by its inability to compete with faster growing bryophytes and particularly with higher plants. As a result lichens can be expected to occur either where low light limits higher plant competitors or in high-light situations where temperature or moisture (or soil ionic characteristics?) limit growth of higher plants. The most frequently occurring environment with good levels of illumination that is extensively utilised by lichens is found where higher plant competition is minimised by low availability of water. This lack of water can be either on a semi-continuous basis or markedly on a seasonal basis and often is in conjunction with very high or low ambient temperatures, typical of desert and arctic environments respectively. Under alpine and arctic conditions there is also an additional limitation of higher plant growth directly by temperature. Within this generalised physical environment it has often

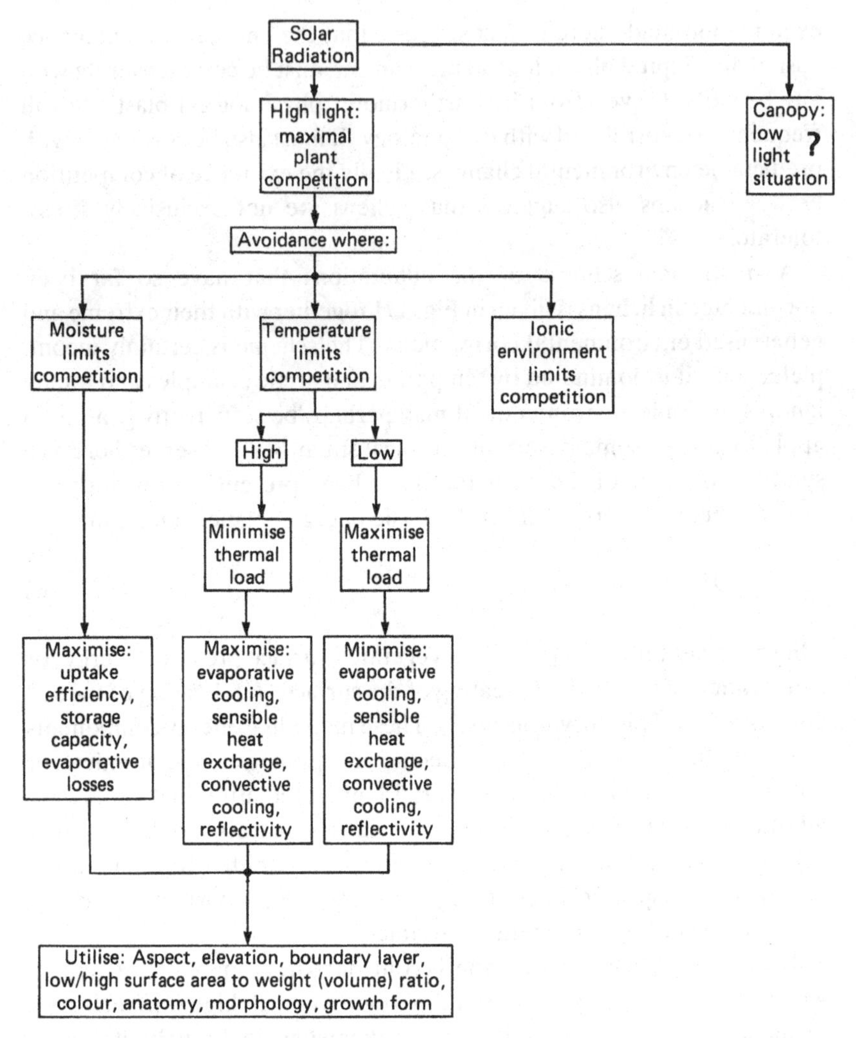

Fig. 171. A provisional scheme to summarise the range of strategies potentially available to lichens which enable them to modify (within limits) their thallus environment, irrespective of the overall latitudinal climate.

been assumed that the central strategy simply involves optimisation of net photosynthesis to correspond with the average ambient temperatures of the latitudinal climate. This is not the case and the scheme presented in Fig. 171 attempts to integrate the morphological variants that can extend the capabilities of a lichen to optimise carbon gain over a much wider range of microclimates. Such a range of microclimates is quite characteristic of any one latitudinal climate. The model also attempts to combine the thermal and moisture requirements of carbon gain and at the same time to recognise the required degree of physiological thermal tolerance that should be commensurate with the maximum thallus operating temperatures. The fact that additional levels of physiological adaptation may also be present is implicit in this argument, and will reflect the potential range and constancy of the thermal operating environment. These concepts have already been discussed previously (Chapter 6) but are again quite relevant here.

Avoidance of competition from higher plants in an environment where there is continuously low light is probably rarely observed, since such conditions would, for example, also usually eliminate lichens from the actual forest floor. In tropical evergreen rainforest the requirement for any screening of corticolous lichen algal symbionts from excessive illumination is unnecessary and as a result both pigmentation and thickness of upper cortex should predictably be minimised. Casual observation suggests that indeed this may be correct. As a direct result, internal diffusive resistances to carbon dioxide via the upper cortex should also be minimal and accordingly photosynthetic response to the degree of thallus hydration in most cases should remain constant until fairly low levels of moisture. It remains to be seen whether these speculations are substantiated experimentally in the future.

Where higher plant competition is limited by moisture there is an equal limitation of carbon gain in lichens. Several potential strategies are available to moderate extreme water shortage or, conversely, to maximise the resource level that is available (Fig. 171). Uptake efficiency has been shown to be quite variable and controlled particularly by the surface area to weight ratio and in some instances by rhizinae (Larson 1981). Lamellae and, in some species, isidia may also be important. Epiphytic species from the genera *Usnea*, *Bryoria*, *Alectoria* and *Ramalina* reflect this strategy which also provides minimal diffusive resistances to carbon dioxide and again presumably a fairly constant photosynthetic response, irrespective of the degree of thallus hydration. Maximisation of storage capacity does not appear to have been utilised extensively as a strategy in lichens and certainly there is no correspondence with 'succulence' as seen in higher plants. The limitation presumably will lie in the ratio between the two bionts, where additional fungal medullary or cortical storage tissue will reduce the free availability of carbon dioxide at the chloroplast surface. This has been discussed at length in Chapter 5 and the general absence of obvious water storage features indicates that only small anatomical or morphological developments have occurred. Even these, however, are reflected in reduced photosynthetic rates at full thallus hydration (see discussion above). It is probable that the most widespread strategy developed has been to use the substrate itself as an additional storage feature to buffer the limited capacity of the thallus. More often, however, the alternative strategy which dominates the ecology of many lichens is that of minimising evaporative losses by utilising a number of mechanisms such as elevation, aspect or the position in the boundary layer, morphology or growth form.

In those ecological situations where higher plant competition is either eliminated or is minimised by temperature, lichens again can form a dominant part of the ground cover. Both the high surface temperatures characteristic of hot deserts and burnt surfaces in low-arctic regions for example, and the low ambient temperatures of the arctic and antarctic regions are involved. The strategies for maximising or minimising thermal load (Fig. 171) involve the albedo of the thallus, with dark colours prevailing where snow-melt contributes an important proportion of the annual total moisture. Additional modification of the thallus thermal environment involves changes to sensible and latent heat exchange as well as to the storage term. Again, efficient energy exchange is controlled particularly by surface area to weight (volume) considerations and, as a result, epiphytic *Bryoria*, *Alectoria* and *Usnea* species can be expected to have moderate temperature optima for photosynthesis as well as a constant response to the degree of thallus hydration. Thallus colour will interact with this ratio to a considerable extent, however, and will certainly modify the thermal characteristics of a thallus significantly. Similarly, elevation, aspect and position of the lichen relative to the surface boundary layer will further control the thermal environment (Fig. 171).

9.4 Differential photosynthetic strategies

Superimposed on the potential range of anatomical and morphological variations and edaphic choices that are open to a lichen (as outlined above) is an additional range of physiological responses determined by the pattern of the thallus operating environment. Accepting that optimising carbon gain is indeed 'success' then a number of net photosynthetic options are available. These options range from a simple, generalised strategy to a complete range of seasonal capacity changes. The former maintains a good level of photosynthesis with a temperature optimum that correlates with the average temperature of the hydrated thallus. The latter strategies allow maximised carbon gain throughout most of the year. Immediately the question arises as to why *all* lichen species do not track environmental changes at least on a seasonal basis and thus continuously maximise their carbon gain.

Two constraints are proposed: the predictability of the seasonal variation in the operating environment, and the maintenance costs of the specific adaptive mechanisms available to lichen phycobionts. The constraint induced by environmental predictability largely determines the extent and level of sophistication of the photosynthetic adaptations. Thus the energetic cost of maintenance must be considerably less than the benefits gained (defined as additional units of carbon fixed). Clearly, a series of photosynthetic adaptations which are rarely used because of the infrequent occurrence of the required environmental combinations, is a poor investment of capital resources: the maintenance costs of the mechanisms could exceed the benefits. Mooney and Gulman (1979) have suggested that resources vary temporally in most habitats and as a result photosynthetic rates will vary with enzyme content, the investment in a specific enzyme being reclaimed and reinvested in a different enzymatic strategy that has a greater return in a different season of the year. This concept is particularly relevant to the desert shrub *Larrea divaricata*, which changes its temperature optimum for net photosynthesis on a seasonal basis. The desert environment, however, varies in a predictable way and accordingly the capital investment in a sophisticated mechanism for optimising net photosynthesis seasonally to ambient temperatures is very worthwhile. Björkman (1981) also questions the energetic costs of maintaining specific photosynthetic mechanisms but again does not directly relate the costs to the constancy or variability of the plant environment. The scheme proposed here (Fig. 172) attempts to develop a series of photosynthetic strategies that are specifically related to the range of variation, but also to the predictability of this variation. These different photosynthetic strategies, although geared to specific latitudinal climates, can then be modified to an extent by the morphological and edaphic scheme above (Fig. 171).

Where the operating environment is more or less constant throughout the year it is suggested that optimal strategies will perhaps involve quite high but constant seasonal photosynthetic rates, sharply maximised within quite narrow temperature, moisture and light values (Fig. 172). Conversely in environmental situations where there is either a large stochastic component to the seasonal fluctuation of the operating environment, or where there is only a limited range of variation, a 'generalised' approach of a constant seasonal photosynthetic response across a much wider range of temperature or light (Fig. 172) offers a good optimal strategy. *Rhizocar-*

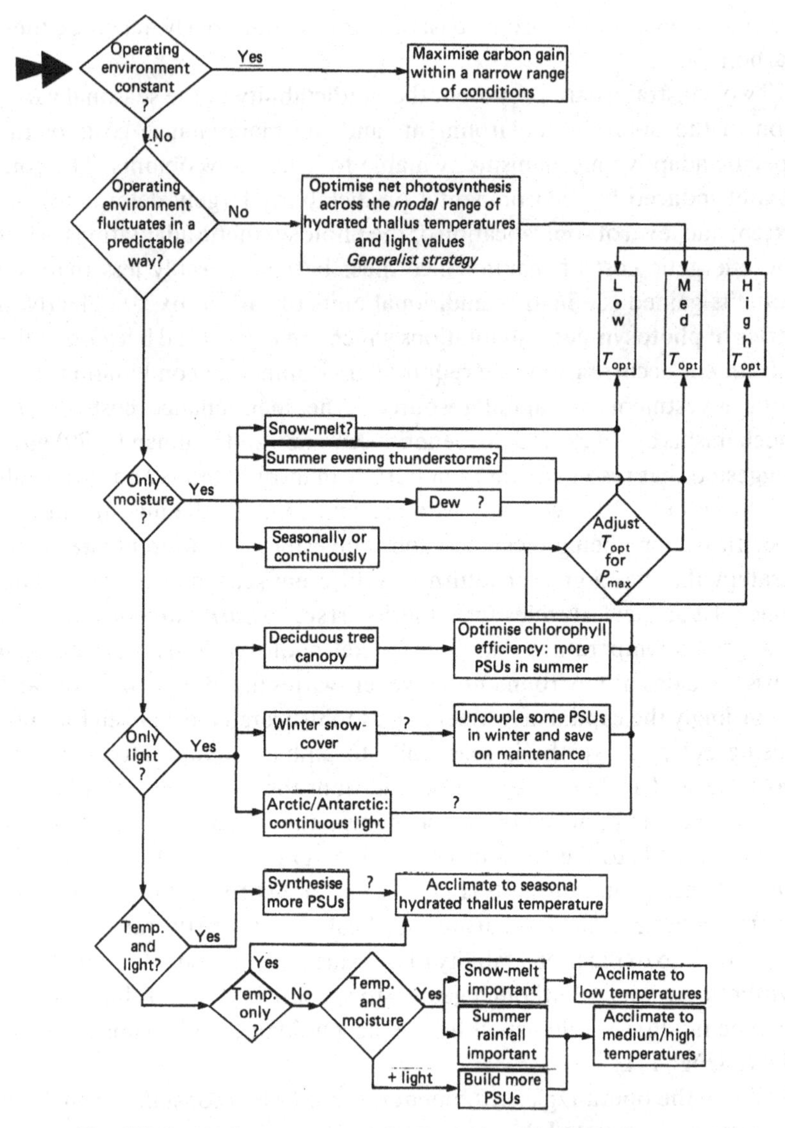

Fig. 172. A tentative summary of the alternative photosynthetic strategies available to a lichen which would enable it to maximise carbon gain within a specific operating environment, but with minimal maintenance costs for the chosen strategy.

pon superficiale is an example of a species with a wide temperature range of net photosynthetic activity reflecting the remarkable variation in thallus microclimate irrespective of season. Thallus hydration periods can occur from snow-melt either in mid-winter during chinook activity, or during mid summer following summer snowstorms; from rainfall throughout spring, summer and autumn coupled with low ambient temperatures; but also from mid-summer thunderstorm activity with concurrent high temperatures (Coxson and Kershaw 1983*b*). As a result there is a very broad range of temperature over which high but almost constant rates of photosynthesis are achieved irrespective of the season (Fig. 173). *Collema furfuraceum* (Kershaw and MacFarlane 1982) similarly has a constant seasonal response which in turn reflects the limited thermal variation of its north-facing corticolous location. The major temperature variation occurs in winter, when, in the absence of water in the liquid phase, no carbon gain at all is possible. Net photosynthetic rates remain fairly high throughout the 5-35 °C range of experimental temperatures examined (Fig. 174) and as a result whatever the ambient temperature throughout early summer to early winter, good rates of carbon gain are achieved without any recourse to seasonal capacity changes with their attendant maintenance costs.

When only moisture as a parameter of the operating environment fluctuates predictably, the available combinations with temperature would suggest that again a constant seasonal response often provides optimal carbon gain with minimal energetic maintenance costs. Of the limited options included in Fig. 172, three of the hydration sources are closely linked with temperature and will result in seasonally constant photosynthetic rates correlated with low to medium temperature optima. This strategy is characteristic of, for example, *Ramalina maciformis* (Lange 1969). The remaining option included under 'predictable thallus moisture' is where there is either a marked seasonality to thallus hydration or continuous rainfall throughout all seasons of the year. Under these conditions there is a correlation between temperature and thallus hydration, and seasonal acclimation of photosynthetic temperature optima is likely to be widespread. Effectively this strategy is part of the temperature-moisture interaction, which is included subsequently in the scheme.

Three major sources of seasonal light variation have been provisionally identified in Fig. 172. The primary factor, a deciduous tree canopy, alters the light environment to a considerable extent. Kershaw and MacFarlane (1980) have shown that there is a large increase in the photosynthetic capacity of *Peltigera praetextata* under summer full-canopy conditions, allowing maintenance of high levels of carbon fixation despite low light

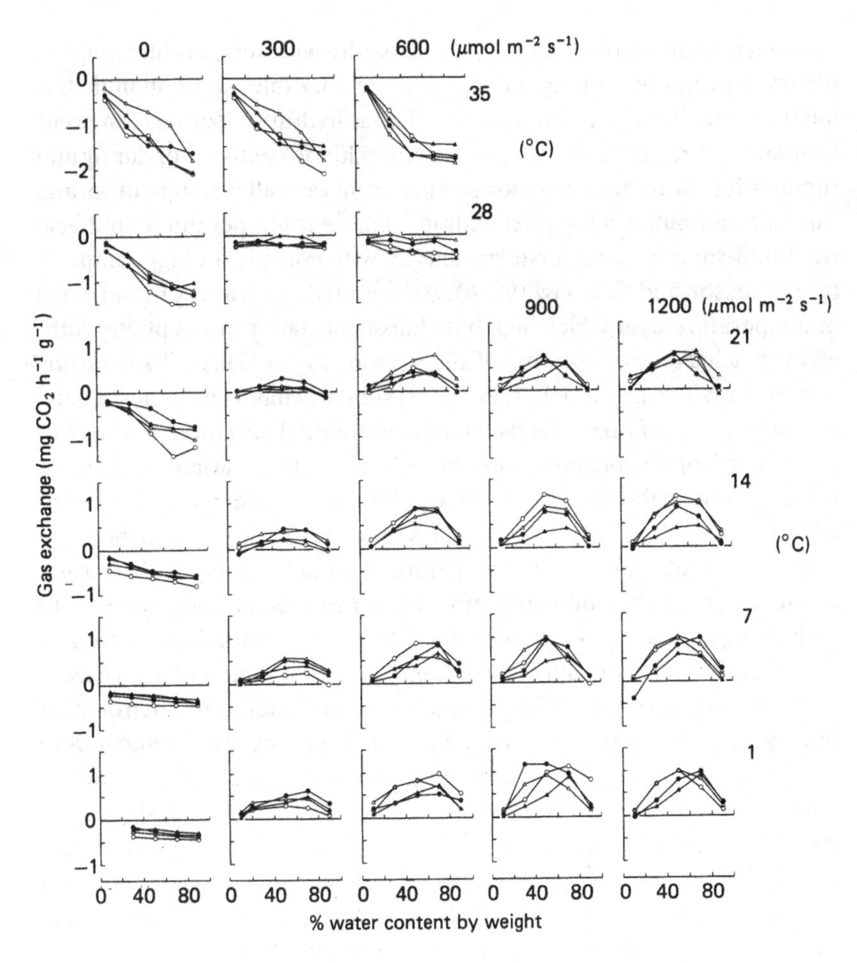

Fig. 173. The seasonal response matrix of *Rhizocarpon superficiale* to temperature, moisture and light. There is a broad photosynthetic temperature optimum, reflecting the wide range of thermal conditions found during periods of thallus hydration. (From Coxson and Kershaw 1983*b*.) ○, 26 December; △, 5 April; ●, 29 May; ▲, 16 July.

levels on the forest floor. This increased photosynthetic capacity results from the synthesis of additional photosynthetic units (PSUs) during the leaf expansion period in early summer. It is probable that other woodland species will employ a similar strategy, either increasing the *number* of PSUs and increasing both efficiency and capacity, or simply synthesising additional chlorophyll as 'extra antennae' and thus optimising photosynthetic efficiency at low illumination levels. This latter strategy has yet to be demonstrated in lichens.

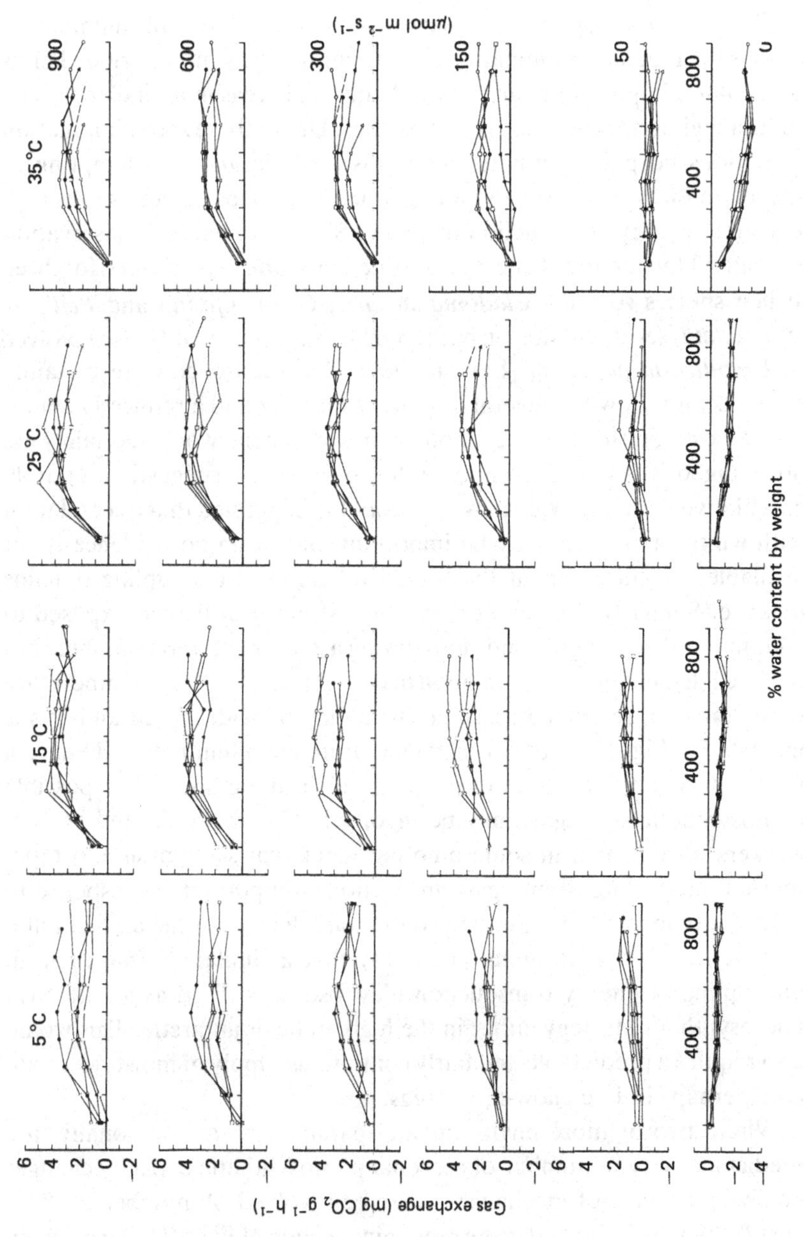

Fig. 174. The seasonal response matrix of *Collema furfuraceum* to temperature, moisture and light. There is an almost constant photosynthetic response to temperature, reflecting the north-facing corticolous habitat of the lichen. □, February (mid-winter); △, April (later winter); ●, May (spring); ▲, June (early summer); ○, July (summer). (From Kershaw and MacFarlane 1982.)

The second major source of seasonal change in ambient illumination level is the regular accumulation of snow during the winter period. This is particularly important in arctic woodlands and areas in alpine and arctic tundra regions that have late snow patches. Under these zero illumination conditions no photosynthetic activity is possible and it is tempting to speculate that one possible strategy which is perhaps advantageous is to uncouple energy transduction in some PSUs as an energy conservation measure. Uncoupling of energy transduction has been confirmed for three lichen species so far – *Cladonia stellaris*, *C. rangiferina* and *Peltigera praetextata* (see discussion on pp. 186–91) – but it is probably also involved in *Umbilicaria papulosa* (Larson 1980). All four species are certainly covered with snow for several months of the year and a reduced maintenance cost for the existing photosynthetic system would certainly be advantageous. At the present time, however, such a suggestion is purely speculative. Similarly Kershaw (1977*a*,*b*) has suggested that a correlation with winter hardening may be important, but again no evidence is yet available. At the moment the true significance of uncoupling remains obscure. Similarly the potential seasonal strategy of lichens exposed to continuous day or night conditions during arctic or antarctic summers has not been documented. Acclimation of the net photosynthetic temperature optimum on a restricted seasonal basis may be widespread and this is indicated in Fig. 172. However, temperature and illumination do have a pronounced diurnal range even in the high arctic and other possible photosynthetic strategies may be involved. Thus Prézelin and her co-workers have shown in some dinoflagellates that sequential and rapid diurnal uncoupling of energy transduction is important, in response to vertical cycling and the resulting continuous changes of the ambient illumination. It is an intriguing possibility that a similar diurnal cycle of uncoupling of energy transduction may also be utilised as an effective photosynthetic strategy either in the high arctic or antarctic. This would also require a predictable and fairly continuous supply of moisture, available perhaps in late snow-melt areas.

Where two or more environmental parameters vary seasonally in a predictable way, a number of seasonal photosynthetic capacity changes involving a range of mechanisms may be involved. A number of these possibilities are indicated in the remaining section of Fig. 172, but no claim is made for completeness nor for their likelihood of occurrence. So far the only documented example involving more than a single seasonal capacity change is *Peltigera praetextata*, and the information there is still incomplete (see discussion in Chapter 6). Thus there is still a considerable requirement for further detailed studies on lichens from a wide range of latitudes and microclimates.

In conclusion it is suggested that many of the morphological and physiological strategies that have evolved in a wide range of lichen species are also available but to a lesser extent within a 'species', either as phenotypic or as genotypic differences. As a result neither the Q_{10} of respiration, the temperature optimum of net photosynthesis, nor the thermal stress limits of a species necessarily bear any general or constant relationship to latitude. However, the discussion above, together with the provisional summary of the potential range of photosynthetic strategies, points to a rather more encouraging situation than was thought possible a few years ago. Thus to examine on a global basis the physiological ecology of lichen species together with their perhaps infinite range of ecotypic and phenotypic variants is truly a daunting prospect. It does now seem more likely that only a limited number of consistent metabolic response patterns which correlate with the specified thallus environment will have evolved. If this is indeed the case then these specific combinations of environment and metabolic responses will number considerably less than the actual numbers of species involved.

References

Ahmadjian, V. (1970). Adaptations of antarctic terrestrial plants. In *Antarctic Ecology*, ed. M.W. Holdgate, pp. 801-11. London, New York & San Francisco: Academic Press.

Ahmadjian, V. & Hale, M.E. (1973). *The Lichens*. London, New York & San Francisco: Academic Press.

Ahti, T. (1959). Studies on the caribou lichen stands of Newfoundland. *Annales Botanici Societatis Zoologicae Botanicae Fennicae 'Vanamo'*, 30, 1-43.

Ahti, T. (1965). Notes on the distribution of *Lecanora conizaeoides*. *Lichenologist*, 3, 91-2.

Alberte, R.S., Hesketh, J.D. & Kirby, J.A. (1976). Comparisons of photosynthetic characteristics of virescent and normal green peanut leaves. *Zeitschrift für Pflanzenphysiologie*, 77, 152-9.

Alexander, V. & Kallio, S. (1976). Nitrogenase activity in *Peltigera aphthosa* and *Stereocaulon paschale* in early spring. *Reports from the Kevo Subarctic Research Station*, 13, 12-15.

Allen, L.H. & Lemon, E.R. (1976). Carbon dioxide exchange and turbulence in a Costa Rican tropical rain forest. In *Vegetation and the Atmosphere*, ed. J.L. Monteith, pp. 265-308. London, New York & San Francisco: Academic Press.

Armstrong, R.A. (1973). Seasonal growth and growth-rate-colony size relationships in six species of saxicolous lichens. *New Phytologist*, 72, 1023-30.

Armstrong, R.A. (1976). Studies on the growth rates of lichens. In *Lichenology: Progress and Problems*, ed. D.H. Brown, D.L Hawksworth & R.H. Bailey, pp. 309-22. London, New York & San Francisco: Academic Press.

Armstrong, R.A. (1977). The response of lichen growth to transplantation to rock surfaces of different aspect. *New Phytologist*, 78, 473-8.

Armstrong, R.A. (1982). Competition between three saxicolous species of *Parmelia* (lichens). *New Phytologist*, 90, 67-72.

Asahina, Y. & Shibata, S. (1954). *Chemistry of Lichen Substances*. Tokyo: Ueno.

Ascaso, C., Galvan, J. & Ortega, C. (1976). The pedogenic action of *Parmelia conspersa*, *Rhizocarpon geographicum* and *Umbilicaria pustulata*. *Lichenologist*, 8, 151-71.

Atkins, C.A. & Pate, J.S. (1977). An IRGA technique to measure CO_2 content of small volumes of gas from the internal atmospheres of plant organs. *Photosynthetica*, 11, 214-16.

Baddeley, M.S., Ferry, B.W. & Finegan, E.J. (1971). A new method of measuring lichen respiration: response of selected species to temperature, pH and sulphur dioxide. *Lichenologist*, 5, 18-25.

Baddeley, M.S., Ferry, B.W. & Finegan, E.J. (1972). The effects of sulphur dioxide on lichen respiration. *Lichenologist*, 5, 283-91.

Badger, M.R., Björkman, O. & Armond, P.A. (1982). An analysis of photosynthetic response and adaptation to temperature in higher plants: temperature acclimation in the desert evergreen *Nerium oleander* L. *Plant, Cell and Environment*, 5, 85-99.

Barkman, J.J. (1958). *Phytosociology and Ecology of Cryptogamic Epiphytes*. Assen, The Netherlands: Van Gorcum.

Beever, R.E. & Burns, D.J.W. (1980). Phosphorus uptake, storage and utilization by fungi. *Advances in Botanical Research*, 8, 127-219.

Berry, J. & Björkman, O. (1980). Photosynthetic response and adaptation to temperature in higher plants. *Annual Review of Plant Physiology*, 31, 491-543.

Billings, W.D. & Mooney, H.A. (1968). The ecology of arctic and alpine plants. *Biological Reviews of the Cambridge Philosophical Society*, 43, 481-529.

Bitter, E. (1901). Über die variabilität einiger Laubflechten und über den Einfluss anderen Bedingungen auf ihr Wachstum. *Jahrbuch für wissenschaftliche Botanik*, 36, 421-92.

Björkman, O. (1972). Photosynthetic adaptation to contrasting light climates. *Carnegie Institute of Plant Biology Year Book*, 70, 82-98.

Björkman, O. (1981). Ecological adaptation of the photosynthetic apparatus. In *Photosynthesis* vol. VI, *Photosynthesis and Productivity, Photosynthesis and Environment*, ed. G. Akoyunoglou, pp. 191-202. Philadelphia: Balaban International Science Services.

Bliss, L.C. (1956). A comparison of plant development in microenvironments of arctic and alpine tundras. *Ecological Monographs*, 26, 303-37.

Bliss, L.C. (1962). Adaptations of arctic and alpine plants to environmental conditions. *Arctic*, 15, 117-44.

Blum, O.B. (1973). Water relations. In *The Lichens*, ed. V. Ahmadjian & M.E. Hale, pp. 381-400. London, New York & San Francisco: Academic Press.

Bone, D.H. (1971*a*). Relationship between phosphates and alkaline phosphatase of *Anabaena flos-aquae* in continuous culture. *Archiv für Mikrobiologie*, 80, 147-53.

Bone, D.H. (1971*b*). Kinetics of synthesis of nitrogenase in batch and continuous culture of *Anabaena flos-aquae*. *Archiv für Mikrobiologie*, 80, 242-51.

Brodo, I.M. (1968). *The Lichens of Long Island, New York: A Vegetational and Floristic Analysis*. Bulletin No. 410, 1-330. New York State Museum, Science Service.

Brodo, I.M. (1973). Substrate ecology. In *The Lichens*, ed. V. Ahmadjian & M.E. Hale, pp. 401-41. London, New York & San Francisco: Academic Press.

Brown, D. & Kershaw, K.A. (1984). Photosynthetic capacity changes in *Peltigera*. II. Contrasting seasonal patterns of net photosynthesis in two populations of *P. rufescens*. *New Phytologist*, 96/1, 447-57.

Brown, D., MacFarlane, J.D. & Kershaw, K.A. (1983). Physiological-environmental interactions in lichens. XVI. A re-examination of resaturation respiration phenomena. *New Phytologist*, 93, 237-46.

Brown, D.H. (1976). Mineral uptake by lichens. In *Lichenology: Progress and Problems*, ed. D.H. Hawksworth & R.H. Bailey, pp. 419-39. Systematics Association Special Volume No. 8. London, New York & San Francisco: Academic Press.

Brown, D.H. & Slingsby, D.R. (1972). The cellular location of lead and potassium in the lichen *Cladonia rangiformis* (L.) Hoffm. *New Phytologist*, 71, 297-305.

Brown, J.S., Alberte, R.S. & Thornber, J.P. (1974). Comparative studies on the occurrence and spectral composition of chlorophyll-protein complexes in a wide variety of plant material. In *Third International Congress of Photosynthesis Research*, pp. 1951-62.

Brown, R.T. & Mikola, P. (1974). The influence of fruticose soil lichens upon the mycorrhizae and seedling growth of forest trees. *Acta Forestalia Fennica*, 141, 1-22.

Buck, G.W. & Brown, D.H. (1979). The effect of desiccation on cation location in lichens. *Annals of Botany*, 44, 265-77.

Burton, M.A.S., Le Sueur, P. & Puckett, K.J. (1981). Copper, nickel, and thallium uptake by the lichen *Cladina rangiferina*. *Canadian Journal of Botany*, 59, 91-100.

Butin, H. (1954). Physiologisch-ökologische Untersuchungen über die Photosynthese bei Flechten. *Biologische Zentralblatte*, 73, 459-502.

Carlisle, A., Brown, A.H.F. & White, E.J. (1966). The organic matter and nutrient elements in the precipitation beneath a sessile oak canopy. *Journal of Ecology*, 54, 87-98.

Carlisle, A., Brown, A.H.F. & White, E.J. (1967). The nutrient content of tree stem flow and ground flora litter and leachates in a sessile oak (*Quercus petraea*) woodland. *Journal of Ecology*, 55, 615-27.

Chambers, S., Morris, M. & Smith, D.C. (1976). Lichen physiology. XV. The effects of digitonin and other treatments on biotrophic transport of glucose from alga to fungus in *Peltigera polydactyla*. *New Phytologist*, 76, 71-83.

Chartier, P. & Catský, J. (1970). Resistances for carbon dioxide diffusion and for carboxylation as factors in bean leaf photosynthesis. *Photosynthetica*, 4, 48-57.

Clarkson, D.T. (1969). Metabolic aspects of aluminium toxicity and some possible mechanisms for resistance. In *Ecological Aspects of the Mineral Nutrition of Plants*, ed. I.H. Rorison, pp. 381-97. Oxford: Blackwell Scientific Publications.

Clegg, M.D., Sullivan, C.Y. & Eastin, J.D. (1978). A sensitive technique for the rapid measurement of carbon dioxide concentrations. *Plant Physiology*, 62, 924-6.

Clymo, R.S. (1962). An experimental approach to part of the calcicole problem. *Journal of Ecology*, 50, 701-31.

Collins, C.A. & Farrar, J.F. (1978). Structural resistances to mass transfer in the lichen *Xanthoria parietina*. *New Phytologist*, 81, 71-83.

Cowles, S. (1982). Preliminary results investigating the effect of lichen ground cover on the growth of black spruce. *Naturaliste canadien*, 109, 573-81.

Cox, R.M. & Fay, P. (1969). Special aspects of nitrogen fixation by blue-green algae. *Proceedings of the Royal Society of London. Series B*, 142, 357-66.

Coxson, D.S. (1983). The ecophysiology of surface cryptograms from alpine tundra and semi-arid grassland of southwestern Alberta. PhD thesis, McMaster University, Hamilton, Ontario.

Coxson, D.S., Brown, D. & Kershaw, K.A. (1983*a*). The interaction between CO_2 diffusion and the degree of thallus hydration in lichens: some further comments. *New Phytologist*, 93, 247-60.

Coxson, D.S. & Kershaw, K.A. (1983*a*). The ecology of *Rhizocarpon superficiale* (Schaer.) Vain. I. The rock surface boundary-layer microclimate. *Canadian Journal of Botany*, 61, 3009–18.

Coxson, D.S. & Kershaw, K.A. (1983*b*). The ecology of *Rhizocarpon superficiale* (Schaer.) *Vain. II. The seasonal response of net photosynthesis and respiration to temperature,* moisture and light. *Canadian Journal of Botany*, 61, 3019–30.

Coxson, D.S. & Kershaw, K.A. (1983*c*). Low temperature acclimation of net photosynthesis in the crustaceous lichen *Caloplaca trachyphylla* (Tuck.) A. Zahlbr. *Canadian Journal of Botany*, 62, 86–95.

Coxson, D.S., Webber, M.R. & Kershaw, K.A. (1984). The thermal operating environment of corticolous and pendulous tree lichens. *Bryologist*, 87 (in press).

Crittenden, P.D. (1975). Nitrogen fixation by lichens on glacial drift in Iceland. *New Phytologist*, 74, 41-9.

Crittenden, P.D. & Kershaw, K.A. (1979). Studies on lichen-dominated systems. XXII. The environmental control of nitrogenase activity in *Stereocaulon paschale* in spruce-lichen woodland. *Canadian Journal of Botany*, 57, 236-54.

Culberson, C.F. (1969). *Chemical and Botanical Guide to Lichen Products*. Chapel Hill: University of North Carolina Press.

Culberson, W.L. (1955). The corticolous communities of lichens and bryophytes in the upland forests of northern Wisconsin. *Ecological Monographs*, 25, 215-31.

Cumming, B.G. & Wagner, E. (1968). Rhythmic processes in plants. *Annual Review of Plant Physiology*, 19, 381-416.

Denmead, O.T. (1964). Evaporation sources and apparent diffusivities in a forest canopy. *Journal of Applied Meteorology*, 3, 383-9.

De Wit, T. (1976). Epiphytic lichens and air pollution in the Netherlands. *Bibliotheca Lichen*, 5, 1-227.

Dibben, M.J. (1971). Whole-lichen culture in a phytotron. *Lichenologist*, 5, 1-10.

Dilks, T.J.K. & Proctor, M.F. (1974). The pattern of recovery of bryophytes after desiccation. *Journal of Bryology*, 8, 97-115.

Drew, E.A. & Smith, D.C. (1967). Studies in the physiology of lichens. VIII. Movement of glucose from alga to fungus during photosynthesis in the thallus of *Peltigera polydactyla*. *New Phytologist*, 66, 389-400.

Dua, R.D. & Burris, R.H. (1963). Stability of nitrogen fixing enzymes and the reactivation of cold labile enzyme. *Proceedings of the National Academy of Sciences, USA*, 50, 169-75.

Englund, B. (1978). Effects of environmental factors on acetylene reduction by intact thallus and excised cephalodia of *Peltigera aphthosa* Willd. In *Environmental Role of Nitrogen-fixing Blue-green Algae and Asymbiotic Bacteria*, ed. U. Granhall, pp. 234-46. Ecological Bulletin (Stockholm) No. 26.

Englund, B. & Meyerson, H. (1974). *In situ* measurement of nitrogen fixation at low temperatures. *Oikos*, 25, 283-7.

Ertl, L. (1951). Über die Lichtverhältnisse in Laubflechten. *Planta*, 39, 245-70.

Fahselt, D. (1976). Growth rates in vegetative and reproductive tissue of *Parmelia cumberlandia* after relocation to a new environment. *Canadian Field-Naturalist*, 91, 134-40.

Farrar, J.F. (1973). Lichen physiology: progress and pitfalls. In *Air Pollution and Lichens*, ed. B.W. Ferry, M.S. Baddeley & D.L. Hawksworth, pp. 238-82. University of Toronto Press.

Farrar, J.F. (1976*a*). The lichen as an ecosystem: observation and experiment. In *Lichenology: Progress and Problems*, ed. D.H. Brown, D.L. Hawksworth & R.H. Bailey, pp. 385-406. London, New York & San Francisco: Academic Press.

Farrar, J.F. (1976*b*). The uptake and metabolism of phosphate by the lichen *Hypogymnia physodes*. *New Phytologist*, 77, 127-34.

Farrar, J.F. (1978). Symbiosis between fungi and algae. In *CRC Handbook Series in Nutrition and Food*, ed. M. Reichigl, pp. 121-39. Cleveland: CRC Press.

Farrar, J.F. & Smith, D.C. (1976). Ecological physiology of the lichen *Hypogymnia physodes*. III. The importance of the rewetting phase. *New Phytologist*, 77, 115-25.

Ferry, B.W., Baddeley, M.S. & Hawksworth, D.L. (1973). *Air Pollution and Lichens*. University of Toronto Press.

Fletcher, A. (1976). Nutritional aspects of marine and maritime lichen ecology. In *Lichenology: Progress and Problems*, ed. D.H. Brown, D.L. Hawksworth & R.H. Bailey, pp. 359-84. London, New York & San Francisco: Academic Press.

Fogg, G.E. & Stewart, W.D.P. (1968). *In situ* determinations of biological nitrogen fixation in Antarctica. *Bulletin of the British Antarctic Survey*, 15, 39-46.

Fogg, G.E. & Than-Tun (1960). Interrelations of photosynthesis and assimilation of elementary nitrogen in a blue-green algae. *Proceedings of the Royal Society of London, Series B*, 153, 111-27.

Follmann, G. & Nakagara, M. (1963). Keimhemmung von Angiospermensamen durch Flechtenstoffe. *Naturwissenschaften*, 50, 696-7.

Foy, R.H., Gibson, C.E. & Smith, R.V. (1976). The influence of day length, light intensity and temperature on the growth rates of planktonic blue-green algae. *British Phycologia Journal*, 11, 151-63.

Fraser, E.M. (1956). The lichen woodlands of the Knob Lake area of Quebec-Labrador. *McGill Subarctic Research Paper*, 1, 3-28.

Galloe, O. (1908). Dansk likenen oekologie. *Danmarks Botanisk Tidsskrift*, 28, 285.

Galun, M. (1963). Autecological and synecological observations on lichen of the Negev, Israel. *Israeli Journal of Botany*, 12, 179-87.

Ganf, G.G. (1974). Rates of oxygen uptake by the planktonic community of a shallow equatorial lake (Lake George, Uganda). *Oecologia (Berlin)*, 15, 17-32.

Garty, J., Galun, M. & Kessel, M. (1979). Localization of heavy metals and other elements accumulated in the lichen thallus. *New Phytologist*, 82, 159-68.

Gauhl, E. (1969). Differential photosynthetic performance of *Solanum dulcamara* ecotypes from shaded and exposed habitats. *Carnegie Institute of Washington, Yearbook*, 67, 482-91.

Gauhl, E. (1970). Leaf factors affecting the rate of light saturated photosynthesis in ecotypes of *Solanum dulcamara*. *Carnegie Institute of Washington, Yearbook*, 68, 633-44.

Geiger, R. (1971). *The Climate near the Ground*. Cambridge, Massachussetts: Harvard University Press.

Gibson, C.E. (1975). A field and laboratory study of oxygen uptake by planktonic blue-green algae. *Journal of Ecology*, 63, 867-80.

Gilbert, O.L. (1965). Lichens as indicators of air pollution in the Tyne Valley. In *Ecology and the Industrial Society*, pp. 35-47. Oxford: Blackwell Scientific Publications.

Gilbert, O.L. (1968). Biological indicators of air pollution. PhD thesis, University of Newcastle-upon-Tyne.

Gilbert, O.L. (1973). Lichens and air pollution. In *The Lichens*, ed. V. Ahmadjian & M.E. Hale, pp. 443-72. New York, London & San Francisco: Academic Press.

Gilbert, O.L. (1976). An alkaline dust effect on epiphytic lichens. *Lichenologist*, 8, 173-8.

Gordon, A.G. & Gorham, E. (1963). Ecological aspects of air pollution from an ion-sintering plant at Wawa, Ontario. *Canadian Journal of Botany*, 41, 1063-78.

Goyal, R. & Seaward, M.R.D. (1981). Metal uptake in terricolous lichens. I. Metal localization within the thallus. *New Phytologist*, 89, 631-45.

Goyal, R. & Seaward, M.R.D. (1982*a*). Metal uptake in terricolous lichens. II. The effects on the morphology of *Peltigera canina* and *Peltigera rufescens*, *New Phytologist*, 90, 73-84.

Goyal, R. & Seaward, M.R.D. (1982*b*). Metal uptake in terricolous lichens. III. Translocation in the thallus of *Peltigera canina*. *New Phytologist*, 90, 85-98.

Granhall, U. & Selander, H. (1973). Nitrogen fixation in a subarctic mire. *Oikos*, 24, 8-15.

Green, T.G.A & Snelgar, W.P. (1981*a*). Carbon dioxide exchange in lichens: partition of total CO_2 resistances at different thallus water contents into transport and carboxylation components. *Physiologia Plantarum*, 52, 411-16.

Green, T.G.A. & Snelgar, W.P. (1981*b*). Carbon dioxide exchange in lichens: Relationship between net photosynthetic rate and CO_2 concentration. *Plant Physiology*, 68, 199-201.

Green, T.G.A., Snelgar, W.P. & Brown, D.H. (1981). Carbon dioxide exchange through the cyphellate lower cortex of *Sticta latifrons* Rich. *New Phytologist*, 88, 421-6.

Greene, S.W. & Longton, R.E. (1970). The effects of climate on Antarctic plants. In *Antarctic Ecology*, ed. M.W. Holdgate, pp. 786-800. London, New York & San Francisco: Academic Press.

Grime, J.P. (1979). *Plant Strategies and Vegetation Processes*. New York: Wiley.

Hale, M.E. (1955). Phytosociology of corticolous cryptogams in the upland forests of southern Wisconsin. *Ecology*, 36, 45-63.

Hale, M.E. (1967). *The Biology of Lichens*. London: Edward Arnold.

Hale, M.E. (1970). Single lobe growth-rate patterns in the lichen *Parmelia caperata*. *Bryologist*, 73, 72-81.

Hale, M.E. (1973). Growth. In *The Lichens*, ed. V. Ahmadjian & M.E. Hale, pp. 473-92. London, New York & San Francisco: Academic Press.

Hale, M.E. (1981). Pseudocyphellae and pored epicortex in the Parmeliaceae: their delimitation and evolutionary significance. *Lichenologist*, 13, 1-10.

Hällgren, J. & Huss, K. (1975). Effects of SO_2 on photosynthesis and nitrogen fixation. *Physiologia Plantarum*, 34, 171-6.

Hampp, R. & Schnabl, H. (1975). Effect of aluminum ions on $^{14}CO_2$–fixation and membrane systems of isolated spinach chloroplasts. *Zeitschrift für Pflanzenphysiologie*, 76, 300-6.

Handley, R. & Overstreet, R. (1968). Uptake of carrier-free ^{137}Cs by *Ramalina reticulata*. *Plant Physiology*, 43, 1401-5.

Harding, L.W., Meeson, B.W., Prézelin, B.B. & Sweeney, B.M. (1981). Diel periodicity of photosynthesis in marine phytoplankton. *Marine Biology*, 61, 95-105.

Harris, G.P. (1969). A study of the ecology of corticolous lichens. PhD thesis, University of London, UK.

Harris, G.P. (1971). The ecology of corticolous lichens. II. The relationship between physiology and the environment. *Journal of Ecology*, 59, 441-52.

Harris, G.P. (1973). Diel and annual cycles of net plankton photosynthesis in Lake Ontario. *Journal of the Fisheries Research Board of Canada*, 30, 1779-87.

Harris, G.P. (1978). Photosynthesis, productivity and growth: the physiological ecology of phytoplankton. *Archiv für Hydrobiologie*, 10, 1-171.

Harris, G.P. & Kershaw, K.A. (1971). Thallus growth and the distribution of stored metabolites in the phycobiont of the lichens *Parmelia sulcata* and *Parmelia physodes*. *Canadian Journal of Botany*, 49, 1367-72.

Hawksworth, D.L. (1973). Mapping studies. In *Air Pollution and Lichens*, ed. B.W. Ferry, M.S. Baddeley & D.L. Hawksworth, pp. 38-76. University of Toronto Press.

Hawksworth, D.L. & Rose, F. (1976). *Lichens as Pollution Monitors*. Institute of Biology's Studies in Biology 66. London: Edward Arnold.

Haynes, F.N. (1964). Lichens. *Viewpoints in Biology*, 3, 64-115.

Haystead, A., Robinson, R. & Stewart, W.D.P. (1970). Nitrogenase activity in extracts of heterocystous and non-heterocystous blue-green algae. *Archiv für Mikrobiologie*, 74, 235-43.

Heatwole, H. (1966). Moisture exchange between the atmosphere and some lichens of the genus *Cladonia*. *Mycologia*, 58, 148-56.

Helfferich, F. (1962). *Ion Exchange*. New York: McGraw-Hill.

Henriksson, E. (1951). Nitrogen fixation by a bacteria-free symbiotic Nostoc strain isolated from *Collema*. *Physiologia Plantarum*, 4, 542-5.

Henriksson, E. & Simu, B. (1971). Nitrogen fixation by lichens. *Oikos*, 22, 119-21.

Henriksson, E., Englund, B., Heden, M.B. & Wass, I. (1972). Nitrogen fixation in Swedish soils by blue-green algae. In *Taxonomy and Biology of Blue-Green Algae*, ed. T.V. Desikachary, pp. 269-73. Madras.

Herron, H.A. & Mauzerall, D. (1972). The development of photosynthesis in a greening mutant of *Chlorella* and an analysis of the light saturation curve. *Plant Physiology*, 50, 141-8.

Hill, D.J. (1971). Experimental study of the effect of sulphite on lichens with reference to atmospheric pollution. *New Phytologist*, 70, 831-6.

Hill, D.J. (1974). Some effects of sulphite on photosynthesis in lichens. *New Phytologist*, 73, 1193-205.

Hill, R. & Bendall, F. (1960). Function of two cytochrome components in chloroplasts: a working hypothesis. *Nature (London)*, 186, 136-7.

Hitch, C.J.B. (1971). A study of some environmental factors affecting nitrogenase activity in lichens. MSc thesis, University of Dundee, UK.

Hitch, C.J.B. & Millbank, J.W. (1975*a*). Nitrogen metabolism in lichens. VI. The blue-green phycobiont content, heterocyst frequency and nitrogenase activity in *Peltigera* species. *New Phytologist*, 74, 473-6.

Hitch, C.J.B. & Millbank, J.W. (1975*b*). Nitrogen metabolism in lichens. VII. Nitrogenase activity and heterocyst frequency in lichens with blue-green phycobionts. *New Phytologist*, 75, 239-44.

Hitch, C.J.B. & Stewart, W.D.P. (1973). Nitrogen fixation by lichens in Scotland. *New Phytologist*, 72, 509-24.

Horne, A.J. (1972). The ecology of nitrogen fixation on Signey Island, South Orkney Islands. *British Antarctic Survey Bulletin*, 27, 1-18.

Horne, A.J. & Goldman, C.R. (1972). Nitrogen fixation in Clear Lake, California. I. Seasonal variation and the role of heterocysts. *Limnology and Oceanography*, 17, 78-92.

Huss-Danell, K. (1977). Nitrogenase activity in the lichen *Stereocaulon paschale*: recovery after dry storage. *Physiologia Plantarum*, 41, 58-61.

Huss-Danell, K. (1978). Seasonal variation in the capacity for nitrogenase activity in the lichen *Stereocaulon paschale*. *New Phytologist*, 81, 89-98.

Huss-Danell, K. (1979). The influence of light and oxygen on nitrogenase activity in the lichen *Stereocaulon paschale*. *Physiologia Plantarum*, 47, 269-73.

Hustich, I. (1951). The lichen woodlands in Labrador and their importance as winter pastures for domesticated reindeer. *Acta Geographica*, 12, 1-48.

Iskander, I.K. & Syers, J.K. (1971). Solubility of lichen compounds in water: pedogenetic implications. *Lichenologist*, 5, 45-50.

Iskander, I.K. & Syers, J,K. (1972). Metal-complex formation by lichen compounds. *Journal of Soil Science*, 23, 255-65.

Jackson, T.A. & Keller, W.D. (1970). A comparative study of the role of lichens and inorganic processes in the chemical weathering of recent Hawaiian lava flows. *American Journal of Science*, 269, 446-66.

Jahns, H.M. (1973). Anatomy, morphology and development. In *The Lichens*, ed. V. Ahmadjian & M.E. Hale, pp. 3-58. London, New York & San Francisco: Academic Press.

James, P.W., Hawksworth, D.L. & Rose, F. (1977). Lichen communities in the British Isles: a preliminary conspectus. In *Lichen Ecology*, ed. M.R.D. Seaward, pp. 295-413. London, New York & San Francisco: Academic Press.

Jarvis, P.G. (1971). The estimation of resistances to carbon dioxide transfer. In *Plant Photosynthetic Production Manual of Methods*, ed. J. Sesták, J. Catský & P.G. Jarvis, pp. 566-631. The Hague: W. Junk.

Jarvis, P.G., James, G.B. & Landsberg, J.J. (1976). Coniferous forest. In *Vegetation and the Atmosphere*, volume 2, ed. J.L. Monteith, pp. 171-236. London, New York & San Francisco: Academic Press.

Johnson, E.A. & Rowe, J.S. (1975). Fire in the wintering ground of the Beverley caribou herd. *American Midland Naturalist*, 94, 1-14.

Jones, K. (1977). The effects of light intensity on acetylene reduction by mats of blue-green algae in sub-tropical grassland. *New Phytologist*, 78, 427-31.

Jones, H.G. & Slatyer, R.O. (1972). Estimation of the transport and carboxylation components of the intracellular limitation to leaf photosynthesis. *Plant Physiology*, 50, 283-8.

Jorgensen, E.G. (1969). The adaptation of plankton algae. IV. Light adaptation in different algal species. *Physiologia Plantarum*, 22, 1307-15.

Jorgensen, E.G. (1970). The adaptation of plankton algae. V. Variation in the photosynthetic characteristics of *Skeletonema costatum* cells grown at low light intensity. *Physiologia Plantarum*, 23, 11-17.

Kallio, S. (1973). The ecology of nitrogen fixation in *Stereocaulon paschale*. *Reports from the Kevo Subarctic Research Station*, 10, 34-42.

Kallio, P. & Heinonen, S. (1971). The influence of short-term lôw temperature on net photosynthesis in some subarctic lichens. *Reports from the Kevo Subarctic Research Station*, 8, 63-72.

Kallio, S. & Kallio, P. (1975). Nitrogen fixation in lichens at Kevo, North Finland. In *Fennoscandian Tundra Ecosystems*, part I. *Ecological Studies*, ed. F.E. Wielgolaski, pp. 292-304. Berlin: Springer-Verlag.

Kallio, P. & Kallio, S. (1978). Adaptation of nitrogen fixation to temperature in the *Peltigera aphthosa* group. In *Environmental Role of Nitrogen-fixing Blue-green Algae and Asymbiotic Bacteria*, pp. 225-53. Ecological Bulletin (Stockholm) No. 26.

Kallio, S., Kallio, P. & Rasku, M.L. (1976). Ecology of nitrogen fixation in *Peltigera aphthosa* (L.) Willd. in North Finland. *Reports from the Kevo Subarctic Research Station*, 13, 16-22.

Kallio, P. & Kärenlampi, L. (1975). Photosynthesis in mosses and lichens. In *Photosynthesis and Productivity in Different Environments*, ed. J.P. Cooper, pp. 393-423. Cambridge University Press.

Kallio, P., Suhonen, S. & Kallio, H. (1972). The ecology of nitrogen fixation in *Nephroma arcticum* and *Solorina* crocea. *Reports from the Kevo Subarctic Research Station*, 9, 7-14.

Kappen, L. (1973). Response to extreme environments. In *The Lichens*, ed. V. Ahmadjian & M.E. Hale, pp. 311-80. London, New York & San Francisco: Academic Press.

Kappen, L., Friedmann, E.I. & Garty, J. (1981). Ecophysiology of lichens in the dry valleys of southern Victoria Land, Antarctica. I. Microclimate of the cryptoendolithic lichen habitat. *Flora*, 171, 216-35.

Kappen, L. & Lange, O.L. (1972). Die Kälteresistenz einiger Makrolichenen. *Flora (Jena)*, 161, 1-29.

Kärenlampi, L. (1970). Distribution of chlorophyll in the lichen *Cladonia stellaris*. *Reports from the Kevo Subarctic Research Station*, 161, 1-29.

Kärenlampi, L. (1971). Studies on the relative growth rate of some fruticose lichens. *Reports from the Kevo Subarctic Research Station*, 7, 33-9.

Kärenlampi, L., Tammisola, J. & Hurme, H. (1975). Weight increase of some lichens as related to carbon dioxide exchange and thallus moisture. In *Fennoscandian Tundra Ecosystems*, Part I, *Plants and Microorganisms*, ed. F.E. Wielgolaski, pp. 135-7. Berlin: Springer-Verlag.

Kelly, B.B. & Becker, V.E. (1975). Effects of light intensity and temperature on nitrogen fixation by *Lobaria pulmonaria*, *Sticta weigelii*, *Leptogium cyanescens* and *Collema subfurvum*. *Bryologist*, 78, 350-5.

Kershaw, K.A. (1963). Lichens. *Endeavour*, 22, 65-9.

Kershaw, K.A. (1972). The relationship between moisture content and net assimilation rate of lichen thalli and its ecological significance. *Canadian Journal of Botany*, 50, 543-55.

Kershaw, K.A. (1974). Dependence of the level of nitrogenase activity on the water content of the thallus in *Peltigera canina*, *P. evansiana*, *P. polydactyla* and *P. praetextata*. *Canadian Journal of Botany*, 52, 1423-27.

Kershaw, K.A. (1975*a*). Studies on lichen-dominated systems. XII. The ecological significance of thallus colour. *Canadian Journal of Botany*, 53, 660-7.

Kershaw, K.A. (1975*b*). Studies on lichen-dominated systems. XIV. The comparative ecology of *Alectoria nitidula* and *Cladina alpestris*. *Canadian Journal of Botany*, 53, 2608-13.

Kershaw, K.A. (1977*a*). Studies on lichen-dominated systems. XX. An examination of some aspects of lichen woodlands in Canada. *Canadian Journal of Botany*, 55, 393-410.

Kershaw, K.A. (1977*b*). Physiological-environmental interactions in lichens. II. The pattern of net photosynthetic acclimation in *Peltigera canina* var. *praetextata* Hue, and *P. polydactyla*. *New Phytologist*, 79, 377-390.

Kershaw, K.A. (1977*c*). Physiological-environmental interactions in lichens. III. The rate of net photosynthetic acclimation in *Peltigera canina* var. *praetextata* and *P. polydactyla*. *New Phytologist*, 79, 391-402.

Kershaw, K.A. (1978). The role of lichens in boreal tundra transition areas. *Bryologist*, 81, 294-306.

Kershaw, K.A. (1983). The thermal operating-environment of a lichen. *Lichenologist*, 15(2), 191-207.

Kershaw, K.A. & Dzikowski, P.A. (1977). Physiological-environmental interactions in lichens. VI. Nitrogenase activity in *Peltigera polydactyla* after a period of desiccation. *New Phytologist*, 79, 417-21.

Kershaw, K.A. & Field, G.F. (1975). Studies on lichen-dominated systems. XV. The temperature and humidity profiles in a *Cladina alpestris* mat. *Canadian Journal of Botany*, 53, 2614-20.

Kershaw, K.A. & Harris, G.P. (1971*a*). Simulation studies and ecology. A simple defined system and model. In *Proceedings of the International Symposium of Statistical Ecology*, vol. III, ed. G.P. Patil, pp. 1-21. Pennsylvania State University Press.

Kershaw, K.A. & Harris, G.P. (1971*b*). Simulation studies and ecology. Use of the model. *Proceedings of the International Symposium of Statistical Ecology*, vol. III, ed. G.P. Patil, pp. 23-42. Pennsylvania State University Press.

Kershaw, K.A. & Harris, G.P. (1971*c*). A technique for measuring the light profile in a lichen canopy. *Canadian Journal of Botany*, 49, 609-11.

Kershaw, K.A. & Larson, D.W. (1974). Studies on lichen-dominated systems. IX. Topographic influences on microclimate and species distribution. *Canadian Journal of Botany*, 52, 1935-46.

Kershaw, K.A. & MacFarlane, J.D. (1980). Physiological-environmental interactions in lichens. X. Light as an ecological factor. *New Phytologist*, 84, 687-702.

Kershaw, K.A. & MacFarlane, J.D. (1982). Physiological-environmental interactions in lichens. XIII. Seasonal constancy of nitrogenase activity, net photosynthesis and respiration, in *Collema furfuraceum* (Am.) DR. *New Phytologist*, 90, 723-34.

Kershaw, K.A., MacFarlane, J.D. & Tysiaczny, M.J. (1977). Physiological-environmental interactions in lichens. V. The interaction of temperature with nitrogenase activity in the dark. *New Phytologist*, 79, 409-16.

Kershaw, K.A., MacFarlane, J.D., Webber, M.R. & Fovargue, A. (1983). Phenotypic differences in the seasonal pattern of net photosynthesis in *Cladonia stellaris*. *Canadian Journal of Biology*, 61, 2169-80.

Kershaw, K.A. & Millbank, J.W. (1970). Nitrogen metabolism in lichens. II. The partition of cephalodial-fixed nitrogen between the mycobiont and phycobionts of *Peltigera aphthosa*. *New Phytologist*, 69, 75-9.

Kershaw, K.A. & Rouse, W.R. (1971*a*). Studies on lichen-dominated systems. I. The water relations of *Cladonia alpestris* in spruce-lichen woodland in northern Ontario. *Canadian Journal of Botany*, 49, 1389-99.

Kershaw, K.A. & Rouse, W.R. (1971*b*). Studies on lichen-dominated systems. II. The growth pattern of *Cladonia alpestris* and *Cladonia rangiferina*. *Canadian Journal of Botany*, 49, 1400-10.

Kershaw, K.A. & Rouse, W.R. (1973). Studies on lichen-dominated systems. V. A primary survey of a raised beach system in N.W. Ontario. *Canadian Journal of Botany*, 51, 1285-307.

Kershaw, K.A. & Rouse, W.R. (1976). *The Impact of Fire on Forest and Tundra Ecosystems*. INA Pub. No. QS-9117-000-EE-A1, Ottawa.

Kershaw, K.A., Rouse, W.R. & Bunting, B.T. (1975). *The Impact of Fire on Forest and Tundra Ecosystems*. INA Pub. No. QS-8038-000-EEE-A1, Ottawa.

Kershaw, K.A. & Smith, M.M. (1978). Studies on lichen-dominated systems. XXI. The control of seasonal rates of net photosynthesis by moisture, light and temperature in *Stereocaulon paschale*. *Canadian Journal of Botany*, 56, 282-30.

Kershaw, K.A. & Watson, S. (1983). The control of seasonal rates of net photosynthesis by moisture, light and temperature in *Parmelia disjuncta* Erichs. *Bryologist*, 86, 31-43.

Kershaw, K.A. & Webber, M.R. (1984). Photosynthetic capacity changes in *Peltigera*. I. The synthesis of additional photosynthetic units in *P. praetextata*. *New Phytologist*, 96, 437-46.

Körner, C.H., Schul, J.A. & Bauer, J. (1979). Maximum leaf diffusive conductance in vascular plants. *Photosynthetica*, 13, 45-82.

Krochko, J.E., Winner, W.E. & Bewley, J.D. (1979). Respiration in relation to adenosine triphosphate content during desiccation and rehydration of a desiccation-tolerant and a desiccation-intolerant moss. *Plant Physiology*, 64, 13-17.

Lamb, I.M. (1970). Antarctic terrestrial plants and their ecology. In *Antarctic Ecology*, ed. M.W. Holdgate, pp. 733-51. London, New York & San Francisco: Academic Press.

Landsberg, J.J. & Ludlow, M.M. (1970). A technique for determining resistance to mass transfer through boundary layers of plants with complex structures. *Journal of Applied Ecology*, 7, 187-92.

Lange, G.E., Reiners, W.A. & Heier, R.K. (1976). Potential alteration of precipitation chemistry by epiphytic lichens. *Oecologia*, 25, 229-41.

Lange, O.L. (1953). Hitze-und Trockenresistenz der Flechten in Beziehung zu ihrer Verbreitung. *Flora (Jena)*, 140, 39-97.

Lange, O.L. (1954). Einige Messungen zum Wärmehaushalt poikilohydrer Flechten und Moose. *Archiv für Meteorologie, Geophysik und Bioklimatologie*, Serie B, 5, 182-90.

Lange, O.L. (1955). Untersuchungen über die Hitzeresistenz der Moose in Beziehung zu ihrer Verbreitung. I. Die Resistenz stark ausgetrockneter Moose. *Flora (Jena)*, 142, 381-99.

Lange, O.L. (1962). Die Photosynthese der Flechten bei tiefen Temperaturen und nach Frostperioden. *Berichte der Deutschen Botanischen Gesellschaft*, 75, 351-52.

Lange, O.L. (1965). Der CO_2-Gaswechsel von Flechten nach Erwärmung im feuchten Zustand. *Berichte der Deutschen Botanischen Gesellschaft*, 78, 441-54.

Lange, O.L. (1969). Experimentell-ökologische Untersuchungen an Flechten der Negev-Wüste. I. CO_2-Gaswechsel von *Ramalina maciformis* (Del.). Borg unter kontrollierten Bedingungen im Laboratorium. *Flora (Jena)*, 158, 324-59.

Lange, O.L. (1980). Moisture content and CO_2 exchange in lichens. *Oecologia (Berlin)*, 45, 82-7.

Lange, O.L., Schulze, E.-D. & Koch, W. (1970*a*). Experimentell-ökologische Untersuchungen an Flechten der Negev-Wüste. II. CO_2-Gaswechsel und Wasserhaushalt von *Ramalina maciformis* (Del.). Borg am natürlichen Standort wärhend der sommerlichen Trockenperiode. *Flora (Jena)*, 159, 38-62.

Lange, O.L., Schulze, E.-D. & Koch, W. (1970*b*). Experimentell-ökologische Untersuchungen an Flechten der Negev-Wüste. II. CO_2 -Gaswechsel und Wasserhaushalt von Krusten-und Blattflechten am natürlichen Standort während der sommerlichen Trockenperiode. *Flora (Jena)*, 159, 525-38.

Lange, O.L. & Tenhunen, J.D. (1981). Moisture content and CO_2 exchange of lichens. II. Depression of net photosynthesis in *Ramalina maciformis* at high water content is caused by increased thallus carbon dioxide diffusion resistance. *Oecologia (Berlin)*, 51, 426-9.

Larcher, W. (1975). *Physiological Plant Ecology*. Berlin: Springer-Verlag.

Larson, D.W. (1978). Patterns of lichen photosynthesis and respiration following prolonged frozen storage. *Canadian Journal of Botany*, 56, 2119-23.

Larson, D.W. (1979). Lichen water relations under drying conditions. *New Phytologist*, 82, 713-31.

Larson, D.W. (1980). Seasonal change in the pattern of net CO_2 exchange in *Umbilicaria* lichens. *New Phytologist*, 84, 349-69.

Larson, D.W. (1981). Differential wetting in some lichens and mosses: the role of morphology. *Bryologist*, 84, 1-15.

Larson, D.W. (1983a). Environmental stress and *Umbilicaria* lichens: the effect of sub-zero temperature pretreatment. *Oecologia*, 55, 268-78.

Larson, D.W. (1983*b*). The pattern of production within individual *Umbilicaria* lichen thalli. *New Phytologist*, 94, 409-19.

Larson, D.W. (1983*c*). Morphological variation and development in *Ramalina menziesii* Tayl. *American Journal of Botany*, 70, 668-81.

Larson, D.W. & Kershaw, K.A. (1975*a*). Studies on lichen-dominated systems. XI. Lichen-heath and winter snow cover. *Canadian Journal of Botany*, 53, 621-6.

Larson, D.W. & Kershaw, K.A. (1975*b*). Measurement of CO_2 exchange in lichens: a new method. *Canadian Journal of Botany*, 53, 1535-41.

Larson, D.W. & Kershaw, K.A. (1975*c*). Acclimation in arctic lichens. *Nature (London)*, 254, 421-3.

Larson, D.W. & Kershaw, K.A. (1975*d*). Studies on lichen-dominated systems. XIII. Seasonal and geographical variation of net CO_2 exchange of *Alectoria ochroleuca*. *Canadian Journal of Botany*, 53, 2598-607.

Larson, D.W. & Kershaw, K.A. (1975*e*). Studies on lichen-dominated systems. XVI. Comparative patterns of net CO_2 exchange in *Cetraria nivalis* and *Alectoria ochroleuca* collected from a raised-beach ridge. *Canadian Journal of Botany*, 53, 2884-92.

Larson, D.W. & Kershaw, K.A. (1976). Studies on lichen-dominated systems. XVIII. Morphological control of evaporation in lichens. *Canadian Journal of Botany*, 54, 2061-73.

Laudi, G., Bonatti, P. & Trovatelli, L.D. (1969). Differenze ultrastrutturali di alcune specie di *Trebouxia* poste in condizioni di illiminazione differenti. *Giornale botanico italiano*, 103, 79-107.

Laundon, J.R. (1967). A study of the lichen flora of London. *Lichenologist*, 3, 277-327.

Laundon, J.R. (1973). Urban lichen studies. In *Air Pollution and Lichens*, ed. B.W. Ferry, M.S. Baddeley & D.L. Hawksworth, pp. 109-23. University of Toronto Press.

Lawrey, J.D (1977*a*). Inhibition of moss spore germination by acetone extracts of terricolous *Cladonia* species. *Bulletin of the Torrey Botanical Club*, 104, 49-52.

Lawrey, J.D. (1977*b*). Adaptive significance of *O*-methylated lichen depsides and depsidones. *Lichenologist*, 9, 137-42.

Lawrey, J.D & Rudolph, E.D. (1975). Lichen accumulation of some heavy metals from acidic surface substrates of coal mine ecosystems in southeastern Ohio. *Ohio Journal of Science*, 75, 113-17.

Lechowicz, M.J. (1978). Carbon dioxide exchange in *Cladonia* lichens from subarctic and temperate habitats. *Oecologia*, 31, 225-37.

Lechowicz, M.J. (1981). The effects of climatic pattern on lichen productivity: *Cetraria cucullata* (Bell.) Ach. in the arctic tundra of northern Alaska. *Oecologia*, 50, 210-16.

Lechowicz, M.J. & Adams, M.S. (1973). Net photosynthesis of *Cladonia mitis* (Sand.) from sun and shade sites on the Wisconsin pine barrens. *Ecology*, 54, 413-19.

Lechowicz, M.J. & Adams, M.S. (1974). Ecology of *Cladonia* lichens. II. Comparative physiological ecology of *C. mitis*, *C. rangiferina*, and *C. uncialis*. *Canadian Journal of Botany*, 52, 411-22.

Lechowicz, M.J., Jordan, W.P. & Adams, M.S. (1974). Ecology of *Cladonia* lichens. III. Comparison of *C. caroliniana*, endemic to southeastern North America, with three northern *Cladonia* species. *Canadian Journal of Botany*, 52, 565-73.

Levitt, J. (1980). *Responses of Plants to Environmental Stresses*, 2nd edn, vol. 1, *Chilling, Freezing and High Temperature Stresses*. Physiological Ecology Series, ed. T.T. Kozlowski. London, New York & San Francisco: Academic Press.

Lewis, M.C. & Callaghan, T.V. (1976). Tundra. In *Vegetation and the Atmosphere*, vol. 2, ed. J.L. Monteith, pp. 399-435. London, New York & San Francisco: Academic Press.

Looman, J. (1964). Ecology of lichen and bryophyte communities in Saskatchewan. *Ecology*, 45, 481-91.

Lounamaa, K.J. (1956). Trace elements in plants growing wild on different rocks in Finland.

A semi-quantitative spectrographic survey. *Annales Botanici Societatis Zoologicae Botanicae Fennicae 'Vanamo'*, 29, 1-196.

MacFarlane, J.D. & Kershaw, K.A. (1977). Physiological-environmental interactions in lichens. IV. Seasonal changes in the nitrogenase activity of *Peltigera canina* var. *praetexta*, and *P. canina* var. *rufescens*. *New Phytologist*, 79, 403-8.

MacFarlane, J.D. & Kershaw, K.A. (1978). Thermal sensitivity in lichens. *Science*, 201, 739-41.

MacFarlane, J.D. & Kershaw, K.A. (1980*a*). Physiological-environmental interactions in lichens. IX. Thermal stress and lichen ecology. *New Phytologist*, 84, 669-85.

MacFarlane, J.D. & Kershaw, K.A. (1980*b*). Physiological-environmental interactions in lichens. XI. Snowcover and nitrogenase activity. *New Phytologist*, 84, 703-10.

MacFarlane, J.D. & Kershaw, K.A. (1982). Physiological-environmental interactions in lichens. XIV. The environmental control of glucose movement from alga to fungus in *Peltigera polydactyla*, *P. rufescens* and *Collema furfuraceum*. *New Phytologist*, 91, 93-101.

MacFarlane, J.D., Kershaw, K.A. & Webber, M.R. (1983). Physiological-environmental interactions in lichens. XVII. Phenotypic differences in the seasonal pattern of net photosynthesis in *Cladonia rangiferina*. *New Phytologist*, 94, 217-33.

MacFarlane, J.D. & Millbank, J.W. (1984). The translocation and partition of fixed nitrogen in *Peltigera aphthosa*. *Canadian Journal of Botany*, (submitted).

MacFarlane, J.D., Maikawa, E., Kershaw, K.A. & Oaks, A. (1976). Environmental-physiological interactions in lichens. I. The interaction of light/dark periods and nitrogenase activity in *Peltigera polydactyla*. *New Phytologist*, 77, 705-11.

Maikawa, E. & Kershaw, K.A. (1975). The temperature dependence of thallus nitrogenase activity in *Peltigera canina*. *Canadian Journal of Botany*, 53, 527-9.

Maikawa, E. & Kershaw, K.A. (1976). Studies on lichen-dominated systems. XIX. The postfire recovery sequence of black spruce-lichen woodland in the Abitau Lake region, NWT. *Canadian Journal of Botany*, 54, 2679-87.

Maquinay, A., Lamb, I.M., Lambinon, J. & Ramant, J.L. (1961). Dosage du zinc chez un lichen calaminaire belge: *Stereocaulon nanodes* Tuck, f. *tryoliense* (Nyl.) M. Lamb. *Physiologia Plantarum*, 14, 284-9.

Mayo, J.M., Despain, D.G. & Van Zinderen Bakker, E.M. (1973). CO_2 assimilation by *Dryas integrifolia* on Devon Island, Northwest Territories. *Canadian Journal of Botany*, 51, 581-8.

Meigs, P. (1966). *Geography of Coastal Deserts*. Arid Zone Research No. 28. Paris: UNESCO.

Millbank, J.W. (1974*a*). Associations with blue-green algae. In *The Biology of Nitrogen Fixation*, ed. A. Quispel, pp. 238-64. Amsterdam & Oxford: North-Holland.

Millbank, J.W. (1974*b*). Nitrogen metabolism in lichens. V. The forms of nitrogen released by the blue-green phycobiont in *Peltigera* spp. *New Phytologist*, 73, 1171-81.

Millbank, J.W. (1976). Aspects of nitrogen metabolism in lichens. In *Lichenology: Progress and Problems*, ed. D.H. Brown, D.L. Hawksworth & R.H. Bailey, pp. 441-55. London, New York & San Francisco: Academic Press.

Millbank, J.W. (1977). The oxygen tension within lichen thalli. *New Phytologist*, 79, 649-57.

Millbank, J.W. (1981). The assessment of nitrogen fixation and throughput by lichens. I. The use of a controlled environment chamber to relate acetylene reduction to nitrogen fixation. *New Phytologist*, 89, 647-55.

Millbank, J.W. (1982). The assessment of nitrogen fixation and throughput by lichens. III. Losses of nitrogenous compounds by *Peltigera membranacea*, *P. polydactyla* and *Lobaria pulmonaria* in simulated rainfall episodes. *New Phytologist*, 92, in press.

Millbank, J.W. & Kershaw, K.A. (1969). Nitrogen metabolism in lichens. I. Nitrogen fixation in the cephalodia of *Peltigera aphthosa*. *New Phytologist*, 68, 721-9.

Millbank, J.W. & Kershaw, K.A. (1973). Nitrogen metabolism. In *The Lichens*, ed. V. Ahmadjian & M.E. Hale, pp. 289-307. London, New York & San Francisco: Academic Press.

Millbank, J.W. & Olsen, J.D. (1981). The assessment of nitrogen fixation and throughput by lichens. II. Construction of an enclosed growth chamber for the use of $^{15}N_2$. *New Phytologist*, 89, 657-65.

Miller, E.V., Greene, R., Cancilla, A.S. & Curry, C. (1963). Antimetabolites in lichens. A preliminary report. *Proceedings of the Pennsylvania Academy of Science*, 37, 104-8.

Miller, E.V. & Schaefers, T. (1964). Antimetabolites in lichens. III. Additional characterizations of the extracts. *Proceedings of the Pennsylvania Academy of Science*, 38, 25-8.

Monteith, J.L. (1964). Evaporation and environment. In *The State and Movement of Water in Living Organisms*. Symposia of the Society for Experimental Biology, No. 19, pp. 205-34. Cambridge University Press.

Monteith, J.L. (1973). *Principles of Environmental Physics*. London: Edward Arnold.

Monteith, J.L. (1975). *Vegetation and the Atmosphere*. London, New York & San Francisco: Academic Press.

Montfort, C. (1950). Photochemische Wokungen des Hohenklimas auf die Chloroplasten photolábiler Pflanzen in Mittel-und Hochgebirge. *Zeitschrift für Naturforschung*, 5B, 221-6.

Mooney, H.A. & Gulman, S.L. (1979). Environment and evolutionary constraints on the photosynthetic characteristics of higher plants. In *Topics in Plant Population Biology*, ed. O.T. Solbrig, S. Jain, G.B. Johnson & P.H. Raven, pp. 316-37. Columbia University Press.

Morgan-Huws, D.I. & Haynes, F.N. (1973). Distribution of some epiphytic lichens around an oil refinery at Fawley, Hampshire. In *Air Pollution and Lichens*, ed. B.W. Ferry, M.S. Baddeley & D.W. Hawksworth, pp. 89-108. University of Toronto Press.

Munding, H. (1952). Untersuchungen zur Frage der Strahlenresistenz des Chlorophylls in den Chloroplasten. *Protoplasma*, 41, 212-234.

Nash, III, T.H. (1975). Influence of effluents from a zinc factory on lichens. *Ecological Monographs*, 45, 183-98.

Nash, III, T.H., Moser, T.J. & Link, S.O. (1980). Nonrandom variation of gas exchange within arctic lichens. *Canadian Journal of Botany*, 58, 1181-86.

Nash, III, T.H., Nebeker, G.T., Moser, T.J. & Reeves, T. (1979). Lichen vegetational gradients in relation to the Pacific coast of Baja California: the maritime influence. *Madronõ*, 26, 149-63.

Neal, M.W. & Kershaw, K.A. (1973). Studies on lichen-dominated systems. IV. The objective analysis of Cape Henrietta Maria raised-beach systems. *Canadian Journal of Botany*, 51, 1177-90.

Nieboer, E.H., Lavoie, P., Sasseville, R.L.P., Puckett, K.J. & Richardson, D.H.S. (1976*a*). Cation-exchange equilibrium and mass balance in the lichen *Umbilicaria muhlenbergii*. *Canadian Journal of Botany*, 54, 720-3.

Nieboer, E.H., Puckett, K.J. & Grace, B. (1976*b*). The uptake of nickel by *Umbilicaria muhlenbergii*: a physicochemical process. *Canadian Journal of Botany*, 54, 724-33.

Nieboer, E. & Richardson, D.H.S. (1980). The replacement of the nondescript term 'heavy metals' by a biologically and chemically significant classification of metal ions. *Environmental Pollution Series B*, 1, 3-26.

Nieboer, E. & Richardson, D.H.S. (1981). Lichens as monitors of atmospheric deposition. In *Atmospheric Pollutants in Natural Waters*, ed. S.J. Eisenreich, pp. 339-388. Ann Arbor, Michigan: American Arbor Science Publications.

Nieboer, E., Richardson, D.H.S., Boileau, L.J.R., Beckett, P.J., Lavoie, P. & Padovan, D. (1982). Lichens and mosses as monitors of industrial activity associated with uranium mining in northern Ontario, Canada. III. Accumulations of iron and titanium and their mutual dependence. *Environmental Pollution Series B*, 4, 181-92.

Nieboer, E., Richardson, D.H.S., Lavoie, P. & Padovan, D. (1979). The role of metal ion binding in modifying the toxic effects of sulphur dioxide on the lichen *Umbilicaria muhlenbergii*. I. Potassium efflux studies. *New Phytologist*, 82, 621-32.

Nieboer, E., Richardson, D.H.S. & Tomassini, F.D. (1978). Mineral uptake and release by lichens: an overview. *Bryologist*, 81, 226-46.

Nielson, A., Rippka, R. & Kunisaura, R. (1971). Heterocyst formation and nitrogenase synthesis in *Anabaena* spp: a kinetic study. *Archiv für Mikrobiologie*, 76, 139-50.

Ochiai, E. (1977). *Bioinorganic Chemistry: An Introduction*. Boston: Allyn & Bacon.

Oke, T.R. (1978). *Boundary Layer Climates*. London: Methuen.

Patterson, D.T., Bunce, J.A., Alberte, R.S. & Van Volkenburgh, E. (1977). Photosynthesis in relation to leaf characteristics of cotton from controlled and field environment. *Plant Physiology*, 59, 384-7.

Pattnaik, H. (1966). Studies on nitrogen fixation by *Westelliopsis prolifica* Janet. *Annals of Botany*, 30, 231-8.

Pearson, L.C. (1970). Varying environmental conditions in order to grow lichens intact under laboratory conditions. *American Journal of Botany*, 57, 659-64.

Pearson, L. & Skye, E. (1965). Air pollution affects pattern of photosynthesis in *Parmelia sulcata*, a corticolous lichen. *Science*, 148, 1600-2.

Penman, H.L. (1963). *Vegetation and Hydrology*. Technical Communication No. 53, Commonwealth Bureau of Soils. Farnham Royal, UK.

Perry, M.J., Larsen, M.C. & Alberte, R.S. (1981). Photoadaptation in marine phytoplankton: response of the photosynthetic unit. *Marine Biology*, 62, 91-101.

Pike, L.H. (1978). The importance of epiphytic lichens in mineral cycling. *Bryologist*, 81, 247-57.

Prézelin, B.B. (1981). Light reactions in photosynthesis. In *Physiological Bases of Phytoplankton Ecology*, ed. T. Platt, pp. 1-43. Ottawa: Department of Fisheries and Oceans.

Prézelin, B.B. & Ley, A.C. (1980). Photosynthesis and chlorophyll *a* fluorescence rhythms of marine phytoplankton. *Marine Biology*, 55, 295-307.

Prézelin, B.B., Meeson, B.W. & Sweeney, B.M. (1977). Characterization of photosynthetic rhythms in marine dinoflagellates. I. Pigmentation, photosynthetic capacity and respiration. *Plant Physiology*, 60, 384-7.

Prézelin, B.B. & Sweeney, B.M. (1978). Photoadaptation of photosynthesis in *Gonyaulax polyedra*. *Marine Biology*, 48, 27-35.

Prézelin, B.B. & Sweeney, B.M. (1979). Photoadaptation of photosynthesis in two bloom-forming dinoflagellates. In *Proceedings of the Second International Conference on Toxic Dinoflagellate Blooms*, ed. C. Ventsch, pp. 101-6.

Prioul, J.L., Reyss, A. & Chartier, P. (1975). Relationships between carbon dioxide transfer resistances and some physiological and anatomical features. In *Environmental and Biological Control of Photosynthesis*, ed. R. Marcelle, pp. 17-28. The Hague: W. Junk.

Prosser, C.L. (1955). Physiological variation in animals. *Biological Reviews*, 30, 229-62.

Puckett, K.J. (1976). The effect of heavy metals on some aspects of lichen physiology. *Canadian Journal of Botany*, 54, 2695-03.

Puckett, K.J., Nieboer, E., Gorzynski, M.J. & Richardson, D.H.S. (1973). The uptake of metal ions by lichens: a modified ion-exchange process. *New Phytologist*, 72, 329-42.

Puckett, K.J., Richardson, D.H.S., Flora, W.P. & Nieboer, E. (1974). Photosynthetic ^{14}C fixation by the lichen *Umbilicaria muhlenbergii* (Ach.) Tuck. following short exposures to aqueous sulphur dioxide. *New Phytologist*, 73, 1183-92.

Puckett, K.J., Tomassini, F.D., Nieboer, E. & Richardson, D.H.S. (1977). Potassium efflux by lichen thalli following exposure to aqueous sulphur dioxide. *New Phytologist*, 79, 135-45.

Pyatt, F.B. (1967). The inhibitory influence of *Peltigera canina* on the germination of graminaceous seeds and the subsequent growth of the seedlings. *Bryologist*, 70, 326-9.

Quispel, A. (1960). Respiration of lichens. *Encyclopedia of Plant Physiology*, 12, 455-60.

Rao, D.N. & LeBlanc, F. (1966). Effects of sulfur dioxide on the lichen alga, with special reference to chlorophyll. *Bryologist*, 69, 69-75.

Rao, D.N. & LeBlanc, F. (1967). Influence of an ion-sintering plant on the epiphytic vegetation in Wawa, Ontario. *Bryologist*, 70, 141-57.

Raven, P.H., Evert, R.F. & Curtis, H. (1976). *Biology of Plants*. New York: Worth Publishers.

Richardson, D.H.S. (1973). Photosynthesis and carbohydrate movement. In *The Lichens*, ed. V. Ahmadjian & M.E. Hale, pp. 249-88. London, New York & San Francisco: Academic Press.

Richardson, D.H.S. & Nieboer, E. (1981). Lichens and pollution monitoring. *Endeavour*, 5, 127-133.

Richardson, D.H.S. & Nieboer, E. (1983). Ecophysiological responses of lichens to sulphur dioxide. *Journal of the Hattori Botanical Laboratory*, 54, 331-51.

Richardson, D.H.S., Nieboer, E., Lavoie, P. & Padovan, D. (1979). The role of metal-ion binding in modifying the toxic effects of sulphur dioxide on the lichen *Umbilicaria muhlenbergii*. II. ^{14}C-fixation studies. *New Phytologist*, 82, 633-43.

Richardson, D.H.S., Nieboer, E., Lavoie, P. & Padovan, D. (1984). Anion accumulation by lichens. I. The characteristics and kinetics of arsenate uptake by *Umbilicaria muhlenbergii*. *New Phytologist*, 96, 71-82.

Richardson, D.H.S. & Puckett, K.J. (1973). Sulphur dioxide and photosynthesis in lichens. In *Air Pollution and Lichens*, ed. B.W. Ferry, M.S. Baddeley & D.L. Hawksworth, pp. 283-98. University of Toronto Press.

Ried, A. (1960a). Thallusbau und Assimilationshaushalt von Laub-und Krustenflechten. *Biologisches Zentralblatt*, 79, 129-51.

Ried, A. (1960*b*). Stoffwechsel und Verbreitungsgrenzen von Flechten. II. Wasser-und Assimilationshaushalt, Entquellungs-und Submersionsresistenz von Krustenflechten benachbarter Standorte. *Flora (Jena)*, 149, 345-85.

Rieman, W. & Walton, H.F. (1970). *Ion Exchange in Analytical Chemistry*. Oxford: Pergamon Press.

Rogers, R.W. (1971). Distribution of the lichen *Chondropsis semiviridis* in relation to its heat and drought resistance. *New Phytologist*, 7, 1069-77.

Rouse, W.R. (1976). Microclimate changes accompanying burning in subarctic lichen woodland. *Arctic and Alpine Research*, 8, 357-76.

Roux, C. (1979). Etude écologique et phytosiologique des peuplements lichéniques saxicoles-calcicoles du sud-est de la France. PhD Thesis, University Pierre et Marie Curie, Paris.

Rudolph, E.D. (1966). Lichen ecology and microclimate studies at Cape Hallet, Antarctica. Biometeorology II. In *Proceedings of the Third International Biometric Congress*, pp. 900-10. Oxford: Pergamon Press.

Rundel, P.W. (1974). Water relations and morphological variation in *Ramalina menziesii* Tayl. *Bryologist*, 77, 23-32.

Rundel, P.W. (1978*a*). Ecological relationships of desert fog zone lichens. *Bryologist*, 81, 277-93.

Rundel, P.W. (1978*b*). Evolutionary relationships in the *Ramalina usnea* complex. *Lichenologist*, 10, 141-56.

Rundel, P.W. (1978*c*). The ecological role of secondary lichen substances. *Biochemical Systematics and Ecology*, 6, 151-70.

Rundel, P.W., Bowler, P.A. & Mulroy, T.W. (1972). A fog-induced lichen community from northwestern Baja, California, with two new species of *Desmazieria*. *Bryologist*, 75, 501-8.

Rutter, A.J. (1967). An analysis of evaporation from a stand of Scots pine. In *Forest Hydrology*, ed. W.E. Sopper & H.W. Hull, pp. 403-417. London, New York & San Francisco: Academic Press.

Sampaio, M.J.A.M., Rai, A.N., Rowell, P. & Stewart, W.D.P. (1979). Occurrence, synthesis and activity of glutamine synthetase in N_2-fixing lichens. *FEMS Microbiological Letters*, 6, 107-10.

Schatz, A. (1963). Soil microorganisms and soil chelation. The pedogenic action of lichens and lichen acids. *Journal of Agriculture and Food Chemistry*, 11, 112-18.

Schell, D.M. & Alexander, V. (1970). Nitrogen fixation in arctic coastal tundra in relation to vegetation and micro relief. *Arctic*, 26, 130-7.

Scholander, P.F., Flagg, W., Walters, V. & Irving, L. (1952). Respiration in some arctic and tropical lichens in relation to temperature. *American Journal of Botany*, 39, 707-13.

Scholander, S.I. & Kanwisher, J.T. (1959). Latitudinal effect on respiration in some northern plants. *Plant Physiology*, 34, 574-6.

Scotter, G.W. (1964). *Effects of Forest Fires on the Winter Range of Barren-ground Caribou in Northern Saskatchewan*. Canadian Wildlife Service, Wildlife Management Bulletin Series 1, No. 18. Ottawa.

Sernander, R. (1912). Studier öfver lafvarnes biologi. I. Nitrofila lafvar. *Svensk Botanisk Tidskrift*, 6. 803-83.

Sernander, R. (1926). *Stockholm's Nature*. Uppsala: Almqvist & Wilksell.

Simon, E.W. (1974). Phospholipids and plant membrane permeability. *New Phytologist*, 73, 377-420.

Simon, J.-P. (1979*a*). Adaptation and acclimation of higher plants at the enzyme level: speed of acclimation for apparent energy of activation of NAD malate dehydrogenase in *Lathyrus japonicus* Willd. (Leguminosae). *Plant, Cell and Environment*, 2, 35-8.

Simon, J.-P. (1979*b*). Adaptation and acclimation of higher plants at the enzyme level: temperature-dependent substrate binding ability of NAD malate dehydrogenase in four populations of *Lathyrus japonicus* Willd. (Leguminosae). *Plant Science Letters*, 14, 113-20.

Skujins, J. & Klubek, B. (1978). Nitrogen fixation and cycling by blue-green algae-lichen-crusts in arid rangeland soils. *Ecological Bulletin (Stockholm)*, 26, 164-71.

Smith, A.L. (1921). *Lichens*. Cambridge University Press.

Smith, D.C. (1960). Studies in the physiology of lichens. III. Experiments with dissected discs of *Peltigera polydactyla*. *Annals of Botany (London) New Series*, 24, 186-99.

Smith, D.C. (1961). The physiology of *Peltigera polydactyla* (Neck.) Hoffm. *Lichenologist*, 1, 209-26.

Smith, D.C. (1962). The biology of lichen thalli. *Biological Reviews*, 37, 537-70.

Smith, D.C. (1980). Mechanisms of nutrient movement between the lichen symbionts. In *Cellular Interactions in Symbiosis and Parasitism*, ed. C. B. Cook, P.W. Pappas & E.D. Rudolph, pp. 197-227. Columbus: Ohio State University Press.

Smith, D.C. & Molesworth, S. (1973). Lichen physiology. XIII. Effects of rewetting dry lichens. *New Phytologist*, 72, 525-33.

Smyth, E.S. (1934). A contribution to the physiology and ecology of *Peltigera canina* and *P. polydactyla*. *Annals of Botany*, 48, 781-818.

Snelgar, W.P., Brown, D.H. & Green, T.G.A. (1980). A provisional survey of the interaction between net photosynthetic rate, respiration rate, and thallus water content in some New Zealand cryptograms. *New Zealand Journal of Botany*, 18, 247-56.

Snelgar, W.P., Green, T.G.A. & Beltz, C.K. (1981*b*). Carbon dioxide exchange in lichens: estimation of internal thallus CO_2 transport resistance. *Physiologia Plantarum*, 52, 417-22.

Snelgar, W.P., Green, T.G.A. & Wilkins, A.L. (1981*a*). Carbon dioxide exchange in lichens: resistances to CO_2 uptake at different thallus water contents. *New Phytologist*, 88, 353-61.

Stålfelt, M.G. (1939). Der Gasaustausch der Flechten. *Planta (Berlin)*, 29, 11-31.

Stewart, W.D.P. (1965). Nitrogen turnover in marine and brackish habitats. I. Nitrogen fixation. *Annals of Botany, New Series*, 29, 229-39.

Stewart, W.D.P. (1974). Blue-green algae. In *The Biology of Nitrogen Fixation*, ed. A. Quispel, pp. 202-37. Amsterdam & Oxford: North-Holland.

Stewart, W.D.P. (1977). Blue-green algae. In *A Treatise on Dinitrogen Fixation*, sec. III, ed. R.W.F. Hardy & W.S. Silver, pp. 63-123. New York: Wiley.

Stewart, W.D.P., Fitzgerald, G.P. & Burris, R.H. (1967). *In situ* studies on N_2 fixation using the acetylene reduction technique. *Proceedings of the National Academy of Sciences, USA*, 58, 2071-78.

Stewart, W.D.P. & Rowell, P. (1977). Modifications of nitrogen fixing algae in lichen symbioses. *Nature (London)*, 265, 371-2.

Stewart, W.D.P., Rowell, P. & Rai, A.N. (1980). Symbiotic nitrogen-fixing Cyanobacteria. In *Nitrogen Fixation*, ed. W.D.P. Stewart & J.R. Gallon, pp. 239-69. London, New York & San Francisco: Academic Press.

Stocker, O. (1927). Physiologische und ökologische Untersuchungen an Laub-und Strauchflechten. *Flora (Jena)*, 21, 334-415.

Tegler, B. & Kershaw, K.A. (1980). Studies on lichen-dominated systems. XXIII. The control of seasonal rates of net photosynthesis by moisture, light and temperature in *Cladonia rangiferina*. *Canadian Journal of Botany*, 58, 1851-8.

Tegler, B. & Kershaw, K.A. (1981). Physiological-environmental interactions in lichens. XII. The seasonal variation of the heat stress response of *Cladonia rangiferina*. *New Phytologist*, 87, 395-401.

Terri, J.A., Patterson, D.T., Alberte, R.S. & Castelberry, R.M. (1977). Changes in the photosynthetic apparatus of maize in response to simulated natural temperature fluctuations. *Plant Physiology*, 60, 370-3.

Thornber, J.P., Alberte, R.S., Hunter, F.A., Shiozawa, J.A. & Kan, K.S. (1977). The organization of chlorophyll in the plant photosynthetic unit. In *Chlorophyll-Proteins, Reaction Centers, and Photosynthetic Membranes*, ed. J.M. Olson & G. Hind, pp. 132-48. Brookhaven Symposium of Biology 28.

Tobler, F. (1925). Zur Physiologie der Farbunterschiede bei *Xanthoria*. *Berichte der Deutschen botanischen Gesellschaft*, 58, 301-5.

Tuominen, Y. (1967). Studies on the strontium uptake of the *Cladonia alpestris* thallus. *Annales Botanici Societis Zoologicae. Botanicae Fennicae 'Vanamo'*, 4, 1-28.

Tuominen, Y. (1968). Studies on the translocation of caesium and strontium ions in the thallus of *Cladonia alpestris*. *Annales Botanici Societatis Zoologicae Botanicae Fennicae 'Vanamo'*, 5, 102-11.

Türk, R. & Wirth, V. (1975). The pH dependence of SO_2 damage to lichens. *Oecologia (Berlin)*, 19, 285-91.

Türk, R., Wirth, V. & Lange, O.L. (1974). CO_2-Gaswechsel-Untersuchungen zur SO_2-Resistenz von Flechten. *Oecologia (Berlin)*, 15, 33-64.

Tysiaczny, M.J. & Kershaw, K.A. (1979). Physiological-environmental interactions in lichens. VII. The environmental control of glucose movement from alga to fungus in *Peltigera canina* var. *praetextata* Hue. *New Phytologist*, 83, 137-46.

Vierling, E. & Alberte, R.S. (1980). Functional organization and plasticity of the photosynthetic unit of the cyanobacterium *Anacystis nidulans*. *Physiologia Plantarum*, 50, 93-8.

Vogel, S. (1955). Nie dere Feusterpflanzen in der südafrikanischen Wüste. Eine ökologische Schilderung. *Beiträge zur Biologie der Pflanzen*, 31, 45-135.

Wainwright, S.J. & Beckett, P.J. (1975). Kinetic studies on the binding of zinc ions by the lichen *Usnea florida* (L.) Web. *New Phytologist*, 75, 91-8.

Walter, H. (1931). *Die Hydratur der Pflanze*. Jena.

Warren Wilson, J. (1957). Observations on the temperatures of arctic plants and their environment. *Journal of Ecology*, 45, 499-531.

Waughman, G.J. (1977). The effect of temperature on nitrogenase activity. *Journal of Experimental Botany*, 28, 949-60.

Weare, N.M. & Benemann, J.R. (1973). Nitrogen fixation by *Anabaena cylindrica*. I. Localization of nitrogen fixation in the heterocysts. *Archiv für Mikrobiologie*, 90, 323-32.

Weber, W.A. (1962). Environmental modification and the taxonomy of the crustose lichens. *Svensk Botanisk Tidskrift*, 56, 293-333.

Index